Role of Giant Honeybees in Natural and Agricultural Systems

Role of Giant Honeybees in Natural and Agricultural Systems provides multidisciplinary perspective about the different facets of giant honeybees. Giant honeybees—*Apis dorsata* and *Apis laboriosa*—are excellent pollinators of crops, fruits, and vegetables in cultivated and natural landscapes. Their large size, long foraging range, and large workforce make them the most spectacular of all honeybee species for crop pollination and honey production. Due to their decline, ecosystems and global food security are being threatened. This book is the first of its kind which deals in detail on varied aspects of giant honeybee biology, management and conservation strategies for protecting biodiversity and enhancing crop productivity. It aims to promote a large, diverse, sustainable, and dependable bee pollinator workforce that can meet the challenge for optimizing food production in the 21st century.

SALIENT FEATURES

1. Covers the latest information on various aspects of biology of giant honeybees and brings the latest advances together in a single volume for researchers and advanced-level students
2. Provides an excellent source of advanced study material for academics, researchers and students and program planners
3. Provides an excellent source of livelihood in mountainous areas and marginal farmers
4. Deals with biology, management and conservation strategies for protecting biodiversity and enhancing crop productivity
5. Excellent pollinator of tropical and subtropical crops, fruits, vegetables, etc., less prone to diseases and enemies

This book will be useful for pollination biologists, honeybee biologists, scientists working in agriculture, animal behavior, conservation, biology, ecology, entomologists, environmental biologists, etc.

Role of Giant Honeybees in Natural and Agricultural Systems

Edited by Dharam P. Abrol

CRC Press
Taylor & Francis Group
Boca Raton London New York

CRC Press is an imprint of the
Taylor & Francis Group, an **informa** business

First edition published 2024
by CRC Press
6000 Broken Sound Parkway NW, Suite 300, Boca Raton, FL 33487–2742

and by CRC Press
4 Park Square, Milton Park, Abingdon, Oxon, OX14 4RN

CRC Press is an imprint of Taylor & Francis Group, LLC

Library of Congress Cataloging-in-Publication Data
Names: Abrol, Dharam P., editor.
Title: Role of giant honeybees in natural and agricultural systems / edited by DP Abrol.
Description: First edition. | Boca Raton, FL : CRC Press, 2024. | Includes bibliographical references and index.
Identifiers: LCCN 2023017059 (print) | LCCN 2023017060 (ebook)
Subjects: LCSH: Honeybee. | Bee culture. | Insect pollinators. | Pollination by insects.
Classification: LCC SF523 .R65 2024 (print) | LCC SF523 (ebook) | DDC 638/.1—dc23/eng/20230628
LC record available at https://lccn.loc.gov/2023017059
LC ebook record available at https://lccn.loc.gov/2023017060

ISBN: 978-1-03227-782-0 (hbk)
ISBN: 978-1-03227-784-4 (pbk)
ISBN: 978-1-00329-407-8 (ebk)

DOI: 10.1201/9781003294078

Typeset in Times
by Apex CoVantage, LLC

Prof. Dr Jerzy Woyke

This book is dedicated to Prof. Dr Jerzy Woyke. Prof. Jerzy Woyke was a distinguished Polish bee scientist and academic. He was born on September 9, 1926 in Malenin and passed away on December 20, 2022. Woyke was a graduate of the University of Warsaw and received his Ph.D. in apiculture from the Institute of Zoology, Polish Academy of Sciences. He held several academic positions throughout his career, including Professor of Apiculture at the University of Agriculture in Lublin, Poland, and Professor of Apiculture at the University of Manitoba, Canada. Woyke was an internationally recognized authority in the field of beekeeping and made significant contributions to the study of honeybee genetics, behavior and ecology. He developed, among others, the method of artificial insemination of the queen bee, conducted research on eye color mutations in bees, and dealt with the history of Polish beekeeping. He was particularly known for his research on the Africanized honeybee, as well as his work on developing methods for the selective breeding of bees with desirable traits. Woyke published numerous scientific articles and books on beekeeping, including "Breeding Bees for Resistance to Diseases" and "Queen Rearing and Bee Breeding." He also served as the Editor of several scientific journals, including the Journal of Apicultural Research *and the* Journal of the International Bee Research Association. *Throughout his career, Woyke received numerous honors and awards for his contributions to the field of apiculture, including the Golden Cross of Merit from the President of Poland and the Apimondia Award for his contributions to international beekeeping.*

Contents

Foreword

In the 21st century, humankind is confronted with the Herculean task of providing food and environmental security to the expanding population, particularly in developing countries. The major challenge before agricultural scientists is to sustain higher productivity in agriculture without compromising the future in terms of resource degradation and depletion. This has necessitated accelerated efforts on the part of the agricultural scientists to sustain higher productivity in agriculture without compromising the future in terms of resource degradation and depletion. The world is facing food deficit coupled with instability of climatic cycles. The growing population pressure has hastened the environmental degradation, ultimately posing a threat to natural resources and fast approaching famine. The human population is expected to increase from 7.8 billion to 9.9 billion in 2050 and 11.2 billion in 2100, thereby doubling the food, feed and crop demands. The situation has further been aggravated due to pollinator decline worldwide, resulting in a pollination crisis adversely affecting crop productivity and due to lack of prophylactic progress in the conservation of biodiversity. Insect pollinators play an important role in producing crops in global agriculture. Pollinator-dependent crops contribute to maintaining a healthy variety in the human diet and often have a high market value, beneficial for local or regional economies. We dwell in the midst of yet another world food shortage that is exacerbated by escalating the prices world over. Coupled with the apparent instability of climate cycles in recent years, one-fourth of our growing human population is fast approaching famine. Concomitantly, this situation includes rampant declines in honeybee populations across three continents, for as yet incompletely resolved reasons, and with no remedy or end in sight. These problems have been further aided and abetted by a lack of prophylactic progress in the conservation of biodiversity and increased agricultural production. Pollinators and pollination are crucial in the functioning of almost all terrestrial ecosystems including those dominated by agriculture because they are in the front line of sustainable productivity through plant reproduction. A growing and increasingly human population is expected to double the demands for food production by 2050. Food production is also a central driver of global environmental change, contributing with negative effects on the climate, water resources, soils and the rich biodiversity the world harbors. Humanity stands in front of a great challenge to tackle these issues and needs to develop the food production system as a part of the solution for a sustainable world. Insect pollination is important for global food nutrition. One often-highlighted ecosystem service is insect pollination. The loss of biodiversity among pollinators has raised questions about whether the pollination services they provide are at risk. Pollination and food production that pollinators provide are threatened by land-use change, agricultural intensification, climate change, pesticide use, pathogens, genetically modified organisms and invasive species. The economic market value of pollination in crop production globally is estimated to be $235 billion to $577 billion annually. Among the various pollinating agents, honeybees play a very important role in pollinating various crops. The honeybee pollination not only results in higher yields; it also gives a better quality of produce, and the efficient pollination of flowers also serves to protect the crops against pests. Of the nine species of *Apis*, the giant honeybees *Apis dorsata* and *A. laboriosa* are the most spectacular, living in the open in huge colonies with exposed positions. They are one of the most important pollinators of agricultural, horticultural and thousands of plants growing in natural landscapes. One of the most outstanding traits of these honeybee species is ability to survive under extreme climatic conditions through their ability to migrate in areas with rich floral abundance and suitable climatic conditions. The greatest advantage of giant honeybees that they are important pollinators of several crops. Their long proboscis, large flight range, large number of field workers and habit of collecting large quantities of pollen and nectar make them the best among the honeybees for crop pollination. They are best honey producers ranging from 50–80 and 55–135 kg/colony per year in case of *A. dorsata* and *A. laboriosa*, respectively. The role of assumes utmost importance in the context of pollinator decline throughout the world, threatening stability of ecosystems and global food security. While *Apis dorsata* have a wide distribution range, covering the Indian subcontinent and Southeast Asia, *Apis laboriosa* is broadly restricted to the Himalayas. These honeybee species live sympatrically with other native honeybee species *A. florea*, *A. cerana* and introduced *A. mellifera.* Giant honeybees provide a source of livelihood in mountainous areas and marginal farmers, are excellent pollinators of tropical and subtropical crops—fruits vegetables, etc.—and are less prone to diseases and enemies.

To promote beekeeping as a sustainable option for rural development, crop production and biodiversity conservation, there is an urgent need to generate information on this important species. This book aims to promote a large, diverse, sustainable and dependable bee pollinator workforce that can meet the challenge for optimizing food production well into the 21st century.

The book *Role of Giant Honeybees in Natural and Agricultural Systems* by Dharam P. Abrol is the first of its kind that deals in details on varied aspects of biology, biogeography, reproductive biology, genetics and genomics, nesting biology. genetic diversity, pests, predators and pathogens, dance language, diversity and distribution, honey harvesting, nesting preferences, floral resources, safety from pesticides, nutritional aspects, population decline, foraging ecology, management and conservation, morphometrics, melittopalynological studies of floral resources. This work encourages utilization of a highly dependable pollinator that can help with food production in tropical and subtropical areas that are increasingly affected by climate change.

I congratulate the author for his efforts to provide this timely contribution.

Prof Dr Michal Woyciechowski
Institute of Environmental Sciences,
Jagiellonian University, Krakow, Poland

Preface

Disproportionately increasing human population at an alarming rate is a major challenge facing agriculture in the 21st century. The population has already crossed the 7 billion mark and expected to rise to 9 billion or more during the middle of this century, requiring raising of food productivity by some 70–100 percent. Production in the developing countries would have to be doubled as compared to present posing a great challenge on the part of the agricultural scientists to develop high yielding production technology and intensification in crop produce practices. Ecological intensification of agriculture is suggested as a way to reach higher crop yields without increasing inputs that may degrade the environment. Increased insect pollination in crops has been suggested to increase yields, but is rarely integrated in crop management. To determine the value of enhanced crop pollination as a means of ecological intensification, reliable estimates of how yield is affected by insect pollination are needed. It needs to be assessed how increasing crop pollination by adding honeybees to crops impacts the wild fauna of flower-visiting insects. One-third of our total diet is dependent, directly or indirectly, upon insect-pollinated plants. Despite good agronomic practices, the level of productivity of most of crops is far below expected. The low productivity of crops can be attributed to various factors such as heavy infestation of pests and diseases. Out of the various factors, one of the most important factors for low production of crops is due to the lack of proper pollination.

The loss of biodiversity among pollinators has raised questions about whether the pollination services they provide are at risk. Pollination and food production that pollinators provide are threatened by land-use change, agricultural intensification, climate change, pesticide use, pathogens, genetically modified organisms and invasive species. The honeybee pollination not only results in higher yields; it also gives a better quality of produce, and the efficient pollination of flowers also serves.

Most of the studies have concentrated on single honeybee species, the European honeybee *A. mellifera*, but recent decline in their populations throughout the world has caused a great concern for global food security. The number of commercial honeybee colonies has declined at an alarming rate. The dependence on single species is extremely dangerous for food security and survival of natural ecosystems. There is a need to promote, conserve and restore the natural native pollinators for sustainable use of pollinator diversity in agriculture and related ecosystems

In Asian countries, more than eight honeybee species are available which need conservation and restoration. Of all these honeybee species, the giant honeybees are the most spectacular of all honeybee species living in the open in huge colonies with exposed position. Rock bees may belong to two distinct species: *Apis dorsata* and *A. laboriosa* Smith. The latter species is the largest among the honeybees in size. It is darker in body color, builds bigger combs and has larger population than *A. dorsata*. It is common in the higher altitudes—between 1200m and 4100m—and is not seen in the tropical plains. *A. dorsata* is common in lower altitudes and in plains, and has a lighter orange brown or tawny body color. The nests of the rock bees are built in open hanging from a support from trees, rocks or buildings, and usually in aggregations.

The Asian giant honeybee, *Apis dorsata* F., and the Himalayan giant honeybee, *Apis laboriosa* S., are not only large in size, but are also distinctive by the extensive size of their colonies and the nests they build. *Apis dorsata* worker bees measure between 1.7cm and 2cm, with colonies that count up to 80,000 bees. Their nests measure up to 1m in height by more than 1.5m in length. Asian giant honeybee colonies frequently nest in dense aggregations, and several dozen nests can be seen on the same tree or cliff. While *Apis dorsata* have a wide distribution range, covering the Indian subcontinent and Southeast Asia, *Apis laboriosa* is broadly restricted to the Himalayas. *Apis laboriosa* Smith is the world's largest honeybee, measuring up to 3cm long. Although very little is known about the biology, *laboriosa* certainly exists under extreme ecological conditions. Each honeybee colony comprises a group of worker and drone bees with one queen, who live together to supply each other's needs and cooperate to raise the off springs. The honey hunters in Nepal harvest honey from nests made on cliffs by *Apis laboriosa*, the world's largest honeybee.

Rock bees are important pollinators of several crops. Its long proboscis, large flight range, large number of field workers and its habit of collecting large quantities of pollen and nectar make it the best among the honeybees for crop pollination. Despite their economic usefulness, biodiversity of giant Asian honeybees *is* suffering precipitous decline and they are threatened with extinction in its entire native habitat. The giant honeybees have considerably declined, due to the loss of food resources and natural nesting sites caused by deforestation and the fast urbanization and landscape fragmentation. Honey-hunting activities and removal of nests from urban habitats has threatened the survival of giant honeybees. During the past four decades, human population has increased more than twofold, exerting a tremendous pressure on the natural resources and the land especially for food, fuel and timber. As a consequence, vast forests have been converted into agricultural land and mountains have become barren due to ruthless cuttings and grazing, thus extensively destroying the food and habitat of several pollinators species. Along with these, use of chemicals has also greatly wiped out the population of natural pollinators, thus resulting in failure of reproduction in several cross-pollinated plant species, including the agricultural crops. The research on giant honeybees has often been ignored by beekeepers and scientists because of their ferocious nature and failure in their domiciliation. The information is available on different aspects but there is no comprehensive book exclusively devoted to *Apis dorsata*. In view of their high honey production potential and pollination of crops in natural and agricultural systems there is need for their conservation as a sustainable option for rural development, crop production and biodiversity conservation. The aim of this book is to fill the gap by providing detailed information on different aspects of giant honeybees leading to food security sustainability and environmental protection. This book deals in details on basic biology, biogeography, reproductive biology, genetics and genomics, nesting biology. genetic diversity, pests, predators and pathogens, dance language, diversity and distribution, honey harvesting, nesting preferences, floral resources, safety from pesticides, nutritional aspects, population decline, foraging ecology, management and conservation, morphometrics and melittopalynological studies of floral resources. Giant honeybees provides sources of livelihood in mountainous areas and marginal farmers, and they are excellent pollinators of tropical and subtropical crops, fruits, vegetables, etc., less prone to diseases and enemies. To promote beekeeping as a sustainable option for rural development, crop production and biodiversity conservation, there is an urgent need to generate information on this important species

The compilation of this book is unique in the sense that in the context of pollinator decline over the world, conservation giant honeybees will be a step for sustaining food security. This book provides a multi-authored and multi-national work. All the contributors deserve special appreciation for writing chapters in their respective fields in great depth with dedication.

I am extremely thankful to Upadhyay, Commissioning Editor and Jyotsna Jangra Editorial assistant, CRC Press, who took great pains and keen interest in publication of this book in a very impressive way. Last but not the least, my sincere thanks are due to my family members for their endurance and help while writing this book.

Jammu **Dharam P. Abrol**

Editor Biography

Dr Dharam P. Abrol, Dean, Faculty of Agriculture, Sher-e-Kashmir University of Agricultural Sciences, has served university in various capacities. He has specialized in pest management, honeybee management and pollination biology. He has authored 22 books, 12 manuals and published over 250 original research papers in various national and international journals. He has completed several externally funded collaborative research projects with international organizations in Poland and Switzerland. Prof Abrol has visited South Korea, Malaysia, Saudi Arabia and several other countries as a special invitee to these countries. In addition, he has received letters of appreciation from different organizations. Prof Abrol is a Fellow of the National Academy of Agricultural Sciences, India, and a Fellow of the Royal Entomological Society London, UK. He is a recipient of the Young Scientist Award (1992) —conferred by the Jammu and Kashmir State Council for Science and Technology —a prestigious state award for his outstanding contributions in the field of agricultural sciences. He is also the recipient of Pran Vohra Award (1993)—a prestigious Young Scientist Award conferred by the Indian Science Congress Association, Calcutta, for his outstanding and innovative research in the field of agricultural sciences. He was also conferred the Prof T. N. Ananthakrishnan Award 1997–1998 —a prestigious national award for his outstanding contributions in the field of entomology by the T. N. Ananthakrishnan Foundation, G.S. Gill Research Institute, Chennai. Prof Abrol won the Dr Rajendra Prasad Puruskar 1999–2000 award —a prestigious national award from the Indian Council of Agricultural Research, New Delhi, for his Hindi book on beekeeping entitled *MadhmakhiPalan-Sidhant Evam Vidhian*. He received the 11th Apicultural Association Award (2010) for outstanding contributions in apiculture. King Saud University conferred on him a gold medal for development of apiculture in Asia.

Contributors

Dharam P. Abrol
Dean, Faculty of Agriculture
Sher-e-Kashmir University of Agricultural Sciences & Technology
Chatha, Jammu –180 009 (J&K)

Tekalign Bergna
Department of Plant Medical
Andong National University
Andong, GB, 36729
South Korea

N.S. Bhat
Department of Apiculture
University of Agricultural Sciences
Bangalore, India

Axel Brockmann
National Centre for Biological Sciences
Tata Institute of Fundamental Research, Bellary Road
Bangalore 560065, India

Nandita Das
Centre for Life Sciences, Vidyasagar University
Midnapore—721102, India

Orawan Duangphakdee
Native Honeybee and Pollinator Research Center
King Mongkut's University of Technology Thonburi Ratchaburi Campus
Bangkok, Thailand

M.S. Engel
Division of Entomology
Natural History Museum and Department of Ecology & Evolutionary Biology
University of Kansas
Lawrence, Kansas 66045, USA
Division of Invertebrate Zoology
American Museum of Natural History
New York, New York 10024, USA

B. Fakrudin
Professor and Head
Dept. of Biotech & Crop Improvement
University of Horticultural Sciences
Post Graduate Centre, GKVK Post
Bengaluru—560 065, Karnataka (State), India

Sampat Ghosh
Agricultural Science and Technology Research Institute
Andong National University
South Korea

Chuleui Jung
Department of Plant Medical
Andong National University
Andong, GB, 36729, South Korea

S. Kahono
Zoology, Research Center for Biology
Indonesian Institute of Sciences (LIPI)
Jl. Raya Jakarta-Bogor Km. 46
Cibinong, Bogor 16911, West Java, Indonesia

Prakash Karmakar
Department of Botany and Forestry
Vidyasagar University
Midnapore, India

Patrick L. Kohl
Department of Animal Ecology and Tropical Biology, Biocenter
University of Würzburg
Am Hubland, 97074 Würzburg, Germany

A.K. Bharath Kumar
Department of Apiculture
University of Agricultural Sciences
Bangalore, India

C. Kumar
Dept. of Biotech & Crop Improvement
University of Horticultural Sciences
Post Graduate Centre, GKVK Post
Bengaluru—560 065, Karnataka (State), India

T.N. Lakshmidevamma
Dept. of Biotech & Crop Improvement
University of Horticultural Sciences
Post Graduate Centre, GKVK Post
Bengaluru—560 065, Karnataka (State), India

J.S.A. Lamerkabel
Agrotechnology Study Program
Faculty of Agriculture
Pattimura University, Jl. Ir. M. Putuhena, KampusPoka
Ambon 97233, Indonesia

Ujjwal Layek
Department of Botany and Forestry
Vidyasagar University
Midnapore, India

S. Mohankumar
Department of biotechnology
Tamil Nadu Agricultural University
Coimbatore, India

Rajib Mondal
Department of Botany & Forestry, Vidyasagar University
Midnapore—721102, India

N. Nagaraja
Department/College UGC
Human Resource Development Centre (HRDC) Jnana Bharathi Campus
Bangalore-560 056, Bangalore University,
India

Saeed Mohamadzade Namin
Agricultural Science and Technology Research Institute
Andong National University
Andong, South Korea

T. Neethu
Department of Apiculture
University of Agricultural Sciences
Bangalore, India

D. Peggie
Zoology, Research Center for Biology
Indonesian Institute of Sciences (LIPI), Jl. Raya Jakarta-Bogor Km. 46
Cibinong, Bogor 16911
West Java, Indonesia

S. Pradeep
Department of Agricultural Entomology
Tamil Nadu Agricultural University
Coimbatore—641 003 India

T. Sonai Rajan
Department of Biotechnology
Tamil Nadu Agricultural University
Coimbatore, India

A.J. Solomon Raju
Department of Environmental Sciences
Andhra University
Visakhapatnam 530003, India

K.B. Rakesh
Dept. of Biotech & Crop Improvement
University of Horticultural Sciences
Post Graduate Centre, GKVK Post
Bengaluru—560 065, Karnataka (State), India

Atsalek Rattanawannee
Department of Entomology
Faculty of Agriculture, Kasetsart University
Yao Chatuchak, Bangkok, Thailand

Preecha Rod-im
Native Honeybee and Pollinator Research Center
King Mongkut's University of Technology Thonburi Ratchaburi Campus
Bangkok, Thailand

Ruchita Thimmarayappa
Dept. of Biotech & Crop Improvement
University of Horticultural Sciences
Post Graduate Centre, GKVK Post
Bengaluru—560 065, Karnataka (State), India

Benjamin Rutschmann
Department of Animal Ecology and Tropical Biology, Biocenter
University of Würzburg, Am Hubland, 97074
Würzburg, Germany

Parul Sharma
Department of Computer Science & IT
University of Jammu, India

B.V. Shwetha
Department of Apiculture
University of Agricultural Sciences
Bangalore, India

M.R. Srinivasan
Professor
Department of Agricultural Entomology
Tamil Nadu Agricultural University
Coimbatore—641 003 India

S. Subiyakto
Indonesian Sweetener and Fibre Crops Research Institute
Ministry of Agriculture Republic of Indonesia
Jl. Raya Karangploso, Malang 65152, East Java, Indonesia

Ratna Thapa
Zoology Department
Tri-Chandra M. Campus Tribhuvan University
Kathmandu, Nepal

J. Ugalat
Dept. of Biotech & Crop Improvement
University of Horticultural Sciences
Post Graduate Centre, GKVK Post
Bengaluru—560 065, Karnataka (State), India

Jerzy Woyke
Agricultural University Apiculture Division
166 Nowoursynowska 02–787 Warsaw, Poland

1 Introduction

Dharam P. Abrol

1.1 INTRODUCTION

Honeybees have been known to humankind since prehistoric times (Singh 1962). The winged creature finds mention in almost all the religious books of the world (Singh 1962). Besides, production of honey and beeswax, giant honeybees are important pollinator of several crops. Their long proboscis, large flight range, large number of field workers and habit of collecting large quantities of pollen and nectar make them the best among the honeybees for crop pollination. The honeybee is one of the most well studied social insects and there is enormous information available mainly in the field of apiculture. However, most of the work has been done on *Apis mellifera*, the European honeybee. The research on giant honeybees has often been ignored by beekeepers and scientists because of their ferocious nature and failure of their domiciliation. The information is available on different aspects but there is no comprehensive book exclusively devoted to *Apis dorsata* and *A. laboriosa*.

The Asian giant honeybee, *Apis dorsata F.,* and the Himalayan giant honeybee, *Apis laboriosa*, are not only large in size but are also distinctive by the extensive size of their colonies and the nests they build (Figure 1.1). *Apis dorsata* worker bees measure between 1.7 and 2 cm with colonies that count up to 80,000 bees. Their nests measure up to 1 m in height by more than 1.5 m in length (Morse and Laigo 1969; Paar et al. 2004). Asian giant honeybee colonies frequently nest in dense aggregations (Koeniger and Koeniger 1980; Lindauer 1955) and several dozen nests can be seen on the same tree or cliff (Oldroyd et al. 2000). While *Apis dorsata* have a wide distribution range, covering the Indian subcontinent and Southeast Asia (Pauly 2015; Ruttner 1988), *Apis laboriosa* is broadly restricted to the Himalayas (Otis 1996; Trung et al. 1996). Nesting at altitudes of 1,200–3,500 m (Woyke et al. 2001), the presence in Southeast Asia of this species is restricted to the mountainous areas in northern Laos, Myanmar and Vietnam. The species is distributed almost continuously over a distance of more than 2,500 km along the pan-Himalaya region from Uttarakhand, India, eastward through Nepal, Sikkim and northern West Bengal (Darjeeling), Bhutan, northeastern India, Yunnan and southern Tibet in China, and the northern portions of Myanmar, Laos and Vietnam, southward along the Arakan Mountains in eastern Arunachal Pradesh, Nagaland, Manipur, and Mizoram (India) to Matupi in west-central Myanmar and the Shillong Hills of Meghalaya (Figure 1.1).

DOI: 10.1201/9781003294078-1

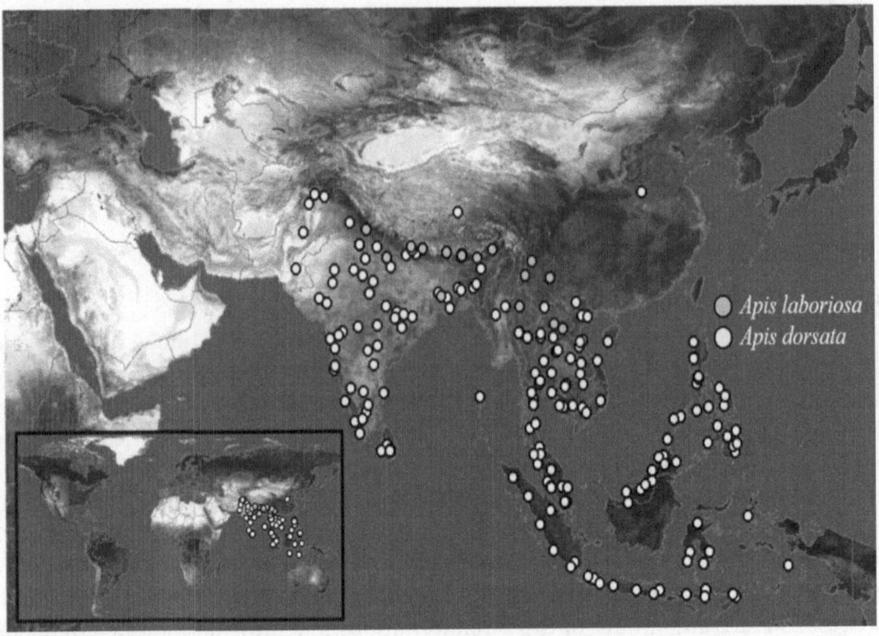

FIGURE 1.1 Global distribution of *Apis laboriosa* in southern Asia which shows some overlap with the related species *Apis dorsata*

(Source: Gregory and Jack 2022).

1.2 THE GIANT HONEYBEES *APIS DORSATA* AND *APIS LABORIOSA*

1.2.1 *APIS DORSATA*

The giant honeybees are certainly the most spectacular of all the honeybee species: an individual bee of the length of a hornet, living in the open in huge colonies, frequently in exposed positions, the motionless bees with spread wings on the surface of the cluster arranged in strict regularity yet ready at any time to launch fierce mass attacks against a supposed enemy within seconds. *A. dorsata* relies on its strength based on a numerous society of large individuals with high defence potential (Seeley et al. 1982). This species is the most ferocious stinging insect in the world (Morse and Laigo 1969) but can be conditioned to live close to humans, nesting on walls of buildings in large towns (Lindauer 1956; Morse and Benton 1967; Reddy 1980).

A. dorsata occurs throughout continental Asia and oceanic Asia, including the Philippines and Sulawesi, Indonesia. In terms of altitudinal distribution, 85% were below 1,000 m and the balance between 1,000 and 3,000 m. *A. laboriosa* mainly occurs between 2,500 and 4,000 m. Both species differ significantly in their altitudinal distribution; however, they are partially sympatric during different seasons (Roubik et al. 1985; Otis 1996). The differences between two giant honeybee species are given in Table 1.1.

TABLE 1.1
Species-Specific Characteristics of Giant Honeybees *A. dorsata* and *A. laboriosa*

Characteristic	*Apis dorsata*	*Apis laboriosa*
Forewing length (mm)	12.5–13.5	14.2–14.8
Cubital index	6.1–9.8	
Tomenta	3–6	3–6
Hind wing: extension of radial vein	Present	Present
Drone	Four pairs of very long thin; cornua short bulb	?
Endophallus	Thick pad of sturdy branched hair	?
Basitarsus 3	Solid	Solid
Behaviour		
Capping of drone cells	Solid	Solid
Nest	Single comb encircling twig to form a "dance floor" fixed with cell bases	Single big comb fixed at bottom side of branch or rock, fixed with midrib
Communication	Sun-oriented dance on platform open to the sky	Sun-oriented dance on vertical comb open to the sky
Commencement of activity	18°C	10°C
Altitudinal variation	1,000–3,000 m above sea level	2,500–4,000 m above sea level
Dorso-ventral defence body twisting (DBT)	*A. dorsata* subspecies by about 10° more. Than *A. laboriosa*	*A. laboriosa* twisted the thorax by 55°
Raising of abdominal tip	Workers *A. dorsata* by about 20° higher than *A. laboriosa*	*A. laboriosa* workers raised the tip of the abdomen by 90°
Twisting of body	*A. laboriosa* twisted the body together with wings folded over the abdomen	*A. dorsata* raised the abdomen between spread wings
Distribution	Sympatric	Sympatric
Hygenic behaviour	*A. dorsata* did not open undamaged sealed cells containing brood killed artificially or naturally.	*A. laboriosa* did not open sealed cells with brood killed by mites and, presumably, some brood diseases.
Honey production kg/colony/year	50–80 kg	55–132 kg

Both *A. dorsata* and *A. laboriosa* are closely related yet they have enormously different dispersal characteristics. *A. dorsata* occurs on all the Philippine islands, Palawan and the Calamian Island groups. Mardan (1989) reported that *A. dorsata* crosses the Strait of Malacca between Sumatra and the Malay Peninsula, which is a distance of about 50 km.

1.2.2 *Apis laboriosa*

The natural range of distribution of *A. laboriosa* barely overlaps that of *A. dorsata*—and then only in different seasons (Otis 1996; Underwood 1990a). However, Joshi et al. (2008) collected both species at the same sites in Nepal, continuous from Uttar Pradesh to northern Laos. *A. laboriosa* is distributed in the lower reaches of the Himalayas and extends from northwestern Nepal along the mountains, through Bhutan, Sikkim, northeastern India, Myanmar and southern China, northern Laos and Vietnam (Sakagami et al. 1980; Roubik et al. 1985; Otis 1996; Takahashi and Nakamura 2003). The climatic zones include subtropical to the east and highland to the west. The vegetation is tropical, moist deciduous forest.

1.2.3 Distribution of *Apis dorsata*

Apis dorsata is found throughout the southern countries of Asia, including Malaysia, Indonesia and the Philippines. Its north–south distribution ranges from southern parts of China to Indonesia. The greatest number of *Apis dorsata* colonies are found in dense forest areas or on cliffs, but nests are occasionally found in urban areas on building ledges. In India, this bee is found up to a height of 1,220 m above sea level (Singh 1962).

1.2.4 The Nest and Nest Site of the Giant Asian Honeybee

The comb of *Apis dorsata* is always attached to the underside of overhanging rocks; suspended from the more or less horizontal branches of tall trees (Verma 1990), or from the eaves of tall buildings (Butler 1954). Deodikar et al. (1977) reported 45% of colonies on terrestrial supports while about 55% were arboreal. Several colonies may be found even on one tree. A hive is more or less semi-circular. The size of a single comb of *A. dorsata*, depending upon the season and stage of development of a colony, measures 1.5–2 m from side to side and 0.6–1.2 m from top to bottom (Deodikar et al. 1977). A single comb may be as much as 5–6 feet long and 3 feet deep (Butler 1954). A comb is made up of two hexagonal wax cell layers fixed back to back. It has a definite zone for storage of honey and brood. The upper portions of the comb store honey and pollen and is generally 10–25 cm thick. Below this storage area is the brood nest (Singh 1962). The colonies are perennial and the development of new colonies takes place by swarming (Butler 1954).

The bees in the nest perform various functions based on the hierarchy of the bees. Some bees construct the comb, take care of the brood and process honey. Most of the bees in a colony (about 80–90%) make a thick multi-layered cover called protective curtain (Morse and Laigo 1969). There is an air space between the two protective layers of the nest, which helps to regulate the temperature of the nest. Bees of the protective curtain remain motionless with their wings spread out. In the lower part of the hive facing the sun, there is an active zone called the "mouth" portion. Here the bees are not uniformly oriented and have their heads directed outwards. Bees regularly undertake landing or taking off flights from the comb for foraging activities. Communication dances are visible in the "mouth" area. It has been found that

TABLE 1.2
Age Related Activities in Honeybees

Duties of Workers	Age of Workers (Days)
Cell cleaning; keeping brood area warm	0–4
Feeding larvae; attending queen	3–14
Fanning and ventilation control (few)	3–22
Polishing cells; packing pollen	4–24
Orientation flights, around midday	5–18
Guard duties; clearing out debris	10–18
Collecting pollen (some bees)	10–26
Collecting nectar	12–35

Source: Brown (1988)

the "mouth" changes its location depending upon the food source and other factors, and can also be utilized to find out the "mood" of the colony. If the "mood" is right, the colony can be handled bare-handed and received no sting, even after a thorough disturbance of the colony and when "out of mood", the colony cannot even be approached, let alone handled, and many a honey collector ignorant of this behaviour learnt about it the hard way. How to detect this "mood", however, remains a mystery. The adult occupants of the nest are thousands of worker bees, a single queen and several hundred drones. The worker bee is light brown in colour. The queen is darker in colour than the workers are and broader by about 2 mm in the region (Singh 1962). The drone is black in colour and has a blunt abdomen without a sting. The drones are not in the hive in the winter. Worker bees are sterile females, arising from fertile eggs, yet are sexually immature because of glandular changes induced with a modified diet after their second day as a larva (Brown 1988), follow a strict division of labour according to their age; starting from cleaning to foraging as described in Table 1.2.

1.2.5 Economic Value of *A. dorsata*

1.2.5.1 Production of Honey and Beeswax

In India, *A. dorsata* accounted for about 70% of the honey in 1951 (Ghatge 1951) and for 80% of the beeswax production in 1961 (Phadke 1961). A colony may yield 40–70 kg of honey (Dutta et al. 1983). Generally, the average honey yield per colony is about 5–10 kg. Singh (1962) reported that a single colony might yield up to 37.3 kg of honey during a year. Many tribal people in India depend either wholly or partially on collection of honey and beeswax for their livelihood.

1.2.5.2 Pollinator Importance of *A. dorsata*

Hundreds of species of agricultural plants in over 40 plant families worldwide are pollinated, at least in part, by bees and other insects (Crane and Walker 1984; Free

1970; McGregor 1976; Southwick and Southwick 1992). Many agricultural crop plants require out-crossing to produce viable fruit and others show hybrid vigour with the production of a better crop (Moritz and Southwick 1992). The yields of many fruit, vegetable, seed and nut crops would drop substantially without pollination by honeybees (Morse 1988; Olmstead and Wooten 1987). It has been estimated that honeybees alone account for as much as 80% or more of all crop insect pollination (Camazine and Morse 1988). *Apis dorsata* is listed as an important pollinator of plants in several reports (Mann and Singh 1983). Pollination efficiency is related to flight range and among the honeybees, *A. dorsata* has the longest flight range (8,500 m) (Singh 1962). Because of migratory activity, the foraging range of *A. dorsata* is large. *A. dorsata* is specialized in exploring rich nectar sources even at distances further than 5 km (Koeniger and Vorwohl 1979). These bees can adapt to extreme climatic conditions. They have been observed (Singh 1962) to begin the days work earlier and stop it later than *A. indica.*

Other uses include curing of arthritis using bee venom. The dreaded bee sting has the mysterious quality of healing muscular and nervous pains and aches of sciatica, rheumatism and arthritis (Singh 1962). Bee venom therapy for multiple sclerosis has been reported by Mraz (1995). However, the use of *A. dorsata* for this purpose has not been reported as yet. Bees have also been used as a weapon of war through the ages. In World War I, infuriated swarms were used to hamper the advance of forces in Belgium. In our country, many a political meeting has ended (Singh 1962) in pandemonium after a stone has been thrown by a mischievous opponent in a colony of wild bees hanging from a branch of a nearby tree.

1.2.5.3 Life Cycle

The life cycle and life stages of *Apis dorsata* are the same as those in all other honeybee species such as the life cycles of *Apis mellifera* and *Apis cerana.* They are Honeybees are holometabolous and undergo four separate life stages (egg, larva, pupa, adult). Compared to *Apis mellifera* workers, *Apis dorsata* seems to live significantly longer, especially during migration swarms, when workers will sometimes travel more than two months to reach a new destination and produce a new generation of bees (Koeniger et al. 2010).

1.2.5.4 Biology

Apis dorsata builds open nests which hang from thick branches of trees or under cliffs of rocks. The nests are single, measuring about 150 cm in length and 70 cm in width. The comb may contain up to 100,000 worker bees. This curtain of comb is of several layers of bees thick, forming a protective covering around the brood to save it from environmental diversities. This curtain protects the nest during storms, wind and rain.

Because of the larger body size, *A. dorsata* has greater flight and foraging range as compared to other honeybee species. *Apis dorsata* colonies undertake seasonal migrations to exploit nectar and pollen resources throughout the year (Oldroyd et al. 2000). Interestingly, the same colony has been observed to return to the exact same branch six months later, even though the bees that knew of the old nesting location would have died long before.

Apis dorsata has been considered as the most dangerous honeybee because of its highly aggressive defensive behaviour—even more than African honeybee (Ellis and Ellis 2009). The bees attack en masse to perceived threat and the stinging is very painful. The stinger is 3 mm long and can easily penetrate the skin and pump the venom into the skin. These bees have evolved a unique method of deterring predators by performing shimmering moments. Shimmering involves a wave which moves across the surface in seconds as the bees raise their abdomens in a sequential manner. The visual display helps to intimidate threats from predatory wasps, birds and mammals.

1.2.5.5 Economic Impact of *A. dorsata*

Apis dorsata is one of the most important as honey producers and for pollination of crops and plants in natural ecosystems. Since the colonies are not transported like *Apis mellifera* and *Apis cerana* for pollination purposes, they are mainly used for harvesting of honey. They are not used for managed or planned pollination, but most of the crops like cotton, mango, coconut, coffee, pepper, star fruit and macadamia are heavily dependent on these bees. *Apis dorsata* is natural host of the *Tropilaelaps* mite, but *A. mellifera* is parasitized where both *Apis dorsata* and *Apis mellifera* are present. This poses a significant threat to the commercial beekeeping industry (Mortensen et al. 2014).

1.2.5.6 Management of *A. dorsata*

Apis dorsata, an open nesting honeybee species, has not been managed so far as other honeybee species. Attempts made to domesticate *Apis dorsata* colonies in wooden Langstroth hives used for *Apis mellifera* and *Apis cerana* have failed (Koeniger et al. 2010) because these bees are not evolved to live in dark cavities. They are instead used to obtain honey, wax and brood. Honey collection from *Apis dorsata* colonies is the common practice in areas wherever this bee is found. Traditional honey collection is done mostly on moonless nights to minimize the number of flying bees once the colony is disturbed. Makeshift ladders or ropes are used to reach the top of the trees or cliffs. Smoking is done to drive away the bees off the comb. Honey hunters normally used to harvest the whole nest to obtain both the honey and the brood, but recently, efforts are being made to just cut away the sections of honeycomb instead of destroying the whole colony. Rafter beekeeping is also being practiced in certain regions to cause less damage to the bees.

1.2.5.7 Biological Characteristics

Foraging is generally a diurnal activity, but in *Apis dorsata*, night time foraging is also reported (Dyer 1985). Besides olfaction, vision is an important part of foraging. The visual organ of honeybee comprises two compound eyes and three simple eyes (in the centre of the head). These simple eyes are called ocelli (sometimes also referred to as dorsal ocelli). They look like small inconspicuous black beads. Honeybees except *A. dorsata* have been domesticated as they are one of the most beneficial insects for man. But all the efforts so far to domesticate the wild honeybees (*A. dorsata*) have failed (Verma 1992). It is only possible to train them for some specific purpose. *A. dorsata* has been trained previously using scented sucrose solution

(100~1/l clove oil, 1.5 M sucrose solution) (Rathore and Wells 1995). Training utilizes the learning capability (visual as well as olfactory) of bees (Frisch 1914). In a typical training experiment, bees associate odour or colour as a marker of reward (e.g., sucrose solution) (Gould and Gould 1988). It is seen that the basic phenomenology of associative learning is remarkably similar in vertebrates and invertebrates (Gould and Towne 1988).

The defensive response is elicited by alerted or stinging workers, which release volatile alarin pheromones from their Koshevnikov's glands and the setose membrane near the sting apparatus (isopentyl acetate) along with mandibular gland secretions (2-heptanone) (Boch and Shearer 1962, 1966; Boch and Rothenbuhler 1974; Koeniger et al. 1979; Southwick and Moritz 1987). The intensity of defensive behaviour is dependent on external environmental factors (aggressiveness is greater under high temperature and high humidity conditions; under cool, overcast and windy conditions, aggressiveness is less), as well as the genetic makeup of the colony (Schua 1952; Crewe 1976; Collins 1981; Southwick and Moritz 1987). Attacking behaviour could also be affected by the time of day (Morse and Laigo 1969).

On comparing the defensive behaviour of three Asian honeybees *(A. dorsata, A. florea* and *A. cerana*), Seeley (1985) pointed out that *A. dorsata* is highly defensive. This results from two obvious reasons: its large size (weighing 1.5–2 times as much as *A. mellifera* and 3–5 times as much as *A. florea* or *A. cerana*) and its easily visible nest which makes it more prone to attack by predators (Baroni Urbani 1979). *A dorsata* often constructs aggregations of nests in particularly favourable nest sites, usually on large-diameter smooth trees, rock overhangs or man-made structures (Seeley et al. 1982). The fact that the *A. dorsata* nest aggregates are within a few metres of each other provides an additional mechanism of cooperative defence against large or persistent predators. There is always a very little response from one isolated bee or a few bees and defensive behaviour is dependent on the number of bees in the group (Southwick and Morse 1985).

Production of hissing sound described as "shimmering behaviour" by Butler (1954) is evoked by various mechanical shocks such as a sudden blow upon the hive, the abrupt opening of the hive lid, etc., but occasionally without any apparent external causes. When an *A. dorsata* nest is smoked gently, there is first a rippling movement across the nest and then a low roar, unlike any noise made by bees in the nest of *A. mellifera* (Morse and Laigo 1969). Lindauer (1956) has observed this first stage in alarm by *A. dorsata*. This bee is more easily aroused under the unfavourable weather conditions or the vernal and autumnal dearths; that is, situations in which *A. mellifera* colonies increase their aggressiveness (Sakagami 1960). Sakagami (1960) found that the repetition of the same shock raises the threshold to evoke this response. With appropriate stimuli, shimmering is usually repeated 4–5 times and the time interval between each shimmering lasts 3–5 seconds, often with a more prolonged delay.

Bees shrug their wings along with the production of sound. The roar, presumably made by bees moving their wings, is the second stage in the alarm system (Morse and Laigo 1969). With the wing stroke, bees push the body forward, simultaneously without any locomotion. This communal reaction appears as wave across the comb. Sakagami (1960) reported a momentary quietness of the comb surface during this reaction as other regular activities like walking, running or dancing ceases. Roepke (1930) and Butler (1954) described similar behaviour in *A. dorsata* when disturbed by intruders such as hornets or men. If *A. dorsata* bees are provoked, unlike other *Apis* species, they may attack in large numbers. Attacking bees fly in a cloud usually 3–6 m in diameter, most of them into open, sunny areas but some close to the ground, even in shady places searching for intruders. The attacking force usually comes from the mouth zone. The maximum attacking force probably includes not more than 10% of a colony population (Morse and Laigo 1969).

1.2.6 *Apis laboriosa*

This species is one of the most important pollinators at higher altitudes of the Himalayas. It has a restricted distribution along the Himalayas and neighbouring mountain ranges of Asia. It is distributed from the eastward mountains of northern Vietnam to southward along the Arakan Mountains to west-central Myanmar, into the Shillong Hills of Meghalaya, India and northwestward in Uttarakhand, India. This species normally occurs at elevations from 1,000–3,000 m above sea level; however, during summer may be found of 850 m above sea level and colonies may maintain their nests throughout the winter. Besides, three regions in Arunachal Pradesh, India, and nine locations in northern Vietnam have been observed where workers of *A. laboriosa* and *A. dorsata* foraging sympatrically. This species stands poorly understood because of nesting at inaccessible cliffs in the Himalaya. (Cronin 1979; Sakagami et al. 1980; Roubik et al. 1985; Underwood 1990a; Joshi et al. 2004; Gogoi et al. 2017).

1.3 SYMPATRIC OCCURRENCE OF *APIS LABORIOSA* WITH *APIS DORSATA*

A. laboriosa was found to forage sympatrically with *A. dorsata* in five different sites in three regions of Arunachal Pradesh in northeastern India. Nyaton Kitnya et al. (2020) added these distribution sites in addition to those reported by Otis (1996). They also added many additional localities for this species in Uttarakhand in northern India, the eastern portion of Nepal, all of Bhutan and much of Arunachal Pradesh in northeastern India, indicating that this species is widespread over that region. Distribution extends to several eastern provinces in northern Vietnam (Trung et al. 1996) and southward for 600 km in the Arakan Mountains which include Patkai Range, Naga Hills, Mizo Hills of Nagaland, Manipur and Mizoram to 21.7° N latitude in the Chin Hills of Myanmar. They also reported occurrence of A. *laboriosa* for the first time from the Shillong Plateau in Meghalaya, India.

Apis laboriosa is notably absent from the western third of Nepal, from 80.5° E to 82.6° E longitude, which may be due to relatively dry climate there. Further studies may reveal the occurrence of this species in some more mountainous areas of Asia. These include northeastern Myanmar (Kumon Range and Gooligong Mountains), eastern Myanmar (much of Shan State), northern Laos (Annam Highlands and Xiangkhoang Plateau) and possibly extreme northern Thailand (Doi Pha Hom Pok

National Park) and the valleys of the Mekong, Yangtze, Yalong and Dadu rivers that extend into the southeastern edge of the Tibetan Plateau and northeastern Punjab, Pakistan and western Jammu, India. *A. laboriosa* is expected to be present in some areas of Pakistan.

Khan et al. (2014) reported several specimens of giant honeybees they collected in Murree, Pakistan (33.92° N, 73.40° E) at an elevation of ~2,300 m above sea level as "*A. dorsata*", despite the general understanding that *A. dorsata* lives below 1,200 m above sea level elevation in Pakistan (Muzaffar and Ahmed 1990). The yellow-rumped honeyguide (*Indicator xanthonotus*), which is generally found associated with *Apis laboriosa* combs (Cronin and Sherman 1976; Underwood 1992; Inskipp et al. 2008), was observed with giant honeybees in Muree which may be indicative of presence of *A. laboriosa* in the region. Mutharaman et al. (2013) analyzed samples of *A. dorsata* from different regions of Jammu division and found that samples collected from Poonch, Jammu, India (33.82° N, 74.12° E), just 60 km to the east of Murree, differed markedly in morphometric analyses from other *A. dorsata* specimens analyzed from Jammu and the rest of India. This finding further needs to be confirmed to determine existence of *Apis laboriosa* in this region which would help to extend its distribution another 400–500 km northwestward.

A. laboriosa shows several unique characteristics that seem to be related to its adaptations to living in mountainous habitats. *A. laboriosa* and *A. dorsata* differ in behaviour such as thermoregulation of thoracic temperature during flight (Underwood 1991; Woyke et al. 2012), dorso-ventral flipping of abdomen to stabilize body temperature (Woyke et al. 2008) and mating flight times. Mating in *A. laboriosa* mostly occurs at early afternoon (Underwood 1990b), compared to after sunset in *A. dorsata* (Tan et al. 1999; Otis et al. 2000). They also differ in their dance communication (Kirchner et al. 1996), pheromonal chemistry (Blum et al. 2000) and body movements related to defensive (Woyke et al. 2008).

1.4 CONCLUSION

Worldwide pollinator declines have increased the urgency to survey abundances of pollinators and to study their biology and ecology for their conservation. Asian honeybee species like *A. laboriosa*, with a restricted distribution in areas difficult to access, are dramatically understudied; there is a need of revised description of the distribution of the Himalayan giant honeybee, *Apis laboriosa*, an important pollinator species in the Himalayas (Batra 1996). Numerous reports on *A. laboriosa* indicate that this honeybee shows specific adaptations to living in high elevation mountainous areas compared to other more tropical honeybee species. Detailed studies on its biology promise to provide interesting insights into the evolutionary history and plasticity of honeybee physiology and social behaviour. Locations where *A. laboriosa* and *A. dorsata* co-occur temporally, like Arunachal Pradesh and Vietnam, are particularly suitable regions for future studies. Likewise, there is a need for conservation of *A. dorsata*, whose population have declined considerably. Moratoriums on destructive harvesting of giant bee nests wherever they exist need to be legally enforced.

REFERENCES

Baroni Urbani, C. 1979. A statistical table for the degree of coexistence between two species. *Oecologia* 44(3): 287–289.

Batra, S.W.T. 1996. Biology of *Apis laboriosa* Smith, a pollinator of apples at high altitude in the great Himalaya range of Garhwal, India (Hymenoptera: Apidae). *Journal of the Kansas Entomological Society* 69: 177–181.

Blum, M.S., Fales, H.M., Morse, R.A., and Underwood, B.A. 2000. Chemical characters of two related species of giant honeybees (*Apis dorsata* and *Apis laboriosa*): Possible ecological significance. *Journal of Chemical Ecology* 26(4): 801–807.

Boch, R., and Rothenbuhler, W.C. 1974. Defensive behavior and production of alarm pheromone in honeybees. *Journal of Apicultural Research* 13(4): 217–221.

Boch, R., and Shearer, D.A. 1962. Identification of geraniol as the active component in the Nassanoff pheromone of the honey bee. *Nature* 194: 704–706.

Boch, R., and Shearer, D.A. 1966. Iso-pentyl acetate in the stings of honeybees of different ages. *Journal of Apicultural Research* 5: 65–70.

Brown, R. 1988. *Honey bees: A guide to management.* Trafalgar Square Publishing, The Crowood Press Ltd. Marlborough, Wiltshire, London.

Butler, C.G. 1954. *The world of the honey bee.* 1st edn. Collins, London.

Camazine, S., and Morse, R.A. 1988. The Africanized honeybee. *American Scientist* 76: 464–472.

Collins, A.M. 1981. Effects of temperature and humidity on honeybees response to alarm pheromones. *Journal of Apicultural Research* 20: 13–18.

Crane, E., and Walker, P. 1984. *Pollination directory for world crops.* International Bee Research Association, London, United Kingdom, 183.

Crewe, R.M. 1976. Aggressiveness of honey bees and their pheromone production. *South African Journal of Science* 72: 209–212.

Cronin Jr, E.W. 1979. *The Arun: A natural history of the world's deepest valley.* Houghton-Mifflin, Boston.

Cronin Jr, E.W., and Sherman, P.W. 1976. A resource-based mating system: The orange-rumped honeyguide. *Living Bird* 15: 5–32. https://eurekamag.com/research/004/619/004619312.php.

Deodikar, G.B., Ghatge, A.I., Phadke, R.P., Mahindre, D.B., Kshirsagar, K.K., Muvel, K.S., and Thakar, C.V. 1977. Nesting behaviour of Indian honeybees. III. Nesting behaviour of *Apis dorsata* Fab. *Indian Bee Journal* 39: 1–12.

Dutta, T.R., Ahmed R., and Abbas, S.R. 1983. The discovery of a plant in Andaman Islands that tranquilises *Apis dorsata*. *Bee World* 64: 158–163.

Dyer, F.C. 1985. Nocturnal orientation by the Asian honey bee, *Apis dorsata*. *Animal Behaviour* 33(3): 769–774.

Ellis, J.D., and Ellis, A. 2009. African honey bee, Africanized honey bee, killer bee, *Apis mellifera scutellata* Lepeletier (Insecta: Hymenoptera: Apidae). *IFAS Extension, Entomology and Nematology, EENY* 429, December 8.

Free, J.B. 1970. *Insect pollination of crops.* Academic Press, London, 544.

Frisch, K.R. 1914. Der Farben- und Formensinn der Bienen. *Zoologische Jahrbücher (Physiologie)* 35: 1–188.

Ghatge, A. 1951. The bees of India. *Indian Bee Journal* 13: 88.

Gogoi, H., Tayeng, M., and Taba, M. 2017. Pan-Himalayan high altitude endemic cliff bee, *Apis laborisa* Smith (Hymenoptera: Apidae): A review. *Proceedings of the Zoological Society* 72: 3–12.

Gould, J.L., and Gould, C.G. 1988. *The honey bee.* W.H. Freeman, New York, 231.

Gould, J.L., and Towne, W.F. 1988. Honey bee learning. *Advances in Insect Physiology* 20: 55–86.

Gregory, M., and Jack, C. 2022. Himalayan giant honey bee, cliff honey bee (suggested common names) *Apis laboriosa* Smith (Insecta: Hymenoptera: Apidae). *UF/IFAS Extension Eeny* 777: 1–7. https://doi.org/10.32473/edis-IN1348-2022.

Inskipp, C., Inskipp, T., Winspear, R., Collin, P., and Robbin, A. 2008. *Bird survey of the Kanchenjunga conservation area*. Report to Critical Ecosystem Partnership Fund. Bird Conservation Nepal and Royal Society for the Protection of Birds. Bird Conservation Nepal, Kathmandu, and Royal Society for the Protection of Birds, Sandy, UK, April. http://himalaya.socanth.cam.ac.uk/collections/inskipp/2008_005.pdf.

Joshi, P.C., Kumar, K., and Arya, M. 2008. Assessment of insect diversity along an altitudinal gradient in Pinderi Forests of Western Himalayas, India. *Journal of Asia-Pacific Entomology* 11: 5–11.

Joshi, S.R., Ahmad, F., and Gurung, M.B. 2004. Status of *Apis laboriosa* populations in Kaski district, Western Nepal. *Journal of Apicultural Research* 43: 176–180.

Khan, K.A., Ansari, M.J., Al-Ghamdi, A., Sharma, D., and Ali, H. 2014. Biodiversity and relative abundance of different honeybee species (Hymenoptera: Apidae) in Murree-Punjab, Pakistan. *Journal of Entomology and Zoology Studies* 2(4): 324–327.

Kirchner, W.H., Dreller, C., Grasser, A., and Baidya, D. 1996. The silent dance of the Himalayan honeybee, *Apis laboriosa*. *Apidologie* 27: 331–339.

Kitnya, N., Prabhudev, M.V., Bhatta, C.P., Pham, T.H., Nidup, T., Megu, K., Chakravorty, J., Brockmann, A., and Otis, G.W. 2020. Geographical distribution of the giant honey bee *Apis laboriosa* Smith, 1871 (Hymenoptera, Apidae). *ZooKeys* 951: 67–81.

Koeniger, G., Koeniger, N., and Fabritius, M. 1979. Some detailed observations of mating in the honeybee. *Bee World* 60(2): 53–57.

Koeniger, N., and Koeniger, G. 1980. Observations and experiments on migration and dance communication of Apis dorsata in Sri Lanka. Journal of Apicultural Research 19(1): 21–34.

Koeniger, N., Koeniger, G., and Tingek, S. 2010. *Honey bees of Borneo: Exploring the centre of 'Apis' diversity*. Natural History Publications (Borneo), Kota Kinabalu, Sabah, Malaysia.

Koeniger, N., and Vorwohl, G. 1979. Competition for food among four sympatric species of Apini in Sri Lanka (*Apis Dorsata*, *Apis Cerana*, *Apis Florea* and *Trigona Iridipennis*). *Journal of Apicultural Research* 18(2): 95–109.

Lindauer, M. 1955. Schwarmbienen auf Wohnungssuche. Journal of Comparative Physiology 37(4): 263–324.

Mann, G.S., and Singh, G. 1983. Activity and abundance of pollinators of plum at Ludhiana (Punjab). *American Bee Journal* 123: 595.

Mardan, M.B. 1989. *Thermoregulation in the Asiatic giant honeybee Apis dorsata (Hymenoptera: Apidae)*. Thesis, University of Guelph.

McGregor, S.E. 1976. *Insect pollination of cultivated crop plants. Agricultural handbook*, Agricultural Research Service, U.S Department of Agriculture, University of Virginia, USA, 496.

Moritz, R.F.A., and Southwick, E.E. 1992. *Bees as superorganisms*. Springer, Berlin, Heidelberg, New York.

Morse, R.A. 1988. Research review. *Gleanings in Bee Culture* 116: 611.

Morse, R.A., and Benton, A.W. 1967. Venom collection from species of honeybees in South-East Asia. *Bee World* 48: 19–29.

Morse, R.A., and Laigo, F.M. 1969. Apis dorsata in the Philippines. Philippine Association of Entomologists, Laguana.

Mortensen, A.N., Burleson, S., Chelliah, G., Johnson, K., Schmehl, D.R., and Ellis, J.D. 2014. Tropilaelaps mite *Tropilaelaps* spp. Delfinado & Baker (Arachnida: Mesostigmata: Laelapidae). *IFAS Extension, Entomology and Nematology EENY* 604, December 2.

Mraz, C. 1995. *Health and the honeybee*. Queen City Publications, Burlington, VT.

Mutharaman, M., Raju, A.J.S., Vijaymkumar, J., Devanesan, S., Abrol, D.P., and Viraktamath, S. 2013. Morphometry of rock bee, *Apis dorsata* Fabricius. In: Viraktamath, S., Fakrudin, B., Vastrad, A.S., and Mohankumar, S. (Eds.), *Monograph on the morphometry and phylogeography of honey bees and stingless bees in India*. Network Project on Honey bees and Stingless bees. Department of Agricultural Entomology. University of Agricultural Sciences, Dharwad, 5–13.

Muzaffar, N., and Ahmed, R. 1990. *Apis* spp. (Hymenoptera: Apidae) and their distribution in Pakistan. *Pakistan Journal of Agricultural Research* 11: 65–69.

Oldroyd, B.P., Osborne, K., and Mardan, M. 2000. Colony relatedness in aggregations of Apis dorsata Fabricius (Hymenoptera, Apidae). Insectes Sociaux 47: 94–95.

Olmstead, A., and Wooten, D. 1987. Bee pollination and productivity growth: The case of Alfalfa. *American Journal of Agricultural Economics* 69(1): 56–63.

Otis, G.W. 1996. Distributions of recently recognized species of honey bees (Hymenoptera: Apidae; Apis) in Asia. Journal of the Kansas Entomological Society 69(4), Supplement: Special Publication Number 2: 311–333.

Otis, G.W., Koeniger, N., Rinderer, T.E., Hadisoesilo, S., Yoshida, T., Tingek, S., Wongsiri, S., and Mardan, M. 2000. Comparative mating flight times of Asian honeybees. *(IBRA/AAA) Proceedings of the 7th International Conference on Tropical Bees: Management and Diversity. 5th Asian Apicultural Association Conference* 137–141.

Paar, J., Oldroyd, B.P., Huettinger, E., and Kastberger, G. 2004. Genetic structure of an *Apis dorsata* population: The significance of migration and colony aggregation. *Journal of Heredity* 95: 119–126.

Pauly, A. 2015. The species of the genus Apis Linnaeus. Atlas Hymenoptera, Mons. www.atlashymenoptera.net/page.aspx?ID=238.

Phadke, R.P. 1961. Some physico-chemical constants of Indian beeswaxes. *Bee World* 42: 149–153.

Rathore, R.R.S., and Wells, H. 1995. Training Asian rock bees (*Apis dorsata*) to forage at specific locations. *Indian Bee Journal* 57: 8–9.

Reddy, C.C. 1980. Studies on the nesting behavior of Apis dorsata F. *International Conference on Apiculture in Tropical Climates* 2: 391–397.

Roepke, W. 1930. Beobachtungen an indischen Honigbienen insbesondere an *Apis dorsata* F. *Meded. Landbouwhoogeschool Wageningen* 34(6): 1–28.

Roubik, W.D., Sakagami, S.F., and Kudo, I. 1985. A note on distribution and nesting of the Himalayan honey bee *Apis laboriosa* Smith (Hymenoptera: Apidae). *Journal of the Kansas Entomological Society* 58: 746–749.

Ruttner, F. 1988. Biogeography and taxonomy of honeybees. Springer-Verlag, Berlin, 284.

Sakagami, S.F. 1960. Preliminary report on the specific difference of behaviour and other ecological characters between European and Japanese honeybees. *Acta Hymenopterologica* 1: 171–198.

Sakagami, S.F., Matsumura, T., and Ito, K. 1980. *Apis laboriosa* in Himalaya, the little known world largest honeybee (Hymenoptera: Apidae). *Insecta Matsumurana* 19: 47–77.

Schua, L. 1952. Untersuchungen über den Einfluss meteorologischer Elemente auf das Verhalten der Honigbiene. *Zeitschrift für Vergleichende Physiologie* 34: 258–277.

Seeley, T.D. 1985. *Honeybee ecology: Study of adaptation in social life*. Princeton University Press, Princeton, NJ08540 USA .

Seeley, T.D., Seeley, R.H., and Akratanakul, P. 1982. Colony defense strategies of the honeybee in Thailand. *Ecological Monographs* 52: 43–63.

Singh, S. 1962. *Beekeeping in India*. Indian Council of Agricultural Research, New Delhi, India.

Southwick, E.E., and Moritz, R.F.A. 1987. Social control of air ventilation in colonies of honey bees, *Apis mellifera*. *Insect Physiology* 33: 623–626.

Southwick, E.E., and Southwick, L. 1992. Estimating the economic value of honey bees as agricultural pollinators in the United States. *Economic Entomology* 85: 621–633.

Takahashi, J.I., and Nakamura, J. 2003. A scientific note on levels of polyandry of 2 queens of the Himalayan giant honeybee, *Apis laboriosa. Apidologie* 34(2): 191–192.

Tan, N.Q., Mardan, M., Thai, P.H., and Chinh, P.H. 1999. Observations of multiple mating flights of *Apis dorsata* queens. *Apidologie* 30(4): 339–346.

Trung, L.Q., Dung, P.X., and Ngan, T.X. 1996. A scientific note on first report of Apis *laboriosa F Smith, 1871 in Vietnam. Apidologie 27: 487–488.*

Underwood, B.A. 1990a. Seasonal nesting cycle and migration patterns of the Himalayan honey bee *Apis laboriosa. National Geographic Research* 6: 276–290.

Underwood, B.A. 1990b. Time of drone flight of *Apis laboriosa* Smith in Nepal. *Apidologie* 21: 501–504.

Underwood, B.A. 1991. Thermoregulation and energetic decision-making by the honey bees *Apis cerana, Apis dorsata* and *Apis laboriosa. Journal of Experimental Biology* 157: 19–34.

Underwood, B.A. 1992. Notes on the orange-rumped honeyguide *Indicator xanthonotus* and its association with the Himalayan honey bee *Apis laboriosa. Journal of the Bombay Natural History Society* 89: 290–295.

Verma, L.R. 1990. *Beekeeping in integrated mountain development.* Oxford and IBH Publishing Company, New Delhi, India, 367.

Verma, L.R. 1992. *Honeybees in mountain agriculture.* International Centre for Integrated Mountain Development, Kathmandu, Nepal, 69.

Woyke, J., Wilde, J., and Wilde, M. 2001. A scientific note on *Apis laboriosa* winter nesting and brood rearing in the warm zone of Himalayas. *Apidologie* 32(6): 601–602.

Woyke, J., Wilde, J., and Wilde, M. 2012. Swarming and migration of *Apis dorsata* and *Apis laboriosa* honey bees in India, Nepal and Bhutan. *Journal of Apicultural Science* 56: 81–91.

Woyke, J., Wilde, J., Wilde, M., Sivaram, V., Cervancia, C., Nagaraja, N., and Reddy, M. 2008. Comparison of defense body movements of *Apis laboriosa, Apis dorsata dorsata* and *Apis dorsata breviligula* honey bees. *Journal of Insect Behavior* 21: 481–494.

2 Biology of Asian Giant Honeybee, *Apis dorsata* Fabricius (Hymenoptera: Apidae)

N. Nagaraja

2.1 INTRODUCTION

The common giant honeybee *Apis dorsata* is a major honeybee species. It is an important source of honey and an effective pollinator of both agricultural crops and wild flora in its distribution range (Chantawannakul et al. 2018; Rajagopal and Nagaraja 1999; Sihag 2014). It differs to a limited extent in behaviour and ecology from other *Apis* species (Morse and Laigo 1969). *A. dorsata* builds single-comb nests on natural and man-made structures. The nests are large and are usually covered by multiple layers of worker bees which generally raise their heads up and abdomens down like a curtain from the branch (Koeniger et al. 2017). They store large amounts of honey, for which they are most frequently harvested. However, unsustainable harvesting methods, deforestation and destruction of suitable nest sites may threaten their local populations. The colonies abandon their nests and migrate to the regions with mass flowering during monsoon season and establish their new colonies (Oldroyd and Wongsiri 2006). They have evolved a unique method of defence to deter predators from attacking their exposed combs. The distinct lifestyle, together with exceptional levels of aggressive behaviour, has limited the domestication of these bee species for commercial honey production (Koeniger et al. 2010; Nagaraja 2012).

The foragers of *A. dorsata* perform dances on its brightest spots of the vertical combs without directing towards a food source, which moves with the position of the sun (Koeniger and Koeniger 1980). They also continue to dance in the dark during nights (Divan and Salvi 1965). *A. dorsata* has the unusual habit of continuing foraging even after sunset on bright moonlight nights (Wongsiri et al. 1991), a nocturnal activity which is not found in other honeybee species. *A. dorsata* foragers can fly at night due to large concentrations of visual pigment in the retinular cells of their ommatidia (Suwannapong and Wongsiri 1999).

2.2 BIOGEOGRAPHY

The nests of *Apis dorsata* are found throughout southern Asia from Pakistan to the Philippines, and from southern China to Indonesia (Oldroyd and Wongsiri 2006;

DOI: 10.1201/9781003294078-2

Hepburn and Radloff 2011). This species is well distributed in plains and hilly regions of up to 1,600 m above the ground level (Nagaraja and Rajagopal 2009), but with a greater number of colonies in dense forest regions. Based on the geographic isolation and morphological differences, *A. dorsata* is further classified in to three sub-species: *Apis dorsata dorsata* F., *Apis dorsata binghami* Cockerell and *Apis dorsata breviligula* Maa. *A. d. dorsata* is widely distributed in southern Asia. However, the nests of *A. d. breviligula* are found in Luzon of the Philippines and *A. d. binghami* in Sulawesi and its surrounding islands of Indonesia (Ruttner 1988; Hepburn and Radloff 2011). The abdomen of these bees is black in colour with white strips, in contrast to brownish orange strips in *A. d. dorsata* and show limited colony aggregation of 3–5 nests on suitable tree branches, in contrast to greater aggregation of *A. d. dorsata* colonies.

2.3 MORPHOLOGICAL FEATURES

The body size is considered as the most important characteristic among honeybee species (Koeniger et al. 2011). *A. dorsata* has the largest individual body size of all honeybee species (Michener 2000). Interestingly, the workers and drones of this species are produced in the cells of similar size. The worker bees are about 17–20 mm long with the forewing length of 12–15 mm and the tongue length of 6–7 mm. The queens are usually darker than the workers, which are yellow in colour. The young *A. dorsata* workers have pale yellow abdomens, but they turn into orange and black by the time they become foragers. In most honeybee species, the queen and drones are larger than worker bees. However, only little difference is found in body size between reproductives and worker bees of *A. dorsata*. The drones of *A. dorsata* are identified by their large eyes, short and round abdomen, and lack of sting. Each colony produces about 1,000 drones seasonally. *Apis dorsata breviligula* has shorter range of body size with a broader metasoma, a short tongue and an intermediate forewing length.

Honeybees possess compound eyes which consist of several thousands of optically isolated ommatidia. The eyes are well suited for orientation and detection of objects. Drones show enlarged compound eyes and are used for detecting small moving objects (Streinzer et al. 2013). Similarly, they have elaborated olfactory and visual systems, as well as flight musculature that is adapted for fast pursuit of flights (Radloff et al. 2003). The facets of queen bees are enlarged due to more number of ommatidia, suggesting that queens trade off special resolution for increased flight sensitivity, a likely adaption for crepuscular mating activity.

2.4 NESTING BIOLOGY

2.4.1 Nesting Sites

Nests are extremely crucial for survival and colony reproduction in most social insects (Hepburn and Radloff 2011). They provide an ideal platform to maintain stable nest temperatures and protect colonies against pests and predators in honeybees. *A. dorsata* build single large-sized vertical nests a suitable substratum. On the

FIGURE 2.1 Aggregation of *Apis dorsata* colonies on a banyan tree in the outskirts of Bengaluru, India.

lower part of the comb, the workers take off and land (Akratanakul 1976). Foragers perform dances on the vertical surface of the comb, where there is a clear view of the sky to observe the exact location of the sun. Its colonies are found on tall trees, multistoried buildings, rock cliffs, water tanks, flyovers, etc. (Reddy and Reddy 1993; Nagaraja 2017), especially at inaccessible elevations (Nagaraja and Rajagopal 2000; Woyke et al. 2012), protected from direct sunrays and rainfall. They tend to nest high in the air, usually from 3–60 m above the ground. Congregation of *A. dorsata* colonies is quite common and the number varies from 50–200 on a tree or rock where the bee forage sources are quite abundant (Rajagopal and Nagaraja 1999; Oldroyd et al. 2000). However, we have recorded more than 300 bee colonies on a banyan tree in the outskirts of Bengaluru for more than six years continuously during January and February months (Figure 2.1). Nesting sites are reoccupied year after year over periods of several decades (Oldroyd et al. 2000). Interestingly, some returning colonies find their way back to exactly the same nesting structure they occupied in the previous season (Neumann et al. 2000; Paar et al. 2000). This is despite the fact that workers probably live for less than two months, so only the queen could have direct information on the previous nest site.

2.5 NEST DISTRIBUTION

Information on colony distribution provides insights on survival and dispersal strategies of honeybee colonies (Winston 1987). *A. dorsata* colonies are found not only

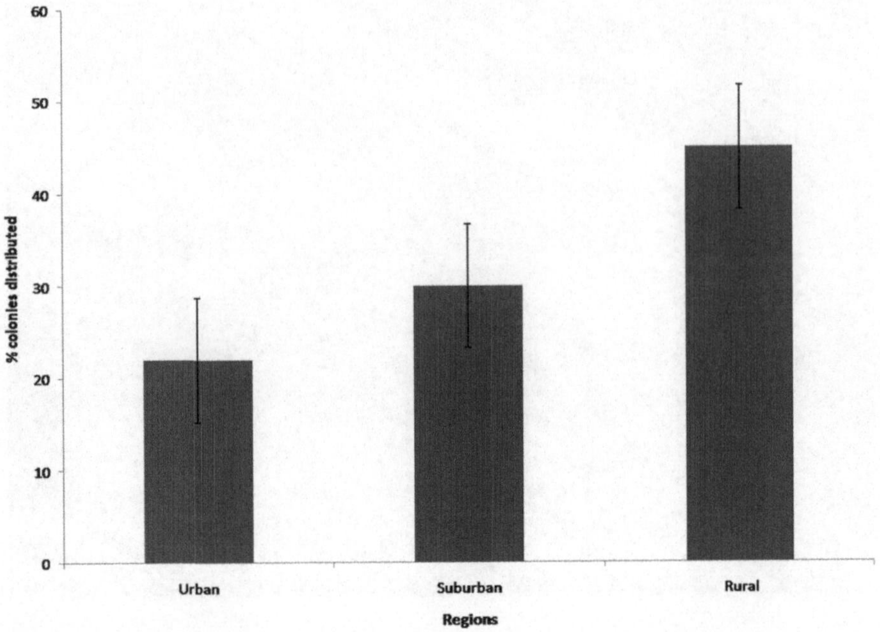

FIGURE 2.2 Distribution of *Apis dorsata* colonies in urban, suburban and rural regions of South Karnataka, India (N = 1,000).

in rural regions, but also are found abundantly in urban and semi-urban regions (Basavarajappa and Raghunandan 2013; Nagaraja 2019). Comparatively, the rural regions recorded greater number of colonies followed by suburban and urban regions (Figure 2.2). The distribution of giant honeybee colonies in rural plains was attributed to continuous flow of pollen and nectar, primarily from agricultural, horticultural and forest bee flora. The rural regions are seasonally cultivated with sunflower, mustard, radish, cucumbers, etc. Similarly, the forest flora such as pongamia, tamarind and eucalyptus supply copious amounts of nectar to *A. dorsata* colonies. The presence of abundant suitable nesting structures in urban regions such as high-rise buildings, water tanks, monumental structures and large trees attract huge number of *A. dorsata* colonies for nesting. The greater protection from wind, slashing rains and sunlight made greater distribution of nests on the buildings in urban regions. Itioka et al. (2001) recorded population fluctuation of *A. dorsata* colonies in tropical lowland forests and found that the density of the colonies was correlated with flowering seasons. Crane (1990) found that the nest sites chosen by *A. dorsata* was usually not directly exposed to wind currents but partially sheltered. However, preference of nests on buildings and trees in suburban regions is attributed to availability of favourable nest sites on both the structures (Nagaraja 2012). Roy et al. (2011) found that bee forage and availability of nesting sites are major factors influencing the distribution of *A. dorsata* colonies. The giant honeybee nests were most abundant in the areas of agro-forestry, in contrast to nests found only in forest fragments (Pavageau et al.

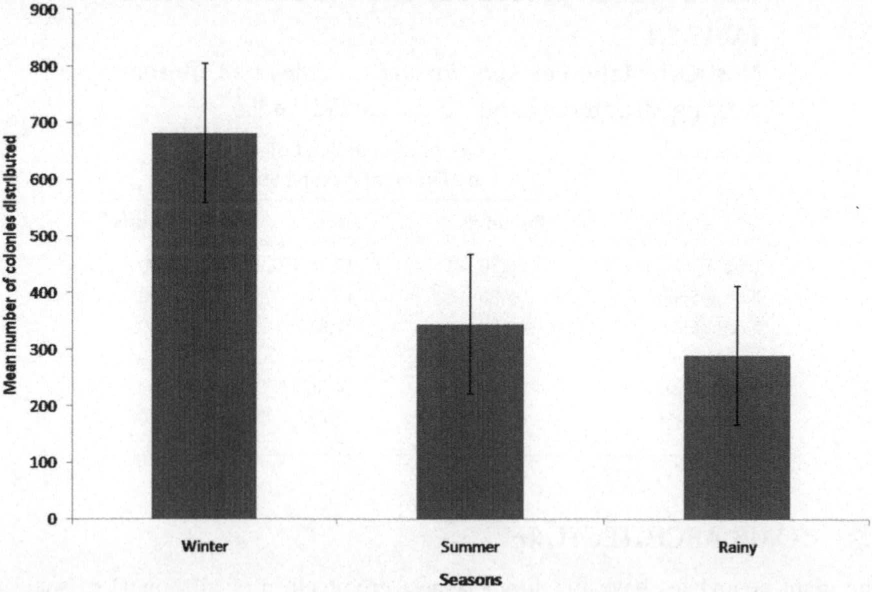

FIGURE 2.3 Fluctuation of *Apis dorsata* colonies during different seasons in and around Bengaluru, India (N = 1,500).

2018). Basavarajappa and Raghunandan (2013) reported huge number of bee nests in winter in and around Mysuru, located in the vicinity of south-western regions of Western Ghats, India. The population of *A. dorsata* colonies showed seasonal fluctuation and the greater number of colonies were found in winter followed by summer seasons (Figure 2.3).

2.6 NEST ORIENTATION

Apis dorsata build nests on manmade and natural nesting structures by orienting towards suitable compass directions. Cole (1994) found that the south-east orientation of ant nests increases the source of solar radiation during morning hours of the day. *A. dorsata* colonies showed differential levels of nest orientation and comparatively, the nests were oriented in more numbers towards north-east and north-west directions on the buildings and north-east and east-west direction on the trees (Table 2.1). However, on the rock cliffs, the nests were oriented towards east-west direction in large numbers (Nagaraja and Yathisha 2015). *A. dorsata* is found to orient its nests in large numbers towards north–south axis by minimizing the exposure to strong wind and bright sunlight (Reddy and Reddy 1993). However, greater orientation of bee nests were found towards north-east direction, irrespective of availability of suitable nest structures on trees in all directions. De Jong (1982) found that the earth's magnetic field was an important cue used by bees in deciding the orientation of their combs during comb building. Lindauer (1971) also reported that bees' choices changed comb orientation according to wind direction.

TABLE 2.1
Nest Orientation of *Apis dorsata* Colonies on Different Nesting Structures in South Karnataka, India

Directions	Orientation of Bee Colonies (%) on Different Nesting Structures		
	Buildings	Trees	Rock cliffs
East–West	06.50	17.18	38.00
North–South	05.00	15.31	08.00
North–East	60.00	21.87	26.00
North–West	17.50	14.48	09.20
South–East	07.00	14.06	10.00
South–West	07.50	15.93	08.80

2.7 COMB ARCHITECTURE

The giant honeybees have the simplest nest architecture of all honeybee species. The combs of *A. dorsata* are more or less semi-circular in shape. They build single vertical combs which measure 1–2 m long and 0.5–1.0 m height (Tan 2007). The nest combs are apically attached to the underside of the nesting substratum. The thickness of the comb cells was varied and the worker, drone and honeycomb cells were found to be 3.3 cm, 3.7 cm and 19.0 cm, respectively. The average cell diameter of the combs was between 5.42 mm and 6.35 mm (Graham 1992). The major portion of the comb was used for brood rearing and the brood cells are more or less similar in both diameter and cell depth. The storage of pollen and honey was found in the upper topmost portion of the nest. Both the diameter and depth of honey storage cells are greater than the brood cells (Lindauer 1957). However, on storage of greater quantities of honey, the depth of the honey cells was increased by the bees for greater storage and for stronger attachment to the substratum. The cells of drone brood and worker brood were hexagonal in shape and horizontal in orientation, but the capping of the drone cell was distinguishable by its dome shaped appearance. The comb was continuously covered by a curtain of many thousands of worker bees and acts as a protective barrier against heavy rain and wind.

Honey storage cells are always attached directly to the support on the upper sides of the comb. However, the drones are reared on the margin of the comb in the newly built cells. This is interesting, because there seems to be no intrinsic reason why drones should be on the comb margins. Indeed, the drones are less sensitive to temperature fluctuations during their development, or the reduction in cognitive abilities due to suboptimal rearing temperatures is less important in drones than workers (Jones et al. 2005).

2.8 COLONY DEVELOPMENT

Brood rearing and survival of adult bees depends on food stores in the colony. The egg laying capacity and population dynamics of *A. mellifera* has been well studied. The brood rearing activity of honeybee colonies is associated with availability of pollen and nectar resources in the vicinity of the nests. Though honeybee colonies rear the brood continuously around the year, the amount of the brood may vary seasonally (Nagaraja 2020). The greater brood rearing activity of *A. dorsata* was observed during September–November followed by February–April. This was due to greater availability of pollen sources, a protein-rich food for the development of brood on availability of plenty of pollen and nectar resources (Rajagopal and Nagaraja 1999). Similarly, the brood production in colonies increase the adult worker bee population. The minimum and maximum brood area recorded in *A. dorsata* was found to be varied from 1,000 cm^2 to 2,000 cm^2 in Bengaluru, India (Nagaraja and Rajagopal 2000).

There was a positive correlation between the size of the colony and the number of queen cells reared. The percentage of drones was limited and varied from 0.1%–17.3% in bee colonies. However, distribution of drone brood was scattered on the brood comb in a random manner (Chuttong et al. 2019). According to Venkatesh and Reddy (1989), the comb size of *A. dorsata* colonies 30 days after settlement on a suitable substratum varied from 80 × 90 cm to 189 × 198 cm. Koeniger and Wijayagunasekera (1976) found a relationship between comb size and the presence of drones, and that drones were present in colonies on if the combs are larger than 0.5 × 1.0 m. Woyke et al. (2007) found many drones in colonies even in smaller combs of 0.5 × 0.8 m or 0.6 × 0.8 m. The greater honey stores were observed from December–March and were attributed to availability and blooming of nectar flora in that season. However, no or limited brood was found during April–July, due to availability of limited floral resources, during which *A. dorsata* colonies migrate to regions with better floral resources.

2.9 COLONY MIGRATION

Migration is a process of seasonal movement of animals from one region to another, usually to exploit resources. Colony migration is a normal part of the seasonal cycle in wild honeybee species such as *A. dorsata* and *A. florea* (Wongsiri et al. 1996). *A. dorsata* undergo seasonal migration between alternate nesting sites to a distance of up to 200 km in search of new nest sites and floral resources (Crane et al. 1993; Dyer and Seeley 1994) by rapid expansion of population size (Itioka et al. 2001). In the north-western region of India, the colonies of *A. dorsata* arrive during October–December, develop their colonies with copious amounts of brood and honey, and left these regions on the onset of summer from May onwards. This migration schedule was in contrast to that found in Bengaluru in South India, wherein the colonies arrive at the beginning of rainy season (May–July). They rear brood and store enough quantities of honey and start migrating from April onwards. The differences in the migration schedule seemed to be due to differences in the local weather and climatic conditions. The north-western region of India represented semiarid environmental conditions, whereas Bengaluru in South India lies in the tropical region. The nonavailability of bee forage supplemented by ambient temperature conditions may cause to induce seasonal migration of the colonies of *A. dorsata*. The giant honeybee is well adapted to high temperatures, during which the bees ventilate the nest and loosen the curtain to cool down the nest at high temperatures. Frequency and the timing of migration depended on the climatic rhythm of the region (Koeniger and Koeniger 1980).

The colonies of *A. dorsata* migrate seasonally between highland and lowland nesting sites, taking advantage of available floral resources at least twice in a year (Robinson 2021; Woyke et al. 2001; Liu et al. 2007), depending on the availability of floral resources and disturbances to colonies if any. However, no brood was found in the combs at least two weeks before migration (Woyke et al. 2005), showing the first-hand information about migration to the queen. Colonies departing after a long stay always left behind barren combs, suggesting that they had left in response to deteriorating quality of resources. Further, the visual, chemical and tactile cues left out at combs may also constitute a navigational map in coordination. These observations support the idea that migration allows colonies to track seasonally varying resources in different regions. Neumann et al. (2000) and Paar et al. (2000) genotyped the migrating *A. dorsata* bees and found that the same swarms return to their previous nesting sites. At stopovers, these bees form combless clusters or bivouacs, and accumulate food reserves for flight and comb construction upon arrival at nest sites (Hepburn 2011). Migration increases colony fitness by improved food availability and enhanced out-breeding (Oldroyd et al. 1996).

2.10 ABSCONDING BEHAVIOUR

Departure of bee colonies without emergence of a new queen/s or overcrowding of bees is generally considered as absconding. Absconding behaviour is a survival strategy of honeybees. It might be seasonal or resource related, or also due to disturbances by predators. The colonies of *A. dorsata* are found to move between habitats during blooming seasons (Crane et al. 1993; Mahindre 2000). Absconding may also helpfully control levels of ectoparasitic mite, *Tropilaelapsclareae*, which needs brood for reproduction in *A. dorsata* (Nagaraja and Rajagopal 2009). Thus, a colony may reduce infestation by this parasite with period of broodless migration (Rinderer et al. 1994). Microsatellite analysis of nests of aggregations showed no colonies were related as mother/daughter, suggesting that rapid increase in number of colonies during flowering periods is most likely to occur by swarms arriving from other areas rather than by *in situ* reproduction. The dances of scout bees and subsequent flights of *A. dorsata* swarms were widely scattered in Sri Lanka and were assumed to be near the end of migrations and seeking nearby nest sites (Koeniger and Koeniger 1980). The traditional honey hunters harvest honey during dark by destroying the entire comb, along with brood. However, the surviving adult population—along with the queen—may abscond and move to new places for nesting. This affects the survival and reproduction of the bee colonies. Similarly, the nests found on buildings, water tanks and trees are sometimes directly exposed to bright sunlight by making the combs soft. As a result, these soft combs could easily be dislodged by strong wind, thereby forcing the adult population of the colony to migrate to alternate nesting sites.

2.11 MASSED FLIGHT ACTIVITY

Colonies of *A. dorsata* perform short daily periodic mass flights of up to six times in a day (Woyke et al. 2004). These flights play a role in the cleansing and orientation of young bees. Similarly, shortly after sunset, the colonies also perform longer dusk mass flights, and drones and queens also fly during such events (Tan et al. 1999). However, Woyke et al. (2005), found that all the castes—such as drones, queen and worker bees—participate in dusk mass flights. Dusk mass flights and diurnal periodic mass flights are performed by different categories of worker bees. There was a strong correlation between the duration of time the colonies stayed in the site and the percentage of drones, which participated in dusk mass flights. The foragers also perform dusk mass flights when drones are absent in all the colonies and even when open brood is absent in the nest.

2.12 DEFENSIVE BEHAVIOUR

Apis dorsata build nests beyond the reach of terrestrial pests and predators, and they are known for their effective defence behaviour. They are quickly alerted on the threat and warn the predators by visual and auditory signals. The workers of these bees twist their bodies in a vertical direction by raising the abdomen and lowering the head whenever a disturbance occurs on their nest curtain (Kastberger et al. 1998; Woyke et al. 2008). They also perform quick runs over the nest curtain by forming clusters at the lower edge of the colony. The colonies respond to disturbance by its enemy with a characteristic defence body twist, while the thorax rotates, the head is lowered and the abdomen and wings are thrust upwards. About three-quarters of the worker population of a colony of *A. dorsata* is engaged in colony defence. While noting this behaviour, the workers of nearby giant bee colonies also emit defence waves on the surface of the nest (Seeley et al. 1982).

2.13 CONCLUSION

Apis dorsata is a wild bee which preferably nests in aggregations. Its defensive strategies in nest protection, coupled with being free from most pests and diseases, made this bee a major honey producer in its distribution range. However, information on biology of Asian honeybees in general—and *A. dorsata* in particular—are fragmentary. Despite *A. dorsata* being a major honeybee species in Indian conditions, the research on its colony development, comb building process and migratory strategies are limited. Indeed, there is greater scope for these studies for conservation of *A. dorsata* for honey production and maintenance of biodiversity with its cross-pollination services in both crop plants and wild flora.

ACKNOWLEDGEMENTS

The author is grateful to University Grants Commission (UGC), New Delhi, India for financial assistance (No. 42–623/2013 [SR]) in conducting part of these studies. I am also indebted to Prof. C. Srinivas, Director, UGC-Human Resource Development Centre, Bangalore University, Bengaluru, for his constant encouragement, and to Bangalore University, Bengaluru, for providing research facilities.

REFERENCES

Akratanakul, P. 1976. Honeybees in Thailand. *American Bee Journal* 116: 120–121.

Basavarajappa, S., and Raghunandan, K.S. 2013. Colony status of Asian giant honeybee, *Apis dorsata* Fabricius in southern Karnataka, India. *African Journal of Agricultural Research* 8: 680–689.

Chantawannakul, P., Williams, G., and Neumann, P. 2018. *Asian beekeeping in the 21st century*. Springer Nature, Singapore, 25.

Chuttong, B., Somana, W., and Burgett, M. 2019. Giant honey bee (*Apis dorsata* F.) rafter beekeeping in Southern Thailand. *Bee World* 96: 66–68.

Cole, B.J. 1994. Nest architecture in the western harvester ant, *Pogonomyrmyx occidentalis* (Cresson). *Insectes Sociaux* 41: 401–410.

Crane, E. 1990. *Bees and beekeeping: Science, practice and world resources*. Heinemann, London.

Crane, E., van Luyen, V., Mulder, V., and Ta, T.C. 1993. Traditional management system for *Apis dorsata* in submerged forests in southern Vietnam and central Kalimantan. *Bee World* 74: 27–40.

De Jong, D. 1982. Orientation of comb building by honeybees. *Journal of Comparative Physiology A* 147: 495–501.

Divan, V.V., and Salvi, S.R. 1965. Some interesting behavioural features of *Apis dorsata* Fab. *Indian Bee Journal* 27(1): 52.

Dyer, F.C., and Seeley, D. 1994. Colony migration in the tropical honeybee, *Apis dorsata* F. (Hymenoptera: Apidae). *Insectes Sociaux* 41: 129–140.

Graham, J.M. 1992. *The hive and the honeybee*. Dadant and Sons Inc., Hamilton, IL, 1324.

Hepburn, H.R. 2011. Absconding, migration and swarming. In: Hepburn, R., and Radloff, S.E. (Eds.), *Honeybees of Asia*. Springer-Verlag, Berlin, 133–158.

Hepburn, H.R., and Radloff, S.E. 2011. Biogeography. In: Hepburn, R., and Radloff, S.E. (Eds.), *Honeybees of Asia*. Springer-Verlag, Berlin, 62–63.

Itioka, T., Inoue, T., Kaliang, H., Kato, M., Nagamitsu, T., Momose, K., Sakai, S. et al. 2001. Six-year population fluctuation of the giant honey bee *Apis dorsata* (Hymenoptera: Apidae) in a tropical lowland dipterocarp forest in Sarawak. *Annals of Entomological Society of America* 94: 545–549.

Jones, J.C., Helliwell, P., Beekman, M., Maleszka, R.J., and Oldroyd, B.P. 2005. The effects of rearing temperature on developmental stability and learning and memory in the honey bee, *Apis mellifera*. *Journal of Comparative Physiology A* 191(12): 1121–1129.

Kastberger, G., Raspotnig, G., Biswas, S., and Winder, O. 1998. Evidence of Nasonov scenting in colony defence of the giant honeybee, *Apis dorsata*. *Ethology* 104: 27–37.

Koeniger, G., Koeniger, N., and Phincharoen, M. 2011. Comparative reproductive biology of honey bees. In: Hepburn, R., and Radloff, S. (Eds.), *Honey bees of Asia*. Springer Verlag, Berlin, 159–206.

Koeniger, N., and Koeniger, G. 1980. Observations and experiments on migration and dance communication of *Apis dorsata* in Sri Lanka. *Journal of Apicultural Research* 19: 21–34.

Koeniger, N., Koeniger, G., and Tingek, S. 2010. *Honeybees of Borneo: Exploring the centre of Apis diversities*. Natural History Publications (Borneo), Kota Kinabalu, Sabah, Malaysia, 262.

Koeniger, N., Kurze, C., Phiancharoen, M., and Koeniger, G. 2017. Up or down that makes the difference: How giant honeybees (*Apis dorsata*) see the world? *PLoS One* 12(11): e0185325.

Koeniger, N., and Wijayagunasekera, H.N.P. 1976. Time of drone flight in the three Asian honeybee species (*Apis cerana, Apis florea, Apis dorsata*). *Journal of Apicultural Research* 15: 67–71.

Lindauer, M. 1957. Communication in swarm bees searching for a new home. *Nature* 179: 63–66.

Lindauer, M. 1971. *Communication among social bees*. Harvard University Press, Cambridge, MA, 161.

Liu, F., Roubik, D.W., He, D., and Li, J. 2007. Old comb for nesting site recognition by *Apis dorsata*? Field experiments in China. *Insectes Sociaux* 54: 424–426.

Mahindre, D.B. 2000. Developments in the management of *Apis dorsata* colonies. *Bee World* 81: 155–163.

Michener, C.D. 2000. *The bees of the world*. The John Hopkins University Press, Baltimore, MD, 913.

Morse, R.A., and Laigo, F.M. 1969. *Apis dorsata in the Philippines*. The Philippine Association of Entomologists, Laguana.

Nagaraja, N. 2012. Asian honeybees: Biology, threats and their conservation. In: Florio, R.M. (Ed.), *Biology, threats and colonies*. Nova Publishers Inc., New York, 99–123.

Nagaraja, N. 2017. Population fluctuation of giant honeybee, *Apis dorsata* in plains of Karnataka. *Journal of Entomological Research* 43(4): 503–508.

Nagaraja, N. 2019. Nesting patterns of giant honeybee, *Apis dorsata* in plains of Karnataka, India. *Journal of Entomological Research* 41(3): 307–310.

Nagaraja, N. 2020. Biology of dwarf honeybee, *Apis florea* Fabricius (Hymenoptera: Apidae). In: Abrol, D.P. (Ed.), *The future role of dwarf honeybee in natural and agricultural systems*. CRC Press, Taylor and Francis Group, New York, 13–23.

Nagaraja, N., and Rajagopal, D. 2000. Foraging and brood rearing activity of rock bee, *Apis dorsata* F. (Hymenoptera: Apidae). *Journal of Entomological Research* 24(3): 243–248.

Nagaraja, N., and Rajagopal, D. 2009. *Diseases, parasites, pests, predators and their management*. MJP Publishers, Chennai, 210.

Nagaraja, N., and Yathisha, V. 2015. Nest orientation of Asian giant honeybee, *Apis dorsata* in plains of Karnataka, India. *Journal of Entomological Research* 39(3): 197–201.

Neumann, P., Moritz, R.F.A., and Mautz, D. 2000. Colony evaluation is not affected by the drifting of drone and worker honeybees (*Apis mellifera* L.) at a performance testing apiary. *Apidologie* 31: 67–79.

Oldroyd, B.P., Osborne, K.E., and Mardan, M. 2000. Colony relatedness in aggregations of *Apis dorsata* Fabricius (Hymenoptera: Apidae). *Insectes Sociaux* 47: 94–95.

Oldroyd, B.P., Smolenski, A.J., Cornuet, J.M., Wongsiri, S., Estoup, A., Rinderer, T.E., and Crozier, R.H. 1996. Levels of polyandry and intracolonial genetic relationships in *Apis dorsata* (Hymenoptera: Apidae). *Annals of Entomological Society of America* 89: 276–283.

Oldroyd, B.P., and Wongsiri, S. 2006. *Asian honey bees: Biology, conservation and human interactions*. Harvard University Press, Cambridge, MA.

Paar, J., Oldroyd, B., and Kastberger, G. 2000. Giant honeybees return to their nest sites. *Nature* 406, 475.

Pavageau, C., Gaucherel, C., Garcia, C., and Guazoul, J. 2018. Nesting sites of giant honeybees modulated by landscape patterns. *Journal of Applied Ecology* 55: 1230–1240.

Radloff, S.E., Hepburn, H.R., and Bangay, L. 2003. Quantitative analysis of intercolonial and intracolonial morphometric variance in honeybees, *Apis mellifera* and *Apis cerana*. *Apidologie* 34: 339–351.

Rajagopal, D., and Nagaraja, N. 1999. Beekeeping status in Karnataka, India. *Asian Bee Journal* 1: 50–59.
Reddy, C.C., and Reddy, M.S. 1993. Studies on the distribution of nests of giant honeybee (*Apis dorsata* F). *Indian Bee Journal* 55: 36–39.
Rinderer, T., Oldroyd, B., Lekprayoon, C., Wongsiri, S., Boonthai, C., and Thapa, R. 1994. Extended survival of the parasitic honeybee mite, *Tropilaelaps clareae* on adult workers of *Apis mellifera* and *Apis dorsata*. *Journal of Apicultural Research* 33: 171–173.
Robinson, W.S. 2021. Surfing the sweet wave: Migrating giant honeybees (Hymenoptera: Apidae: *Apis dorsata*) display spatial and temporal fidelity to annual stop over site in Thailand. *Journal of Insect Science* 21(6): 1–12.
Roy, P., Leo, R., Thomas, S.G., Varghese, A., Sharma, K., Prasad, S., Bradbear, N., Roberts, S., Potts, S.G., and Davida, P. 2011. Nesting behaviour of rock bee, *Apis dorsata* in the Nilgiri biosphere reserve in India. *Tropical Ecology* 52: 285–291.
Ruttner, F. 1988. *Biogeography and taxonomy of honeybees*. Springer-Verlag, Berlin.
Seeley, T.D., Seeley, R.H., and Aktratanakul, P. 1982. Colony defence strategies of the honeybees in Thailand. *Ecological Monographs* 52: 43–63.
Sihag, R.C. 2014. Phenology of migration and decline in colony numbers and crop hosts of giant honeybee (*Apis dorsata* F.) in semiarid environment of northwest India. *Journal of Insects* 639467.
Streinzer, M., Brockmann, A., Nagaraja, N., and Spaethe, J. 2013. Sex and caste specific variation in compound eye morphology of five honeybee species. *PLoS One* 8(2): e57702.
Suwannapong, G., and Wongsiri, S. 1999. Ultrastructure of the compound eyes of the giant honey bee queens, *Apis dorsata* Fabricius, 1793. *Journal STREC* 7(1–2): 60–68.
Tan, N.Q. 2007. Biology of *Apis dorsata* in Vietnam. *Apidologie* 38: 221–222.
Tan, N.Q., Mardan, M., Thai, P.H., and Chinh, P.H. 1999. Observations on multiple mating flights of *Apis dorsata* queens. *Apidologie* 30: 339–346.
Venkatesh, G., and Reddy, C.C. 1989. Rates of swarming and absconding in the giant honey bee, *Apis dorsata* F. *Proceedings of Indian Academy of Sciences (Animal Sciences)* 98(6): 425–430.
Winston, M.L. 1987. *The biology of the honeybee*. Harvard University Press, Cambridge, UK.
Wongsiri, S., Rinderer, T.E., and Sylvester, H.A. 1991. *Biodiversity of honeybees in Thailand*. Prachachon, Bangkok, 50–63.
Wongsiri, S., Thapa, R., Oldroyd, B., and Burgett, D.M. 1996. A magic bee tree: Home to *Apis dorsata*. *American Bee Journal* 136: 796–799.
Woyke, J., Kruk, C., Wilde, J., and Wilde, M. 2004. Periodic mass flights of the giant honey bee *Apis dorsata*. *Journal of Apicultural Research* 43: 180–185.
Woyke, J., Wilde, J., Reddy, C.C., and Nagaraja, N. 2005. Periodic mass flights of giant honeybee, *Apis dorsata* performed in successive days at two environmental conditions. *Journal of Apicultural Research* 44(4): 180–189.
Woyke, J., Wilde, J., and Wilde, M. 2001. *Swarming, migration and absconding of 'Apis dorsata' colonies*. Proceedings of the 7th International Conference on Apiculture in Tropical Climates, Chiang Mai, 183–188.
Woyke, J., Wilde, J., and Wilde, M. 2008. Comparison of defence body movements of *Apis laboriosa*, *Apis dorsata dorsata* and *Apis dorsata breviligula* honeybees. *Journal of Insect Behaviour* 21: 481–494.
Woyke, J., Wilde, J., and Wilde, M. 2012. Swarming and migration of *Apis dorsata* and *Apis laboriosa* honey bees in India, Nepal and Bhutan. *Journal of Apicultural Science* 56: 81–91.
Woyke, J., Wilde, J., Wilde, M., Reddy, M.S., Nagaraja, N., and Sivaram, V. 2007. Presence or absence of drones in drone dusk mass flights performed by *Apis dorsata breviligula* worker bees. *Journal of Apicultural Research* 46(1): 40–49.

3 Reproductive Biology of Asian Giant Honeybee, *Apis dorsata* Fabricius (Hymenoptera: Apidae)

N. Nagaraja

3.1 INTRODUCTION

The social organization in bees is a result of the long evolutionary changes in their social life. Honeybee colonies are characterized by reproductive division of labour with overlapping generations (Wilson 1971). Honeybees are often referred to as superorganisms, since the colony itself can be considered as a unit for reproduction (Hölldobler and Wilson 2008; Moritz and Southwick 1992). A honeybee colony normally consists of one queen and tens of thousands of worker bees, along with worker brood and small amount of drone brood. However, a few hundreds of drones are produced during reproductive seasons (Nagaraja and Abrol 2020). The queen depends on her workers to maintain routine colony activities, including colony defence. Honeybees show a haplo-diploid type of sex determination system in which males and females develop from unfertilized and fertilized eggs, respectively (Palmer and Oldroyd 2000; Collison 2004). The information on reproductive biology is crucial in management of *A. dorsata* colonies for honey production and crop pollination. The queens and drones are reproductives and the mated queen propagates the colony by laying thousands of eggs daily. Queens mate at the beginning of their life, typically during one or very few nuptial flights (Boomsma et al. 2005). Drones on sexual maturity fly and congregate at identified mating yards during afternoon hours of the day where mating takes place with virgin queens (Koeniger and Koeniger 1991). Absence of remating in bees shows that mating does not interfere with normal colony life. Worker sterility is the characteristic of eusociality and is an environmentally sensitive process that continues in honeybee worker bees (Ronai et al. 2017).

3.2 THE QUEEN AND DRONE BEES

The queen and drones are larger than the worker bees in honeybee colonies. The queen is the most important member in a colony and have a shortest development period than other bee castes. The queen bee continuously monitors the activities of worker bees by releasing pheromones (Cobey 2007; Naumann et al. 1990). Similarly,

DOI: 10.1201/9781003294078-3

the queen maintains reproductive division of labour by emitting specific pheromones which induce the workers to remain sterile in her presence (Holman 2018). During queen rearing, worker bees cooperate to rear queens of similar reproductive potential (Tarpy et al. 2004). Furthermore, queens invest substantial metabolic resources into maintaining viable spermatozoa within their spermathecae. *A. dorsata* colonies produce natural queen cells during breeding season having huge adult bee population. The queen cells are most often elongated but are broader in shape. The queen cells are larger in size than drone and worker cells, different in shape and vertical in orientation. In general, the number and position of queen cells are not much different from other honeybee species.

Rearing drones is costly to the bee colonies, as they do not participate in any types of colony activities except mating with virgin queens. They are reared during the reproductive season, which coincides with availability of potential amounts of floral resources and huge worker population (Rangel et al. 2013; Smith et al. 2014). Production of drones is initiated by the construction of drone cells that are comparatively larger than worker cells (Smith et al. 2014). The percentage of drone pupae in a colony varies from 6%–17% of the total worker pupae and is scattered on the brood comb in a banding pattern (Chuttong et al. 2019). Drones mature at the age of 6–8 days and they interact, fed and groomed by worker bees (Goins and Schneider 2013; Collison 2004). They perform orientation flights to learn the local landmarks and precise location of the nest before leaving to drone congregation areas (DCAs) for mating (Galindo-Cardona et al. 2015).

3.3 REPRODUCTIVE SYSTEM OF THE QUEEN

The reproductive system of *A. dorsata* queens is more or less similar to the queens of other *Apis* species. It consists of a pair of ovaries and associated accessary structures. Each ovary consists of a bundle of more than 150 ovarioles. At the end of each ovariole, there is a germinal tissue from which the nurse cells, follicle cells and true egg cells are formed. The ovarioles open in to the two lateral oviducts and join the median oviduct. Eggs are moved from the ovarioles into the oviducts and then to the spermatheca. The queen stores the spermatozoa in its spermatheca, and they are viable for many years. When the queen lays an egg into a worker cell, a few spermatozoa are released out of the spermatheca and fertilize the egg. However, when the queen lays eggs into a drone cell, no spermatozoa are released and eggs are not fertilized. The fertilized eggs laid by the queen are developed into queens and worker bees, and the unfertilized eggs into drones (Winston 1991). However, the queens of honeybee species mate several times during single mating flight. Mating signs have been reported in hive honeybee species, but no mating signs were noticed protruding from the sting chamber of *A. dorsata* queens.

3.4 REPRODUCTIVE SYSTEM OF THE DRONE

Honeybee drones mature sexually at the age of around 8–12 days. Spermatozoa are produced and matured in the testes, which are composed of bundles of tubules. They have large testes at the time of emerge from the cell and the contents of each testis are pass through the vas deferens to the seminal vesicle, after which the testes are completely shrunk. The seminal vesicles are large since they contain semen suspended with spermatozoa. The endophallus is the copulatory organ and is differentiated into a broad vestibulum with cornua, a slender cervix and a thick bulb with its lobe. In the drones of *A. dorsata*, endophallus is elongated with four long curved cornua and is covered by an orange-coloured sticky secretion. Elongation of the endophallus is mainly caused by the extended cervix. During copulation, the endophallus is everted successively and introduced into the queen. After copulation, the eversion of irreversible endophallus results in the paralysis of the drone (Koeniger and Koeniger 1991).

3.5 SEX PHEROMONE COMMUNICATION

The chemical communication systems have evolved in the context of mating and social behaviours in eusocial insects (Ayasse et al. 2001). The pheromones produced by the queen mandibular gland signal the presence of a fertile queen in the colony (Princen et al. 2019). The complex and phase-dependent chemical compositions of secretions of queen mandibular glands (Plettner et al. 1997; Free 1987) show that a single component or a mixture of components are dealt differently in between queen and drone bees. The major component of queen mandibular pheromone is found to be 9-oxo-2- decenoic acid (9ODA), which is an effective attractant over large distances and elicits highly predictable responses in flying drones (Brockmann et al. 2006). Interestingly, Nagaraja and Brockmann (2009) found 10-hydroxy-2- decenoic acid (10-HDA) also acts as a sex pheromone in dwarf honeybees, *Apis florea*. It was found that mandibular extracts of queen containing 9-ODA successfully attracted drones of *A. dorsata* (Koeniger et al. 1994). Further investigations suggest that other components of the secretion might play a role in the communication between queen and drones of *A. dorsata* (Koeniger and Koeniger 2000; Naumann et al. 1990).

3.6 DRONE CONGREGATION AREAS

Drone congregation areas (DCAs) are identified locations where drones of *Apis* species meet and mate with virgin queens. Honeybee drones are known to fly to DCAs where they mate with visiting virgin queens (Koeniger et al. 2005). Similarly, the virgin queens fly to the vicinity of DCAs after a few minutes of arrival of drones. DCAs are unique locations with a diameter of 30–200 m and are formed 10–35 m above ground level, where mating of the queen takes place with multiple drones. Each DCA has a highly mixed drone population from many surrounding colonies of up to a distance of 5 km radius. The drone congregations may contain thousands of drones from more than 200 different colonies (Koeniger et al. 2005). However, Baudry et al. (1998) found molecular evidence that drones from about 238 colonies were present at a single congregation area in Germany. Kraus et al. (2005) found a high diversity of drone genotypes in DCAs of *A. dorsata* for four consecutive days. The DCAs are perineal and are reported to exist even for many decades. The influence of environmental cues, if any, for gathering at a given DCA are not well understood (Koeniger et al. 2005). However, there is possibility that the density of vegetation may affect the navigation of flights of drones (Galindo-Cardona et al.

2015). The mating period of *A. dorsata* is of an extremely short duration of about 20–30 minutes (Koeniger and Wijayagunasekera 1976). The temporal sequence of mating periods is strictly correlated with size and it begin with the smallest (*A. andreniformis*) and ends with the largest sympatric species i.e. *A. dorsata*. The drones of *A. dorsata* congregate under the canopy of tall emergent trees and these treetops seem to serve as visual landmarks. They flew under the branches of the trees at a height of 20–25 m. Therefore, the distribution of drones resulted in a clear spatial separation without any overlap between the three species, viz. *A. dorsata*, *A. florea* and *A. cerana*.

3.7 MATING BEHAVIOUR

The mating behaviour of honeybees is affected by eusociality in several ways. The sex ratio is strongly male biased whereby the colonies produce many hundreds of drones and only a few queens. Similarly, honeybee drones perform multiple nuptial flights, while a queen flies only once or twice for mating. The mating of queen and drones takes place in DCAs at a short distance from the colony. Furthermore, a considerable number of queens do not return to the colony from their flights showing risk in mating. The drones are monogamous because they die after mating and queens are polyandrous, mating with multiple drones. The information on mating behaviour of *A. dorsata* is meagre compared to other honeybee species.

3.8 MATING FLIGHTS

The drones of honeybees perform mating flights on maturity at the age of 8–12 days. These flights begin around noon or in the early afternoon hours of the day with maximum solar elevation and light intensities in most of the *Apis* species. However, the drones of *A. dorsata* perform mating flights during sunset. They showed mating flights during 18:00–18:45 hours in Sri Lanka (Koeniger and Wijayagunasekera 1976), 18:13–18:45 hours in southeast Thailand (Rinderer et al. 1993), 18:15–19:05 hours in Borneo, Indonesia (Koeniger et al. 1994) and 8:15–18:35 hours in Nepal (Woyke et al. 2001) (Table 3.1). The presence of large-facet diameters of *A. dorsata* drones is quite effective in maximizing contrast sensitivity in low light environments that visually limit visual contrast. The facet and ocelli diameters of drones of *A. dorsata* were found similar to a few some nocturnal bee species, thus making mating flights

TABLE 3.1
Drone Flight Periods of *Apis dorsata* in Different Asian Countries

Period (hrs)	Country from Where Reported	Reference
18.00–18.45	Sri Lanka	Koeniger and Wijayagunasekera (1976)
18.13–18.45	Thailand	Rinderer et al. (1993)
18.15–19.05	Indonesia	Koeniger et al. (1994)
18.15–18.35	Nepal	Woyke et al. (2001)

around sunset (Streinzer et al. 2013). There was a similar correlation between facet and eye size and mating flight time in queen.

The unmated queens of *A. dorsata* fly to DCAs after one or two short orientation flights, where they copulate with several drones. The number of mating flights performed by *A. dorsata* queens is comparatively higher than other honeybee species as genetically diversified colonies easily adapt to their environment (Moritz et al. 1995). The duration of mating flights of queens of *A. dorsata* was 15.4 ± 4.3 minutes, which is shorter than *A. florea*, where it was from 18–30 minutes. Body size might have been a pre-adaption which allows a shift of mating flights to lower weight levels.

3.9 MATING STRATEGIES

The mating system in honeybees is of male aggregation type, as the males reach the congregation areas before arrival of females (Boomsma et al. 2005; Paxton 2005). Male aggregations leads to rapid mating of queens, with many drones during one successful mating flight of short duration (Tan et al. 1999). However, female mate choice is also possible, as successful copulation depends on queens opening their sting chamber to allow a male to insert his genitalia (Winston 1991; Strassmann 2001). Furthermore, the queens need to contract their bursa to press the endophallus in order to transfer sperm into their lateral oviducts (Koeniger and Koeniger 1991). Drones follow a virgin queen in a comet-like swarm, engage in competition, approach and mate with the queen. The virgin queen in flight responds to drones in the vicinity of a DCA by holding open the entrance to her sting chamber. One of the drones flying succeeds in seizing her between his legs. The first drone inserts his endophallus into the bursa copulatrix of the queen and ejaculates his spermatozoa into the oviducts of the queen. Similarly, each consecutive drone inserts his endophallus beneath the mating sign of his predecessor and removes it by a special organ on the endophallus (Koeniger 1986).

According to Koeniger (1991), there are two mechanisms of sperm transfer from drones to queen in *A. dorsata*. First, the sperm is transferred from drones to oviduct of the queen and after that, only 5%–10% of them reach the spermatheca as in hive honeybees; second, drones transfer sperm directly into the spermatheca as in *A. florea* and *A. andreniformis*. Further study on sperm transfer mechanism in *A. dorsata* is strongly suggested, but no mating signs were noticed protruding from the sting chamber and queen was permitted to re-enter the colony in *A. dorsata* (Tan et al. 1999) since mating signs may be too small to see without catching the returning queens for closer examination. Woyke et al. (2001) inseminated 2–3 mm^3 of *A. mellifera* semen into the virgin queens of *A. dorsata*. Though the queen laid the eggs, only 3% of the eggs hatched and developed into larvae and pupae, but they failed to emerge as adults.

3.10 MATING FREQUENCY

Polyandry in social insects has been shown to play an important role in natural selection. The queens of honeybees are extremely polyandrous (Tarpy et al. 2004) and are known to mate with tens of drones over two to three successive mating flights.

However, the degree of polyandry varies substantially among different species of honeybees. Cornuet (1986) reported a good number of 12.4 matings in *A. mellifera* populations. Similarly, Koeniger et al. (1989) found on the basis of semen counts that *A. florea* mates with fewer than five drones. Woyke (1975) reported high degree of polyandry in *A. cerana* and found semen of up to 27 drones in one queen returning from a mating flight. The most extreme level of polyandry was found in *A. dorsata*, in which queens mate with an average of 30–44 drones over up to six consecutive days (Moritz et al. 1995; Tan et al. 1999; Wattanachaiyingcharoen et al. 2003). It is good for the queens to mate with as many drones as possible since better colony performance was associated with mating with a greater number of drones (Delaplane et al. 2015).

3.11 SPERM STORAGE

The sperm numbers are biological data which are taken as a basis to understand several aspects of honeybee mating biology. The sperm numbers of one-week-old drones are comparatively lower than the drones of more than two weeks old. The dynamics of sperm storage and sperm use is most interesting aspect in honeybees. To maintain a better reproductive success of the colony for many years, a honeybee queen requires a large storage of sperm in the spermatheca ranging from 4–6 million.

The percentage of drone spermatozoa stored in the spermatheca differs among honeybee species. The total number of spermatozoa in the queen's spermatheca divided by the effective number of matings indicates the number of spermatozoa contributed by each drone. The rate of sperm movement into the spermatheca is nearly constant during the first 24 hours after copulation, at least for large ejaculates. The sperm storage process lasts for about 40 hours after mating, during which most of the previously acquired sperm gets lost as it flows back into the bursa copulatrix to be eventually expelled through the vagina. Extreme sperm dumping is not only found in *A. mellifera* but also in *A. dorsata* (Oldroyd et al. 1996). The average number of spermatozoa per drone in *A. dorsata* was varies from 1.2×10^6–2.4×10^6 (Koeniger et al. 1990) and the number of spermatozoa per individual drone varied from 0.22×10^6 to 2.65×10^6. The average number of sperm in spermatheca of a newly mated *A. dorsata* queen was found to be 5.5 ± 0.9 million (Tan et al. 1999). After mating, the sperm is stored in the spermatheca, where it keeps its viability for years together (Pamilo 1991).

3.12 REPRODUCTIVE SWARMING

Honeybee queens generally found their colonies through reproductive swarming during breeding season (Winston 1991). During swarming, the old queen and about half of the colony's workers leave the colony and form a temporary cluster in the surrounding vegetation (Seeley 1985). From these temporary clusters, about 5% of the bees search the surrounding environment for new nest sites (Seeley et al. 1979). On return to the temporary cluster, the scouts indicate the locations found using the waggle dance. Once a new site has been decided on, the scout bees coordinate lift-off and guide the swarm to the new site. Scouts guide the swarm by flying swiftly in the direction of the site they have located (Beekman et al. 2006). In *A. mellifera*, the mother queen leaves the hive before emergence of virgin queens, whereas the newly developed queens stay behind. While several virgin queens develop simultaneously, they fight with each other until only one remains in the colony (Pflugfelder and Koeniger 2003).

The success of a swarm depends upon number of workers present in colony. Large swarms are more successful in establishing their colonies in the following year. The average distance those swarms move away from the natal nest is unknown, but on at least one occasion such a swarm travelled more than 500 m (Lindauer 1957). Similarly, the queen and a group of workers separated from the nest and start a new colony a short distance from the mother colony without forming an intermediate cluster. Reproductive swarming is influenced by favourable environmental conditions. For survival, a new swarm needs more or less immediate access to nectar and pollen for comb building and brood rearing. Otherwise, the natural mortality of workers cannot be compensated, and the swarm is later reduced to beyond the critical threshold.

In *A. dorsata*, during swarming, a small number of dances are performed by scouts before commencement of orientation flights. However, the scouts repeatedly left the swarm surface during the decision-making process. Differences in the number of circuits per dance for different locations suggest that *A. dorsata* makes some sort of assessment of site quality, with higher numbers of circuits per dance indicating sites of higher quality. Similar to *A. florea*, but in contrast to *A. mellifera*, *A. dorsata* scouts do not reduce the duration of their dance after repeated returns from scouting flights. Many scouts that dance for a non-preferred location switch preference during the decision-making process after following dances for the consensus direction in which the swarm eventually departed. Therefore, the consensus-building process of *A. dorsata* swarms relies on the interaction of scout bees on the swarm rather than the process of dance attenuation as occurs in the consensus building process of *A. mellifera* swarms (Makinson et al. 2014).

3.13 NEST-SITE SELECTION

The nest-site requirements of *A. dorsata* are intermediate between *A. mellifera* and *A. florea*. *A. dorsata* colonies are migratory, following the flow of nectar through the environment, but due to the large size of their combs, they prefer to nest in large congregations (Oldroyd and Wongsiri 2006). The differences in nesting biology have the potential to strongly influence the decision-making process. Upon relocating forest patches, the swarms either start rarely for a specific location in the surrounding canopy or migrate on to a location with better forage conditions. Alternatively, these swarms may behave in a fashion similar to those of *A. florea*, deciding where to land on the wing and then testing the suitability of their roosting spot for a few days or weeks before relocating again if it proves not to be ideal (Makinson et al. 2011). In contrast, during migration, bees select a single direction in which to fly, but information with respect to distance is highly variable in *A. dorsata* (Makinson et al. 2014).

3.14 CONCLUSION

Social insects have evolved unique methods of reproduction for their successful establishment on the earth. Despite the detailed reproductive biology of western honeybee, *A. mellifera*, is well documented; information on the reproductive biology of Asian honeybees in general—and *A. dorsata* in particular—is fragmentary. However, these studies are limited to a particular extent for the difficulties in rearing them in apiary conditions. As *A. dorsata* is the major honeybee species in India, research on its mating flights, drone congregation areas and swarming mechanisms are needed. Indeed, there is a greater scope for these studies for conservation of *A. dorsata* for honey production and utilization of these species for maintenance of biodiversity and pollination of crops and wild plants.

ACKNOWLEDGEMENTS

The author is grateful to Prof. C. Srinivas, Director, UGC-Human Resource Development Centre, Bangalore University, Bengaluru, India for his constant encouragement, and to Bangalore University, Bengaluru for providing facilities.

REFERENCES

Ayasse, M., Paxton, R.J., and Tengo, J. 2001. Mating behaviour and chemical communication in the order hymenoptera. *Annual Review of Entomology* 46: 31–78.

Baudry, E., Solignac, M., Garnery, L., Gries, M., Cornuet, J.M., and Koeniger, N. 1998. Relatedness among honeybees (*Apis mellifera*) of a drone congregation. *Proceedings of Royal Society of London B* 2009–2014.

Beekman, M., Fathke, R.L., and Seeley, T.D. 2006. How does an informed minority of scouts guide a honeybee swarm as it flies to its new home? *Animal Behaviour* 71: 161e171.

Boomsma, J.J., Baer, B., and Heinze, J. 2005. The evolution of male traits in social insects. *Annual Review of Entomology* 50: 395–420.

Brockmann, A., Dietz, D., Spaethe, J., and Tautz, J. 2006. Beyond 9-ODA: Sex pheromone communication in the European honeybee, *Apis mellifera* L. *Journal of Chemical Ecology* 32: 657–667.

Chuttong, B., Somana, W., and Burgett, M. 2019. Giant honey bee (*Apis dorsata* F.) rafter beekeeping in Southern Thailand. *Bee World* 96: 66–68.

Cobey, S.W. 2007. Comparison studies of instrumentally inseminated and naturally mated honeybee queens and factors affecting their performance. *Apidologie* 38: 390–410.

Collison, C.H. 2004. *Basics of beekeeping*. The University of Pennsylvania, University Park.

Cornuet, J.M. 1986. Population genetics. In: Rinderer, T.E. (Ed.), *Bee genetics and breeding*. Academic Press, Orlando, FL.

Delaplane, K.S., Pietravalle, S., Brown, M.A., and Budge, G.E. 2015. Honeybee colonies headed by hyperpolyandrous queens have improved brood rearing efficiency and lower infestation rates of parasitic *Varroa* mites. *PLoS One* 10: e0142985.

Free, J.B. 1987. *Pheromones of social bees*. Cornell University Press, Ithaca, NY.

Galindo-Cardona, A., Monmany, A.C., Diaz, G., and Giray, T.A. 2015. Landscape analysis to understand orientation of honeybee (Hymenoptera: Apidae) drones in Puerto Rico. *Environmental Entomology* 44: 1139–1148.

Goins, A., and Schneider, S.S. 2013. Drone quality and caste interactions in the honeybee, *Apis mellifera* L. *Insectes Sociaux* 60: 453–461.

Hölldobler, B., and Wilson, E.O. 2008. *The superorganisms: The beauty, elegance, and strangeness of insect societies*. W. W. Norton, New York, 522.

Holman, L. 2018. Queen pheromones and reproductive division of labour: A meta-analysis. *Behavioural Ecology* 29(6): 1199–1209.

Koeniger, G. 1986. Diversity in *Apis* mating systems. In: Smith, D.R. (Ed.), *Diversity in the genus Apis*. Westview Press, Boulder, 199–212.

Koeniger, G. 1991. Diversity in *Apis* mating systems. In: Smith, D.R. (Ed.), *Diversity in the genus Apis*. Westview Press, Boulder, CO, 199–212.

Koeniger, G., Koeniger, N., Mardan, M., Punchihewa, R.W.K., and Otis, G.W. 1990. Numbers of spermatozoa in queens and drones indicate multiple mating of queens in *Apis andreniformis* and *Apis dorsata*. *Apidologie* 21: 281–286.

Koeniger, G., Koeniger, N., Pechhacker, H., Ruttner, F., and Berg, S. 1989. Assortative mating in a mixed population of European honeybees, *Apis mellifera ligustica* and *Apis mellifera carnica*. *Insectes Sociaux* 36: 129–138.

Koeniger, G., Koeniger, N., and Tingek, S. 1994. Mating flights, number of spermatozoa, sperm transfer and degree of polyandry in *Apis koschevnikovi* (Buttel-Reepen, 1906). *Apidologie* 25: 224–238.

Koeniger, N., and Koeniger, G. 1991. An evolutionary approach to mating behavior and drone copulatory organs in *Apis*. *Apidologie* 23: 581–590.

Koeniger, N., and Koeniger, G. 2000. Reproductive isolation among species of the genus *Apis*. *Apidologie* 31: 313–339.

Koeniger, N., Koeniger, G., and Pechhacker, H. 2005. The nearer the better? Drones (*Apis mellifera*) prefer nearer drone congregation areas. *Insectes Sociaux* 52: 31–35.

Koeniger, N., and Wijayagunasekera, H.N.P. 1976. Time of drone flight in the three Asian honeybee species (*Apis cerana, Apis florea, Apis dorsata*). *Journal of Apicultural Research* 15: 67–71.

Kraus, F.B., Koeniger, N., Tingek, S., and Moritz, R.F.A. 2005. Temporal genetic structure of a drone congregation area of the giant Asian honeybee (*Apis dorsata*). *Naturwissenschaften* 92: 578–581.

Lindauer, M. 1957. Communication in swarm bees searching for a new home. *Nature* 179: 63–66.

Makinson, J.C., Oldroyd, B.P., Schaerf, T.M., Wattanachaiyingcharoen, W., and Beekman, M. 2011. Moving home: Nest-site selection in the red dwarf honeybee (*Apis florea*). *Behavioural Ecology and Sociobiology* 65: 945–958.

Makinson, J.C., Schaerf, T.M., Rattanawannee, A., Oldroyd, B.P., and Beekman, M. 2014. Consensus building in giant Asian honeybee, *Apis dorsata* swarms on the move. *Animal Behaviour* 93: 191–199.

Moritz, R.F.A., Kryger, P., Koeniger, N., Estoup, A., and Tingek, S. 1995. High degree of polyandry in *Apis dorsata* queens detected by DNA microsatellite variability. *Behavioural Ecology and Sociobiology* 37: 357–363.

Moritz, R.F.A., and Southwick, E.E. 1992. *Bees as superorganisms: An evolutionary reality*. Springer, Berlin, Heidelberg and New York.

Nagaraja, N., and Abrol, D.P. 2020. Reproductive biology of *Apis florea*. In: Abrol, D.P. (Ed.), *The future role of dwarf honeybee in natural and agricultural systems*. CRC Press, Taylor and Francis Group, New York, 25–35.

Nagaraja, N., and Brockmann, A. 2009. Drones of dwarf honeybee, *Apis florea* are attracted to (2E)-9-oxodecinoic acid and (2E)-10-hydroxydecinoic acid. *Journal of Chemical Ecology* 35: 653–655.

Naumann, K., Winston, M.L., Wyborn, M.H., and Slessor, K.N. 1990. Effects of synthetic honey bee (Hymenoptera: Apidae) queen mandibular gland pheromone on workers in packages. *Journal of Economic Entomology* 83: 1271–1275.

Oldroyd, B.P., Smolenski, A.J., Cornuet, J.M., Wongsiri, S., Estoup, A., Rinderer, T.E., and Crozier, R.H. 1996. Levels of polyandry and intracolonial genetic relationships in *Apis dorsata* (Hymenoptera: Apidae). *Annals of Entomological Society of America* 89: 276–283.

Oldroyd, B.P., and Wongsiri, S. 2006. *Asian honey bees: Biology, conservation, andhuman interactions*. Harvard University Press, Cambridge, MA.

Palmer, K., and Oldroyd, B.P. 2000. Evolution of multiple mating in the genus *Apis*. *Apidologie* 31: 235–248.

Pamilo, P. 1991. Evolution of sterile caste. *Journal of Theorotical Biology* 149: 75–95.

Paxton, R.J. 2005. Male mating behaviour and mating systems of bees: An overview. *Apidologie* 36: 145–156.

Pflugfelder, J., and Koeniger, N. 2003. Fight between virgin queens (*Apis mellifera*) is initiated by contact to the dorsal abdominal surface. *Apidologie* 34: 249–256.

Plettner, E., Otis, G.W., Wimalaratne, P.D.C., Winston, M.L., Slessor, K.N., Pankiw, T., and Punchihewa, P.W.K. 1997. Species- and caste-determined mandibular gland signals in honey bees (*Apis*). *Journal of Chemical Ecology* 23: 363–377.

Princen, S.A., Van Oystaeyen, A., Petit, C., Van Zweden, J.S., and Wenseleers, T. 2019. Cross activity of honeybee queen mandibular pheromone in bumble bees provides evidence for sensory exploitation. *Behavioural Ecology* 31(2): 303–310.

Rangel, J., Keller, J.J., and Tarpy, D.R. 2013. The effects of honeybee (*Apis mellifera* L.) queen reproductive potential on colony growth. *Insectes Sociaux* 60: 65–73.

Rinderer, T.E., Oldroyd, B.P., and Sheppard, W.S. 1993. Africanized bees in the US. *Scientific American* 269: 52–58.

Ronai, I., Allsopp, M.H., Tan, K., Dong, S., Liu, X., Vergoz, V., and Oldroyd, B.P. 2017. The dynamic association between ovariole loss and sterility in adult honeybee workers. *Proceedings of Royal Society B* 284: 20162693.

Seeley, T.D. 1985. *Honeybee ecology*. Princeton University Press, Princeton, NJ.

Seeley, T.D., Morse, R.A., and Visscher, P.K. 1979. The natural history of the flight of honey bee swarms. *Psyche* 86: 103–113.

Smith, M.L., Ostwald, M.M., Loftus, J.C., and Seeley, T.D. 2014. A critical number of workers in a honeybee colony triggers investment in reproduction. *Naturwissenschaften* 101: 783–790.

Strassmann, J. 2001. The rarity of multiple mating by females in the social hymenoptera. *Insectes Sociaux* 48: 1–13.

Streinzer, M., Brockmann, A., Nagaraja, N., and Spaethe, J. 2013. Sex and caste-specific variation in compound eye morphology of five honeybee species. *PLoS One* 8: e57702.

Tan, N.Q., Mardan, M., Thai, P.H., and Chinh, P.H. 1999. Observations on multiple mating flights of *Apis dorsata* queens. *Apidologie* 30: 339–346.

Tarpy, D.R., Nielsen, R., and Nielsen, D.I. 2004. A scientific note on the revised estimates of effective paternity frequency in *Apis*. *Insectes Sociaux* 51: 203–204.

Wattanachaiyingcharoen, W., Oldroyd, B.P., Wongsiri, S., Palmer, K., and Paar, R. 2003. A scientific note on the mating frequency of *Apis dorsata*. *Apidologie* 34: 85–86.

Wilson, E.O. 1971. *The insect societies*. Harvard University Press, Cambridge.

Winston, M.L. 1991. *The biology of the honeybee*. Harvard University Press, Cambridge, 294.

Woyke, J. 1975. Natural and artificial insemination of *Apis cerana* in India. *Journal of Apicultural Research* 14: 153–159.

Woyke, J., Wilde, J., and Wilde, M. 2001. *Apis dorsata* drone flights, collection of semen from everted endophalli and instrumental insemination of queens. *Apidologie* 32: 407–416.

4 Insights into the Genetics and Genomics of *Apis dorsata* Fabricius *and A. laboriosa* Smith

S. Mohankumar and T. Sonai Rajan

4.1 INTRODUCTION

Honeybee species *Apis dorsata* Fabricius and *Apis laboriosa* Smith are considered as the indispensible components of the ecosystem because of their pollinating potential, rich source of honey, and medicinal and antibacterial properties against bacterial pathogens (Partap 2011; Crane 1999; Joshi et al. 2004). The studies on phylogenetic analysis of honeybee species demonstrated that these species were grouped into a clade of "giant honeybees" (Chhakchhuak et al. 2016). Thus, these giant honeybee species are considered as an important source for the production and productivity of several economically important agricultural and horticultural crops because of their pollinating potential and other characters. Studies demonstrated that the deforestation, honey hunting, loss of nest sites, use of pesticides, and climate change are significantly reducing the population of honeybee species (Oldroyd and Nanork 2009). Thus, the conservation of *A. dorsata* and *A. laboriosa* is crucial for both ecological and economic reasons. Recently, the genetics and genomics of honeybee species have demonstrated as an effective tool for better understanding of bee biology, evolutionary process, distribution of honeybee species, and identification and selection of desired trait against pests and disease for the establishment of conservation strategy of honeybees (Fouks et al. 2021). In addition, these approaches were also reported to provide the information related to population structure, gene flow, effective population size, density of the colony and foraging behaviour, and historical events of honeybee species (Zyed 2009; Mohankumar et al. 2020), and these approaches are also employed for rapid detection of subspecies of honeybees, for solving taxonomic questions, and for indicating bee populations in decline. Integrating the genetic approaches with ecological studies also reported to provide greater insight into bee conservation. The genetics and genomics of *A. mellifera* provide opportunities to dissect out the genetic architecture and functions of numerous genes in various honeybee species, including *A. dorsata*. In this chapter, we discussed about the genetics and genomics of *A. dorsata* and *A. laboriosa*.

DOI: 10.1201/9781003294078-4

4.2 KARYOTYPE OF *A. DORSATA*

Since 1901, the details on chromosome numbers of honeybee species are widely used for the speciation of honeybees. In general, the genetic system of honeybee species is called as haplodiploidy; with this, males arise from unfertilized haploid eggs (n = 16) while females are developed from fertilized diploid eggs (2n = 32) (Aamidor et al. 2018). The identification of chromosome configuration is considered as useful key for genetically conserving strategy, as the heritable substance is organized in nucleus chromosomes (Asadi et al. 2010). In this context, only two studies have reported the chromosome details of *A. dorsata*. Earlier, Fahrenhorst (1977) reported that the species *A. dorsata* has haploid number of chromosomes (n = 16) but none of the studies have explained the chromosome nature of *A. dorsata* in detail. Recently, Rajak and Sekarappa (2021) recorded the chromosomes in *A. dorsata* populations collected from different geographical regions of southern Karnataka, India and observed a diploid (2n = 32) number of chromosomes across the samples with varied lengths, but there was no detailed information on the structure and position of chromosome.

4.3 MITOCHONDRIAL GENOME OF *A. DORSATA* AND *A. LABORIOSA*

Mitochondrial genomic data play a major role for determining the evolutionary process and phylogeny of honeybee species, and also facilitate to assess the origin of honeybees and their dispersal in different geographical regions (Tihelka et al. 2020). In this context, a few studies have been reported the mitochondrial genome of *A. dorsata* from different parts of the world (Chhakchhuak et al. 2016; Takahashi et al. 2018; Yang et al. 2017). In India, Chhakchhuak et al. (2016) sequenced and characterized the near complete mitochondrial genome of *A. dorsata* using next generation sequencing approach and reported that the mitochondrial genome was 15,076 bp long with 13 protein coding genes (PCGs), 21 tRNA genes and two ribosomal RNA genes and one AT-rich region. The phylogenetic analyses demonstrated that the resultant phylogenetic tree placed *A. dorsata* with other honeybee species of *A. mellifera scutellata*, *A. mellifera intermissa*, *A. mellifera ligustica*, *A. mellifera syiriaca*, *A. cerana*, *A. andreniformis*, and *A. florea* belonging to the tribe Apini with open and cavity nesting behaviour (Chhakchhuak et al. 2016). Similarly, in Thailand, Takahashi et al. (2018) analyzed the complete mitochondrial genome of *A. dorsata* using Illumina's NextSeq 500 technology sequencing method with the reference sequence of *A. laboriosa* (AP018039) and reported that the mitochondrial genome was assembled to 15,279 bp long and reported that the mitochondrial genome of *A. dorsata* represents a specific hymenopteran mitochondrial genome. Further phylogenetic analyses revealed that the Thailand *A. dorsata* was more closely related to the Chinese *A. dorsata*, and that species of *A. laboriosa* is a sister species of *A. dorsata*. Similarly, in southwest China, Yang et al. (2017) sequenced the complete mitochondrial genome of *A. dorsata* and recorded a length of 15,933 bp consisting of 13 protein coding genes, 21 transfer RNA genes (tRNA), two riposomal RNA genes (rRNA), and one control region (D-loop) with high level of AT base content and also reported that the genome has comparable codon usage and organization of genes reported in other species of honeybees and hymenopteran insects. All these *A. dorsata* mito-genome displayed the same orientation and arrangement as those reported in other honeybee species and hymenopterans. Chhakchhuak et al. (2016) studied the complete mitochondrial genome of *A. laboriosa* collected from Mizoram, India, and reported that the length of the sequences was 15,266 bp, and also demonstrated that the genome arrangement of *A. laboriosa* was similar to other *Apis* species. In Nepal, Takahashi et al. (2018) observed the average AT content of the *A. laboriosa* mitochondrial genome is 84.7%. Start codons ATG and ATT were found in three and ten genes, respectively, while stop codons TAA and TAG were observed in 12 and one gene, respectively. Recently, Mohankumar et al. (2013) assessed the genetic diversity of *A. dorsata* collected from different geographical regions of India using mitochondrial deoxyribonucleic acid (DNA) (COI) and demonstrated that the sequencing of *COI* region yielded 12 discrete haplotypes (h = 12) with the haplotype diversity of 0.849 and a variance of haplotype diversity of 0.0006, and indicated the prevalence of huge genetic diversity across the population of *A. dorsata*. Overall, the mitochondrial genome analysis of *A. dorsata* revealed that this species is originated from Asian country, and further research is needed from other parts of the country for assessing the evolutionary status of this bee species. A phylogenetic analysis inferred from the 13 mitochondrial PCGs, based on maximum likelihood, indicates that *A. laboriosa* and *A. dorsata* are very closely related. Takahashi et al. (2018) found that the genetic distance between *A. laboriosa* and *A. dorsata* is 0.197, indicating that while they are genetically similar enough to be considered sister species, they are indeed two distinct species.

4.4 GENOME INFORMATION AND GENE DISCOVERY OF *A. DORSATA*

With the recent advent of next generation sequencing technology, numerous genome sequences have been developed, providing an opportunity for better understanding of the evolutionary and biological process of insect pests, especially social insects. Honey Bee Genome Sequencing Consortium (2006) was established during 2006, and the genetic architecture of several bee species have been explored and a number of gene annotation versions were also developed and made available in NCBI. Oppenheim et al. (2020) performed the whole genome sequencing and assembly of *A. dorsata* using a hybrid Oxford Nanopore and Illumina approach and reported that 224 Mb genome has an N50anf 35 kb with the largest scaffold of 302 kb and using MAKEr identified 13,517 protein coding genes. More than 80% of the predicted genes had matches from other hymenopteran insects. Mohankumar et al. (2013) assessed the genetic diversity of *A. dorsata* collected from different geographical regions of India using COI and microsatellites, and demonstrated that the sequencing of *COI* region yielded 12 discrete haplotypes (h = 12) with the haplotype diversity of 0.849 and a variance of haplotype diversity of 0.00056, and indicated the prevalence of huge genetic diversity across the population of *A. dorsata*. The draft genome of *A. laboriosa* based on the de novo assembly is 226.1 Mbp in length with a scaffold N50 size of 3.34 Mbp, a GC content of 32.2%, a repeat content of 6.86%, and a gene family number of 8,404. Comparative genomic analysis revealed that the genes in the *A.*

laboriosa genome have undergone stronger positive selection (2.5 times more genes) and more recent duplication/loss events (6.1 times more events) than those in the *A. dorsata* genome (Lin et al. 2021).

Recently, Fouks et al. (2021) sequenced, assembled, and annotated the genomes of *A. dorsata* and produced high quality genome assembles with the length and GC content of 230 Mb, N50:732kb, GC: 32.5% with size of scaffolds: 200 bp–3.6 Mb, total count 4040. In addition, they also mentioned about the repetitive sequences of 17.5% with 40.4 Mb data. A total of 584 gene sets were also reported to be recorded in the *A. dorsata* genome assembly. Gene ontology analyses demonstrated that the *A. dorsata* and *A. mellifera* shared similarly functional categories involved in cellular ion exchange. Further, the overlap analyses of genes identified from the *A. dorsta* genome assemble with other *Apis* species revealed that there was no significant overlap with other *Apis* species. Though, they were reported that the overlap of positively selected genes with genes present in quantitative trait locus (QTL) studies. Enrichment analysis through comparing GO term annotations of the genes specific to lineage demonstrated that *A. dorsata* genome contained two lineage specific genes related to vision, *gelsolin-like* and *calphotin-like*.

Identification of chemosensory genes and their diversification analysis is crucial for evolutionary process of insect pests (Brand et al. 2020). In this context, Fouks et al. (2021) annotated the genes through manual annotation and identified varied number of genes from chemosensory families including odorant receptors (161 nos), gustatory receptors (22 nos), ionotropic receptors (10 nos), odorant binding proteins (20 nos), and chemosensory genes (6 nos). The number of chemosensory genes identified in the *A. dorsata* was reported to be similar to the other species of *A. mellifera* and *A. florea*. Comparative genomic analysis of *A. dorsata* identified the molecular signature of adaptations to long distance migration in *A. dorsata* lineage. In addition, night foraging behaviour of *A. dorsata* was also justified that this behaviour might be explained by homologs of specific genes involved in the movement towards light (i.e. phototaxis) viz., *gelsolin*-like and *calphotin*-like (Stocker et al. 1999; Yang and Ballinger 1994). In addition, several other genes *viz*., Forkhead box protein P1-like, Arrestin domain-containing protein 17-like, Intersectin-1-like, Dynamin and Debiquitinase receptor 1, and major royal jelly protein gene (MRJP), have also been identified in *A. dorsata*.

4.5 MAJOR ROYAL JELLY PROTEIN GENE (MRJP) IN *A. DORSATA*

The protein-rich substances secreting from the hypopharyngeal gland of worker honeybees is called royal jelly and it plays a major role in the nutrition of larvae and development of the queen caste (Fujita et al. 2013). In addition, it is also considered as a natural source of protein, vitamins, minerals, amino acids, lipids, sugars, fats, and other nutrients (Schmitzova et al. 1998). Hanes and Simuth (1992) identified the first royal jelly protein and termed as major royal jelly protein gene (MJPR). Later, nine other proteins of royal jelly were also observed and the encoding genes (MJPR–MJPR 9) were identified in the honeybee genome (Honey Bee Genome Sequencing Consortium 2006). The prevalence of these genes was also reported to be varied across the bee species (Helbing et al. 2017). Earlier, Albert and Schmitz (2002) characterized the MRJP-like sequences in *A. dorsata* and identified the homologues of MRJP3 and MRJP5 termed as *adMRJP3* and *adMRJP5*, and they also identified the polymorphic nature of *adMRJP3* and demonstrated that the *adMRJP3* repeat is highly informative locus for the population studies of *A. dorsata*. Later, Helbing et al. (2017) analyzed the MRJP gene in three species of *Apis* including *A. dorsata* with long amplicon sequencing and confirmed the arrangement of MRJP3–5. The MRJP genes 3, 1, 4, 6 and 5 displayed an identical location at the 5' end of the cluster and MRJP8 and MRJP9 were found at the 3' end. *A. dorsata* genome sequence reported to be comprised nine full-length *mrjp* genes. Similar to *A. florea*, the MRJP amino acid based phylogenetic reconstruction revealed the prevalence of new member MRJP10 and the expression and function of MRJP10 needs to be explored for better understanding in *A. dorsata*. Helbing et al. (2017) concluded that the comparative analyses of MRJP gene expression pattern in *A. dorsata* would assist to infer the expression pattern of genes and or to assess the functional redundancy of gained or lost genes.

4.6 GENE RESPONSIBLE FOR BODY COLOUR IN *A. DORSATA*

Body colour of the honeybee species is considered as an important and most distinct character. The body colouring pattern of honeybees is reported to be varied between and within the species, as well as between the castes of queen, workers, and drones in the same nest or colony. Studies have demonstrated that the minimum of seven different loci of the genes may alter the colouring pattern of honeybee species (Roberts and Mackensen 1951). In *A. dorsata*, Woyke (1997) reported that the body colour of worker bees is yellow, whereas the queens and drones are brown in colour. The thoracic area of queens and drones are dark brown but the scuttelum is brown in colour. The abdominal portion of *A. dorsata* is light brown with dark brown bands. In addition, Woyke (1997) demonstrated that the gene responsible for the body colour of *A. dorsata* was *Do*.

4.7 SEX DETERMINATION IN *A. DORSATA*

Sex determination is considered a fundamental process of life and the mechanism of sex determination was reported to be varied from species to species (Bull 1983). Earlier, Dzierzon (1845) discovered that the haplo-diploid nature of sex is present in honeybee species and other hymenopteran insects including ants, wasps, and some parasitoids, in which males develop from unfertilized embryos and females of either queen or worker bees develop from fertilized embryos, and it is indicated that the sexual characteristics of honeybee is determined by fertilization or ploidy level (Nachtsheim 1916). Later, studies demonstrated that the allelic composition of single locus determines the sex (Woyke 1963; Crozier 1971); such a mechanism of sex determination is called complementary sex determination (*csd*). Lechner et al. (2013) have assessed the molecular control behind the determination of sex in honeybees, finally identifying and tracing back a detailed number of *csd* alleles to develop knowledge about the variability of the *csd* gene over evolutionary periods. They looked at a data set of 244 csd sequences from all the caste of bee species and showed that the total number of *csd* alleles ranged from 53 to 87, which was reported to be much higher than the earlier report. Using an evolutionary model, they also extrapolated the presence of total 116–145 *csd* alleles globally and deciphered the minimum number of mutations leading to the heterozygous *csd*. They also assessed the evolutionary period of the *csd* and found that the novel *csd* function affecting the sex determination arises about every 400,000 years. Overall, the genetics and genomics base of sex determination studies provides the detailed view of genetic diversity and the evolutionary pressures shaping the sex determination in bees, as well as how changes in *csd* affect the colony fitness of honeybee species. Most of these studies have used *A. mellifera* as a test insect to assess the sex determination and very limited information is available with other species of honeybee. Liu et al. (2012) assessed the sex determination gene in *A. dorsata* through sequencing analysis and reported that the *csd* allele is prevalent in *A. dorsata* populations collected from different provinces of China, and also demonstrated very little or no clustering according to geographical origin of each *csd* allele and concluded that there is no founder effect in the *csd* gene due to high polyandrous nature of *A. dorsata* queens.

4.8 POPULATION STRUCTURE AND GENETIC DIVERSITY OF *A. DORSATA*

Studies demonstrated that the seasonal migration, absconding to alternative nest sites, nesting behaviour (i.e. aggregation and solitary nesting), mating behaviour and frequencies, unsustainable methods of harvesting or hunting, widespread use of insecticides, destruction of appropriate nesting sites, or deforestation may the reason for the genetic variation and population structure of *A. dorsata* populations across the world (Paar et al. 2004; Sahebzadeh et al. 2012; Rattanawannee et al. 2012). Several molecular markers have been employed for assessing the population structure and genetic diversity of *A. dorsata* (Paar et al. 2004). Among them, the nuclear (i.e. SSR) and mitochondrial markers (COI, II, and others) are often used for genetic diversity analysis. Paar et al. (2004) assessed the genetic structure of *A. dorsata* population collected from northeastern parts of India using microsatellite markers and demonstrated that the significant population structuring occurs between the geographical areas and also reported that the level of structuring caused by aggregation exceeds the differentiation attributable to geographic areas. Cao et al. (2012) studied the genetic structure of *A. dorsata* populations collected from different geographical regions of China based on DNA microsatellite markers and results revealed that the populations from various geographical regions are significantly differentiated and also demonstrated that the population from Hainan Island very probably diverged from that of the mainland of China and also exhibits comparatively lower level of genetic diversity, which is probably the result of founder effect and genetic drift. It is remarkable that the *A. dorsata* populations from Hainan have lesser level of genetic differentiation than their continental populations (Estoup et al. 1996; Widmer et al. 1998). Similarly, significant differentiation was also reported from the populations of *A. dorsata* collected from the mainland and islands of Thailand (Insuan et al. 2007). This clearly shows that the samples of *A. dorsata* collected from different geographical regions the country demonstrate significant genetic structure and it is indicated that the seasonal migration of this bee species may be the possible reason for greater genetic differentiation within the country. The genetic diversity analysis of *A. dorsata* samples collected from islands shows a lower level of genetic diversity. Genetic variation of *A. dorsata* by DNA sequencing of the mtDNA (mitochondrial DNA) (COI) region have been conducted by Tanaka et al. (2003), Insuan et al. (2007), and Mohankumar et al. (2013), which determined a huge amount of genetic variation in different geographical populations of *A. dorsata*. Recently, Mohankumar et al. (2013) assessed the genetic diversity of *A. dorsata* collected from different geographical regions of India and demonstrated Bayesian clustering analyses of population structure using microsatellite genotypes suggested that the sampled individuals are derived from five hypothetical populations (K = 5) and suggested that the populations sampled from different locations were geographically clustered in nature.

4.9 CONCLUSION

The studies on genetics and genomics of *A. dorsata* have provided knowledge on genetic architecture, cytological background, sexual background, and functional characterization of several genes of *A. dorsata*, and it provides an opportunity to understand about the biology of *A. dorsata*. In addition, studies of population genetic analyses of *A. dorsata* demonstrated the historical events and dispersal pattern of *A. dorsata* and the genetic variation and population structure of *A. dorsata* from different geographical regions. Several aspects, especially the global gene expression profiling and SNP (single-nucleotide polymorphism) genotyping of *A. dorsata*, are yet to be studied, and future research on this area may be helpful for better understanding of *A. dorsata*. Recent research has removed *A. laboriosa* from inclusion within *A. dorsata*, as a separate species, with supporting evidence including a significant region of sympatry. Population-level genetic studies are very scant in *A. laboriosa*. In the future, more in-depth comparative and population-level studies with *A. laboriosa* will help to describe the landscape of local adaptation between these two species, thereby facilitating the development of feasible strategies for protecting these important pollinators. Overall, the emerging field of genetics and genomics of *A. dorsata* has considered as potential tool for the development of effective conservation strategies for *A. dorsata*.

REFERENCES

Aamidor, S.E., Yagound, B., Ronai, I., and Oldroyd, B.P. 2018. Sex mosaics in the honeybee: How haplodiploidy makes possible the evolution of novel forms of reproduction in social Hymenoptera. *Biological Letter* 14: 20180670. http://dx.doi.org/10.1098/rsbl.2018.0670.

Albert, S., and Schmitz, J. 2002. Characterization of major royal jelly protein like DNA sequences in *Apis dorsata*. *Journal of Apicultural Research* 41(3–4): 75–82.

Asadi, N., Ghafari, S.M., Gharahdaghi, A.A., Tahmasebi, G.H.H., and Khederzadeh, S. 2010. Karyotype and C- banding analysis of haploid male chromosome of *Apis florea* F. *African Journal of Biotechnology* 9(27): 4471–4474.

Brand, P., Hinojosa-Díaz, I.A., Ayala, R., Daigle, M., Yurrita Obiols, C.L., Eltz, T., Ramírez, S.R. 2020. The evolution of sexual signaling is linked to odorant receptor tuning in perfume-collecting orchid bees. *Nature Communication* 11: 244.

Bull, J.J. 1983. *Evolution of sex determining mechanisms*. Benjamin, Cummings Publishing Company Inc., Menlo Park, CA, 316.

Cao, L.F., Zheng, H.Q., Liang, H., and Hepburn, H.C. 2012. Genetic structure of Chinese *Apis dorsata* population based on microsatellite. *Apidologia* 43: 643–651.

Chhakchhuak, L., Mandal, S.D., Gurusubramanian, G., Sudalaimuthu, N., Gopalakrishnan, C., Mugasimangalam, R.C., and Vanramliana, K.N.S. 2016. Complete mitochondrial genome of the Himalayan honey bee, *Apis laboriosa*. *Mitochondrial DNA Part A* 27(5): 3755–3756.

Crane, E. 1999. *The world history of beekeeping and honey hunting*. Routledge, New York, 720.

Crozier, R.H. 1971. Heterozygosity and sex de-termination in haplo-diploidy. *The American Naturalist* 105: 399–412.

Dzierzon, J. 1845. Gutachten ueber die von Herrn Direktor Stoehr in ersten und zweiten Kapitel des General-Gutachtens aufgestellten Fragen. *Bienenzeitung* 1: 109–113.

Estoup, A., Solignac, M., Cornuet, J.M., Goudet, J., and Scholl, A. 1996. Genetic differentiation of continental and island populations of *Bombus terrestris* (Hymenoptera: Apidae) in *European Molecular Ecology* 5: 19–31.

Fahrenhorst, H. 1977. Nachweis übereinstimminder chromosomen-Zahlen (n=16) bei allen 4 Apis-Arten. *Apidiologie* 8: 89–100.

Fouks, B., Brand, P., Nguyen, H.N., Herman, J., Camara, F., Ence, D., Hagen, D.E., Hoff, K.J. et al. 2021. The genomics and basis of evolutionary differentiation among honey bees. *Genome Research* 31(7): 1–13.

Fujita, H., Kozuka-Hata, H., Ao-Kondo, T., Kunieda, M., and Oyama, T.K. 2013. Proteomic analysis of the royal jelly and characterization of the functions of its derivation glands in the honeybee. *Journal of Proteome Research* 12: 404–411.

Hanes, J., and Simuth, J. 1992. Identification and partial characterization of the major royal jelly protein of the honey bee (Apis mellifera L.). *Journal of Apicultural Research* 31: 22–26.

Helbing, S., Lattorff, H.M., Moritz, R.F.A., and Buttstedt, A. 2017. Comparative analyses of the major royal jelly protein gene cluster in three Apis species with long amplicon sequencing. *DNA Research* 24(3): 279–287.

Honey Bee Genome Sequencing Consortium. 2006. Insights into social insects from the genome of the honeybee *Apis mellifera*. *Nature* 443: 931–949. https://doi.org/10.1038/nature05260.

Insuan, S., Deowanish, S., Klinbunga, S., Sittipraneed, S., Sylvester, H.A., and Wongsiri, S. 2007. Genetic differentiation of the giant honey bee (Apis dorsata) in Thailand analyzed by mitochondrial genes and microsatellites. *Biochemical Genetics* 45: 345–361.

Joshi, S., Ahmad, F., and Gurung, M.B. 2004. Status of *Apislaboriosa* populations in Kaski district, Western Nepal. *Journal of Apicultural Research* 43(4): 176–180.

Lechner, S., Ferretti, L., Schöning, C., Kinuthia, W., Willemsen, D., and Hasselmann, M. 2013. Nucleotide variability at its limit? Insights into the number and evolutionary dynamics of the sex-determining specificities of the honey bee *Apis mellifera*. *Molecular Biology and Evolution* 31(2): 272–287.

Lian-Fei, C., Zheng, H.Q., Hu, F.L., and Hepburn, H. 2012. Genetic structure of Chinese Apis dorsata population based on microsatellites. *Apidologie* 43(6): 643–651.

Lin, D., Lan, L., Zheng, T., Shi, P., Xu, J., and Li, J. 2021. Comparative genomics reveals recent adaptive evolution in himalayan giant honeybee *Apis laboriosa*. *Genome Biology and Evolution* 13(10): evab227. https://doi.org/10.1093/gbe/evab227.

Liu, Z., Wang, Z., Yan, W., Wu, X., Zeng, Z., and Huang, Z. 2012. The sex determination gene shows no founder effect in the giant honey bee, *Apis dorsata*. *PLoS One* 7.

Mohankumar, S., Preetha, B., Fakrudin, B., Omkar-Babu, K., Chapalkar, S., Lazar, K.V., and Vikramnath, S. 2013. Genetic diversity and phylogeography of *A. dorsata* Fabricius. *Monograph on Morphometry and Phylogeography of Honey Bees and Stingless Bees in India* 66–100.

Mohankumar, S., Sonai Rajan, T., Praghadeesh, M., Saravanan, P.A., and Srinivasan, M.R. 2020. The genetics and genomics of the dwarf honeybee *Apis florea* fabricius. In: Abrol, D.B. (Ed.), *The future role of dwarf honeybee in natural and agricultural systems*. CRC Press, Boca Raton, 37–45.

Nachtsheim, H. 1916. Zytologische Studien ueber die Geschlechtsbestimmung bei der Honigbiene (*Apis mellifera*). *Arch Zellforsch* 11: 169–241.

Oldroyd, B.P., and Nanork, P. 2009. Conservation of Asian honey bees. *Apidologie* 40(3): 296–312.

Oppenheim, S., Cao, X., Rueppel, O., Krongdang, S., Phokasem, P., DeSalle, R., Goodwin, S., Xing, J., Chantawannakul, S., and Rosenfeld, J.A. 2020. Whole genome sequencing and assembly of the Asian honey bee *Apis dorsata*. *Genome Biology and Evolution* 12(1): 3677–3683. https://doi.org/10.1093/gbe/evz277.

Paar, J., Oldroyd, B.P., Huettinger, E., and Kastberger, G. 2004. Genetic structure of an *Apis dorsata* population: The significance of migration and colony aggregation. *Journal of Heredity* 95: 119–126.

Partap, U. 2011. The pollination role of honeybees. In: Hepburn, R., and Radloff, S.E. (Eds.), *Honeybees of Asia*. Springer-Verlag, Berlin, 227–256.

Rajak, B., and Sekarappa, B. 2021. Record of chromosomes in giant honeybee, Apis dorsata Fabricius population from different geo-locations of Southern Karnataka, India. *Journal of Entomology and Zoology Studies* 9(3): 22–25.

Rattanawannee, A., Chanchao, C., and Wongsiri, S. 2012. Geometric morphometric analysis of giant honeybee (*Apis dorsata* Fabricius, 1793) populations in Thailand. *Journal of Asia-Pacific Entomology* 15: 611–618.

Roberts, W.C., and Mackensen, O. 1951. Breeding improved honey bees. II: Heredity and variation. *The American Bee Journal* 91: 328–330.

Sahebzadeh, N., Mardan, M., Ali, A.M., Tan, S.G., Adam, N.A., and Lau, W.H. 2012. Genetic relatedness of low solitary nests of *Apis dorsata* from Marang, Terengganu, Malaysia. *PLoS One* 7(7): e41020.

Schmitzova, J., Klaudiny, J., Albert, S., Hanes, J., Schroder, W., Schrockengost, V., Judova, J., and Simuth, J. 1998. A family of major royal jelly proteins of the honeybee *Apis mellifera* L. *CMLS: Cell Molecular Life Science* 54: 1020–1030.

Stocker, S., Hiery, M., and Marriott, G. 1999. Phototactic migration of dictyostelium cells is linked to a new type of gelsolin-related protein. *Molecular Biology of the Cell* 10: 161–178. https://doi.org/10.1091/mbc.10.1.161.

Takahashi, J., Ilyasov, R.A., Park, J., and Kwon, H. 2018. Phylogenetic uniqueness of honeybee Apis cerana from the Korean Peninsula inferred from the mitochondrial, nuclear, and morphological data. *Journal of Apicultural Science* 62: 189–214.

Tanaka, H., Suka, T., Kahono, S., and Rousik, D. 2003. Mitochondrial variation and genetic differentiation in hone bees (*Apis cerana*, *A. koschernikovi* and *A. dorsata*) of Borneo. *Tropics* 13(2): 107–117.

Tihelka, E., Cai, C., Pisani, D., and Donoghue, P. 2020. Mitochondrial genomes illuminate the evolutionary history of the Western honey bee (*Apis mellifera*). *Scientific Reports* 10(1): 14515.

Widmer, A., Schmid-Hempel, P., Estoup, A., and Scholl, A. 1998. Population genetic structure and colonization history of Bombus terrestris s.l. (Hymenoptera: Apidae) from the Canary Islands and Madeira. *Heredity* 81: 563–572.

Woyke, J. 1963. What happens to diploid drone larvae in a honeybee colony. *Journal of Apicultural Research* 2: 73–76.

Woyke, J. 1997. Expression of body and hair color in three adult castes of the red honeybee Apis koschevnikovi von Buttel-Reepen, 1906 in Sabah, Borneo. *Apidologie* 28: 275–286.

Yang, J., Xu, J., He, S., and Wu, J. 2017. The complete mitochondrial genome of wild honeybee *Apis florea* (Hymenoptera: Apidae) in south-western China. *Mitochondrial DNA B* 2: 845–846.

Yang, Y., and Ballinger, D. 1994. Mutations in calphotin, the gene encoding a Drosophila photoreceptor cell-specific calcium-binding protein, reveal roles in cellular morphogenesis and survival. *Genetics* 138: 413–421. https://doi.org/10.1093/genetics/138.2.413.

Zyed, A. 2009. Bee genetics and conservation. *Apidologie* 40(3): 237–262.

5 Nesting Biology of Giant Honeybees *Apis dorsata* and *Apis laboriosa*

Jerzy Woyke

5.1 INTRODUCTION

Prof. Jerzy Woyke studied the nesting biology of Giant honeybees *Apis dorsata* and *A. laboriosa* in Nepal and India. Details of observations made by him are reproduced in his words in what follows. Data presented in this chapter concerning biology of nesting behaviour of *A. dorsata* and *A. laboriosa* in Nepal and India were collected when I investigated other problems of the biology of these bees with co-authors (Woyke et al. 2001a, 2001b, 2003, 2004a, 2004b, 2005). Observations were made under the following headings.

5.2 NEST CONSTRUCTION

Figure 5.1 shows Jerzy Woyke with a removed *A. dorsata* nest. Inside the nests of *Megapis* bees is a comb. Figure 5.2 shows bee brood in the bottom and honey at the top. In nature, the comb is covered with protective curtain of worker bees (Figure 5.3).

5.3 LOCALIZATION

Megapis bees *Apis dorsata* and *Apis laboriosa* are open nesting colonies. I investigated the following *Megapis* bees: *Apis dorsata*, *Apis dorsata breviligula and Apis laboriosa*. I conducted the investigations in Nepal in 1998 and 1999, in India 2002, in the Philippines in 2004 and in Bhutan in 2008. I counted the number of active bee colonies and the number of fresh recently abandoned combs at each nesting site. I counted also the number of remains of queen cells at the lower margin of abandoned combs, when the cells were visible. In Chitwan, where *A. dorsata* bees covered the combs, I smoked the workers from the lower margin of the combs. At six nesting sites occupied by *A. dorsata* and *A. laborios*a in Nepal and India, I conducted also the whole day uninterrupted observations. Data concerning Bhutan were collected in 2008.

I conducted four types of observation:

1. Sporadic observations.
2. Whole-day observations.
3. Ten-day consecutive observations.
4. Observations during different seasons of the year.

DOI: 10.1201/9781003294078-5

FIGURE 5.1 J. Woyke with removed *A. dorsata* nest.

FIGURE 5.2 *A. dorsata* comb with brood at the bottom and honey at the top.

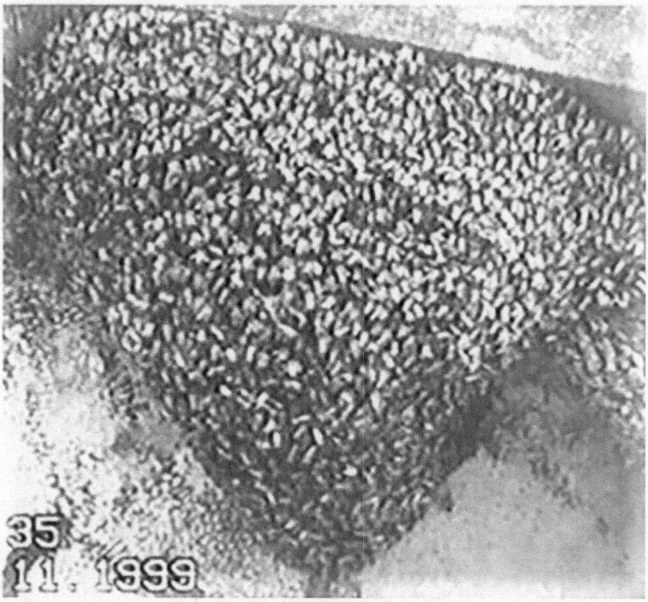

FIGURE 5.3 *A. dorsata* comb covered with protective curtain of worker bees.

Characteristics of a number of nesting sites and number of active colonies of *Apis dorsata* and *Apis laboriosa* investigated in Nepal. India and Bhutan during 1998–2008 years are presented in Table 5.1.

Data presented in Table 5.1 show that I investigated the behaviour of 42 nesting sites at which 659 nests of *Megapis* bees were present.

I found nests of the following bees: *A. dorsata*, *A. laboriosa* and *A. d. breviligula*.

A. dorsata occupied eight nesting sites with 263 nets.

A. laboriosa occupied 18 mountain cliffs with 495 nests.

A. d. brevuligula occupied two sites with five nests.

Of *A. dorsata* colonies in Nepal, I found one nesting site in Chitwan National Park with four nests. One nest was under a large limb of a tree (Figure 5.4).

I found 132 colonies nesting under two water tanks on two towers (Figure 5.5) and on two houses on the campus of the Tribhuvan University in Rampur, and on a nearby water tank tower in Bharatpur. I found 65 nests under the bottom of the water tank on the tower in Rampur (Figure 5.6).

5. *dorsata* colonies in India were observed in Bangalore in March 2002. The colonies occupied two nesting sites. One, with 29 colonies, was on the polytechnic building in the centre of the city (Figure 5.7) The other, with 98 active colonies, was on a banyan tree at the campus of the Agricultural University (Figure 5.8).

I investigated together 42 nesting sites at which 659 colonies nested.

A. dorsata occupied nine nesting sites with 261 nests.

A. laboriosa occupied 31 nesting sites with 391nests.

TABLE 5.1
Characteristics of Number of Nesting Sites and Number of Active Colonies of *Apis dorsata* and *Apis laboriosa* Investigated in Nepal, India and Bhutan during 1998–2008

Species	Country	Year	Place	No. of Sites	No. of Nests	Data	Period
Apis dorsata	Nepal	1998	Chitwan	1	4	20.9	1 day
Apis dorsata	Nepal	2000	Tribhuvan University Rampur	5	132	11.10-15.01	45 days
Apis dorsata	India	2002	Banglore Politechnik Banyan tree	2	29	03–14.05	10 days
Apis laboriosa	Nepal	1998	Bhote Koshi ryversite Annapurna	3	61	27.11-02.12	9 days
Apis d. breviligu	Philippines	2004	L. Banios Alfonso	1 1	4 1	28.02-08.03	9 days
Apis laboriosa		2008	Douchula pass, Tang valley		92	25–29.09	5 days
Total				**42**	**659**		**90 days**

FIGURE 5.4 *A. dorsata* nest under limb of a tree in Cheetwan National Park in Nepal, 1998.

FIGURE 5.5 Two water tank towers with 86 *A. dorsata* nests in Rampur, Nepal, 1999.

FIGURE 5.6 Water tank on tower in Nepal in Rampur with 65 *A. dorsata* nests.

FIGURE 5.7 *A. dorsata* nests under the balcony of the polytechnic building in Bangalore India, 2002.

FIGURE 5.8 *Ficus bangalense* with as many as 96 *A. dorsata* nests at the Agriculture University in Bangalore, India, 2002.

FIGURE 5.9 *Apis dorsata* nests under cliffs of Annapurna mountain in Nepal 1999.

A. dorsata breviligula occupied two nesting sites with five nests.

A. dorsata and *A. laboriosa* nests in Nepal were examined at 19 nesting cliffs in 1998 and 1999 in the Himalayas. I observed 61 nests in 1998 and 238 nests in 1999. Three nesting cliffs were situated at Bhote Koshi riverside along the highway from Kathmandu to Kodari at the Nepal–Tibet border. The first cliff was located in Chale (alt. 1178 m), near the Chaku village. The second Kodari nesting site was further north near the Kodari waterfall (alt. 1475 m) and the third in Tatopani, near the bridge to China (alt. 1520 m). I examined those nesting cliffs in spring, 28 March 1988, and in the winter, 5–6 December 1999. The fourth nesting cliff was on the Annapurna slope at the Modi Khola riverside near Landrung (alt. 1250 m) (Figure 5.9).

5.4 PERIODIC MASS FLIGHTS OF *APIS DORSATA*

Young *Apis dorsata* worker bees perform periodic mass flights (PMFs). During those flights, the bees defecate and made orientation flights of locations of their nests. PMFs last in *Apis mellifera* about 30 minutes and in *Megapis* bees about 5 minutes. This description presents results of investigations which I conducted in Rampur, Chitwan, Nepal during five months from October–April 2000. I conducted the whole day continuous observations from 08:00–19:00 hours. I observed six nests of *Apis dorsata* in different seasonal conditions.

5.5 PERIODIC MASS FLIGHTS PERFORMED BY *A. DORSATA* IN DIFFERENT SEASONS

I conducted investigations on PMFs of *Apis dorsata* in Rampur, Chitwan, Nepal, during October–April 2000. I observed during whole days *A. dorsata* nests from 08:00–19:00 hours in different seasonal conditions. At the beginning of the investigations in Nepal (19 October), worker bees from six nests performed the PMF. Fourteen days later, at the end of observations (2 April), the bees only from five nests made PMFs. During the day, *A. dorsata* workers made 0–6 PMFs, each of which lasted about five minutes. In cool conditions (Nov. 15, 23.7°C), worker bees from six nests performed 20 PMFs. The index of PMF; (No. PMF per No of nests = 11 PMF per six nests = 11 / 6 = **1.8**). The PMF activity lasted from 10:00–18:00 hours during the day.

In warm conditions (15 Nov. 23.7°C), worker bees from six nests performed 19 PMFs. The index = 19 PMF / 6 nests = **3.2**. In hot conditions (2 April 33.6° C), worker bees from five nests performed 11 PMFs. The index 11 PMF per five nests = **2.2**. Thus, the index in the three conditions—cool, warm and hot—was; **1.8, 3.2** and **2.2**. The bees performed proportionally the highest No. of PMFs in warm conditions. The PMF activity in the hot condition lasted 11 hours, between 8:00 hours in the morning to 19:00 hours in the evening. However, PMFs were performed not during the whole day. Worker bees made PMFs only in the morning, 8:00–10:00 hours, and in the evening, 16:00–19:00 hours. Lack of six hours of no PMFs occurred between. Worker bees evidently protected the bee brood from being overheated.

Worker bees performed the highest number of PMFs in worm conditions, index 3.3. There was not risk to chill the bee brood in the morning or to overhead it at the evening.

Such distribution of PMFs during the day protected the bee brood from being chilled in low temperature or overheated at high temperature. Thus, the index PMF/number of nests is a quite good characteristic of the nest's performance. The index shows two pieces of information: the total number of PMFs and the mean PMFs per one nest.

The number of PMFs performed by *A. dorsata* bees in successive days differed very much. However, I found that after performing low number of PMFs per day (2, 3, 4), the bees made high number of PMFs the next days (4, 5, 6); the reverse was also true. After performing a high number of PMFs (4, 5, 6), the bees made a low number of PMFs (3, 2, 1). Thus, *A. dorsata* bees performed PMFs in cycles of 2–3 PMF per day.

5.6 THE NUMBER OF PMFS IN CONSECUTIVE DAYS

At the time, I investigated PMFs of *A. dorsata* bees nesting under the balcony of the polytechnic in Bangalore India (2002); many plants were in bloom in the parks, gardens and street avenues, particularly *Ailantus excelsia*, *Azadrichta indica*, *Bombax* sp., *Cassia fistula*, *Ceiba pentandra*, *Jacaranda mimosifolia* and *Mangifera indica*. In contrast, the banyan tree site at the Agriculture University was characterized by harvested fields with scattered flowering trees of *Azadrichta indica*, *Mangifera indica* and *Eucalyptus* sp.

5.7 NUMBER OF PMFS AND DIFFERENT FOOD AVAILABILITY

The relation of the index of PMF activity in the city of Bangalore and the harvested agriculture fields was (5 March, 32 / 6 = **5.3**. Thus the activity of worker bees in the centre of Bangalore was 5.3 times higher than at the University of Agriculture, where the fields were harvested. The observations on PMFs performed within 10 days by *A. dorsata* worker bees nesting on banian tree at the Agriculture University in Bangalore were recorded during 2002. The Index PMF in the Agriculture University 10 March was 19 / 5 = **3.8**.

Thus, the relation of PMF activity of worker bees nesting in polytechnic and banian tree (5.3 / 3.8 = 1.4) was 1.4 times higher than of those nesting on the banyan tree at Agriculture University. I suggest that both the short duration of five minutes of PMF activity, as well as the distribution of PMFs during the day, are strategies to protect the bee brood from thermal damages.

5.8 OPEN-AIR-NESTING *MEGAPIS* DIFFER IN HYGIENIC BEHAVIOUR FROM THE CAVITY-NESTING *APIS* BEES

The brood of honeybees is subject to be infested for different reasons. It is infected by bacteria and viruses and by parasitic mites. Among the most dangerous bacterial diseases is European foul brood (*Streptococcus pluton*), which cause the death of larvae (unsealed brood) and American foul brood (*Baccillus larvae*), which kills prepupae and pupae in sealed cells. Two mites, *Varroa destructo*r and *Tropilaelaps clareae*, parasitize pupae and prepupae in sealed cells (Woyke 1987). The mite *Tropilaelaps clareae* coms out from sealed cells, together with emerging young worker bees. Next, the mite enters into other cells with bee brood. The mite reproduces there on bee larvae or preapupae. *V. destructor* can feed on adult bees and survive for several months. However, *T. clareae* is unable to pierce the intersegmental membrane of worker bees. It is not able to feed also on adult bees. Without bee brood in the colonies, the mite die within a few days (Woyke 1984, 1985a). *T. clareae* is dangerous in those world areas where bee brood enjami interruption does not occur. However, *T. clareae* is not dangerous in world areas where bee brood rearing does not occur for ten days or more (Woyke 1985b). Honeybees have been known for long time to maintain clean and hygienic conditions in their colonies. This important social behaviour encompasses all aspects of detection, uncapping and removal of dead, diseased and infested larvae and pupae from their combs. This display of hygiene behaviour, including its characteristic features, was first reported in *Apis mellifera* (Rothenbuhler 1964; Spivak and Gilliam 1998; Woodrow 1941). Subsequently, it was reported also in the Asian hive bee *Apis cerana* (Peng et al. 1987) and the giant wild bee *Apis dorsata* (Woyke 1984, 1996). Although both, the hive (cavity-) nesting bees and the open-air-nesting ones display hygiene behaviour, it differs significantly from species to species with respect to both quality and quantity. Although *A. cerana* opens cells infested by *V. destructor*, it has been reported that *A. cerana* do not remove drone brood infested by *V. destructor* mites (Koeniger et al. 1983; Rath and Dresher 1990). Instead, that bee plugs the pore in the cappings of the dead drone brood (Boecking 1999). Boecking suggests that the hard cap structure of *A. cerana* drone cells prevents the hygiene behaviour of the nurse bees. Woyke (1984, 1996) found sealed cells with *A. dorsata* pupae killed by *T. clareae* within the brood area from which the workers had already emerged. He suggested that *A. dorsata* workers do not open sealed cells with brood killed by *T. clareae*. Koeniger et al. (1993, 2002) found a number of capped cells in combs left by colonies which had migrated, and he suggested that this may represent a mechanics of mite elimination. It seems that brood hygiene behaviour of the open-air-nesting *Megapis* bees *A. dorsata* and *Apis laboriosa* differs from that of the cavity-nesting *A. mellifera* and *A. cerana*. The purpose of the present investigation was to confirm and describe comprehensively the nature and extent of hygiene behaviour of *A. dorsata*. I investigated also whether similar hygiene behaviour occurs in the closely related open-air-nesting bee *A. laboriosa*. The whole description is divided into two parts, concerning *Apis dorsata* and *Apis laboriosa*.

5.9 HYGIENIC BEHAVIOUR OF *APIS DORSATA*

The investigations of *A. dorsata* were conducted in March 2002 on three colonies in Bangalore, India. Two assays of hygienic behaviour were conducted: freeze and pin killing.

in freeze killing, three pieces of combs with sealed bee brood were carefully cut out and removed from three *A. dorsata* nests. The pieces were shaped as equilateral triangles of 7 cm and 10 cm sides. The wax triangles with bee brood were frozen in a freezer at 20°C for 24 hours. This killed the bee prepupae and pupae in the triangle comb pieces. Next, the pieces were refrozen. Afterwards, they were inserted into three *A. dorsat*a nests.

in pins killing, two pins of different diameters—0.75 mm and 0.30 mm—were used. One of the pins was inserted through the lids into the cell containing bee prepupae or pupae of *A. dorsata*.

The number of unopened lids on cells containing bee brood killed by either method was recorded daily throughout the observation.

For control purposes, the sealed cells were counted that remained in all three colonies within the brood area of 400 cells from which worker bees emerged.

Observations were also conduct on sealed brood left behind in combs deserted by *Apis dorsata*. The combs in investigated nests were observed with the aid of 16 × 50 binoculars. They were also photographed and video-recorded (zoom 24×).

5.10 HYGIENIC BEHAVIOUR OF *APIS LABORIOSA*

The investigations were conducted in Nepal in 1998. *A. laboriosa* pieces of combs with bee brood were cut from nests in the Himalayan cliff in Chale opposite to Totapani near the border with China. The comb pieces with bee brood were cut from two *A. laboriosa* nests. Those combs were brought to Dabur Apicultural Centre in Jugedi, Chitwan. Here the cappings were opened and examined under a stereo microscope. The contents of 400 sealed brood cells was examined.

5.11 RESULTS CONCERNING TWO MITES

The effect of hygiene behaviour of *A. dorsata* workers toward brood killed naturally: Out of 156 pupae killed with the thick pin (0.75 mm) and 193 killed with the thin pin (0.30 mm), only 26.4% and 63.0%, respectively, were still found in their cells the next day. The proportion of pupae remove from cells pinned with the thick pin was significantly higher than that from those treated with the thin pin. The number of removed brood did not change during the next five days. Cappings on pin-treated cells were not damaged or even scratched by the bees throughout the rest of the observations. The next day after the insertion of freeze-killed bee brood, there still remained 100–310 undamaged lids on sealed pupae per freeze-killed replicate. After four days, 95% of pupae remained in colonies 2 and 3, and a few percent less in colony 1. In the central portion of the bee brood pieces, only one pupa was removed in each of colonies 1, 2, 3 and 7. There was no change in the number of pupae left in colonies 1 and 2 on the next (fifth) day. Undamaged lids remained intact enja the end of the observations. The cells were not opened by the bees, even after all the worker bees emerged from brood area around the inserted piece of frozen brood.

Highly significantly, more brood was left in comb cells with freeze-killed pupae than with killed by pins of either size. Within the control areas of 400 cells from which worker bees emerged, 0.75% (N = 300), 0.75% (N = 300), and 1% (N = 4) cells remained unopened in combs of colonies 1, 2, and 3, respectively. Thus, all the results show that *A. dorsata* workers do not open undamaged sealed cells containing artificially killed bee brood.

The effect of hygiene behaviour of *A. dorsata* workers toward artificially and naturally killed bee brood: Results of artificially killed bee brood (frozen or pinned) were just presented. A banyan tree was growing in Bangalore India. I found 98 *A. dorsata* colonies nesting on it. Additionally, I found there 45 recently abandoned combs. During the next 10 days of my observations, 11 colonies absconded in my presence. I was able to record the number of sealed cells without any outside interference. In most combs, there were only a few (10–40) unopened cells, which displayed an infestation level below 1% of the number of brood cells. However, two colonies, which absconded in my presence, left combs with many sealed cells. Both combs were relatively small. Probably, many sealed cells with dead brood disabled the development of the colonies.

All the results show that *A. dorsata* do not open undamaged sealed cells containing brood killed artificially or naturally.

5.12 HYGIENIC BEHAVIOUR OF *APIS LABORIOSA*

The investigations were conducted in the Himalayas in Nepal in 1998. A rock cliff with 53 *A. laboriosa* nests was located in Chale opposite Totapani near the Tibetan border. There was found also rock cliff with 15 *A. laboriosa* nests, 8 km apart, in the direction of Tibetan border. Additionally, there was one recently deserted comb with 40% of sealed brood cells. The results show that *A. laborios*a leaves sealed brood cells in abandoned combs.

Comb pieces with bee brood were cut out from two *A. laboriosa* nests. Those combs were brought to Dabur Apicultural Centre in Jugedi, Chitwan. Here the cappings were opened and examined under a stereo microscope. The contents of 400 sealed brood cells was examined. In the comb from the first colony there were 30% sealed cells (N = 120) containing dead brood. The cappings of the cells were not damaged. Dead mites, both *T. clareae* as well as *Tropilaelaps koenigerum*, were present in most cells. However, inside some cells with dead brood, I did not find any mites. It looked as if the brood had died due to some disease. Within 300 sealed cells in the comb from the second colony only 10% (N = 30) contained dead brood. Thus, *A. laboriosa* did not open sealed cells with brood killed by mites and possibly also by some brood diseases.

5.13 DISCUSSION CONCERNING HYGIENIC BEHAVIOUR OF *MEGAPIS* AND *APIS* BEES

Three types of eliminating mites and diseases from *A. dorsata* and *A. laboriosa* nests may be presented.

First mechanism of elimination. *A. dorsata* actively grooms and hunts for *T. clareae* outside sealed brood cells (Koeniger et al. 2002; Koeniger and Muzafar 1988; Rath and Delfinado-Baker 1991; Woyke et al. 2004a, 2004b).

Second mechanism of elimination. *A. dorsata* and *A. laboriosa* do not open cells with dead brood. As a result, the mites cannot escape from infected cells. A high percentage of sealed cells with dead brood and lack of adequate space for bee reproduction may cause the colonies to migrate. However, migration of bees does not eliminate from the colonies mites left behind in deserted combs in sealed cells with dead brood. Those cells would not have been opened and the mites would not have come out, even when the bees did not migrate.

Third mechanism of elimination. Migration of *A. dorsata* and *A. laboriosa* indirectly eliminates large number of mites. Migration of colonies causes the cessation of egg-laying by the queen, which diminishes the amount of brood and, finally, results in absence of mites (Woyke et al. 2001b). Even mites emerging along with worker bees from sealed brood do not survive for more than a few days on adult bees (Aggarwal 1988; Koeniger and Muzafar 1988; Woyke 1984). As a result, a heavy fall in the number of mites occurs from *A. dorsata* colonies shortly before migration (Koeniger et al. 2002; Rath and Delfinado-Baker 1991). Therefore, mites are absent in new arriving colonies (Kavinseksan et al. 2003). The new swarms arrive at a new site within a period of several weeks or even several months (Woyke et al. 2001a). They may become infested again by mites from established colonies due to bee drifting or robbing (Paar et al. 2002).

Final Two Questions

8. What advantages may be achieved by *A. dorsat*a and *A. laboriosa* from not opening sealed cells with dead brood?

Both the *Megapis* bees, *A. dorsata* and *A. laboriosa*, are open-air-nesting and migratory species. It may be beneficial for them to disuse temporarily part of the comb cells in exchange for arresting the mites there and thus reducing the increase of their population. When the number of disused cells becomes too high, and the bee reproducing area is insufficient, the bees abscond.

2. What advantages may be achieved by *A. mellifera* and *A. cerana* from opening sealed cells and removing infected or infested brood?
   In the nest, there are no present sealed cells with dead brood, which cannot be used for brood rearing of *Apis* bees.

5.14 TRIAL TO DOMESTICATE THE OPEN NESTING GIANT HONEYBEE *APIS DORSATA*

About 70% of honey produced in India originates from free living giant bees *Apis dorsata*. That bee nests in difficult accessible places, like under limbs of big trees and under different construction built by people, like water tanks and other towers. Management of *Apis dorsata* nests would be easy if the bees could be kept in hives. Forager worker bees nesting in hives, communicate about the food sources (nectar and pollen) by special dances. *Megapis* bees (*A. dorsata* and *A. laboriosa*) perform also recruitment dances. However, the recruitment dances performed by free nesting bees and bees in the hives differ.

The observations of the nest of *Apis dorsata* in the hive with three walls show that the behaviour of the bees was normal. The forager bees collected nectar and pollen. The queen laid the eggs and the nursery bees fed the bee larvae. Emigration occurs when no more food is available during the monsoon time, which starts in June. Thus, the behaviour of *Apis dorsata* bees do not permit keeping them in stationary apiaies during the year around. However, it should be possible to keep that bee in three-wall hives, in migratory apiaries, if food resources (nectar and pollen) would be available throughout the whole year. Thus, keeping *Apis dorsata* bees in hives is not excluded. The same concerns also *Apis laboriosa*.

REFERENCES

Aggarwal, K. 1988. Incidence of *Tropilaelaps clareae* on three *Apis* species in Hisar India. In: Needham, G.R., Page, R.E., Delfinado-Baker, M., and Bowman, C.E. (Eds.), *Africanized honey bees and bee mites*. Ellis Horwood Lim, Chichester, 396–403.

Boecking, O. 1999. Sealing up and non-removal of diseased and *Varroa jacobsoni* infested drone brood cells is part of the hygienic behaviour in *Apis cerana*. *Journal of Apicultural Research* 38: 159–168.

Kavinseksan, B., Wongsiri, S., De Guzman, L.I., and Rinderer, T.E. 2003. Absence of *Tropilaelap*s infestation from recent swarms of *Apis dorsata* in Thailand. *Journal of Apicultural Research* 42: 49–50.

Koeniger, G., Koeniger, N., Lekprayoon, C., and Tingek, S. 2002. Mites from debris and brood cells of *Apis dorsata* colonies in Sabah (Borneo) Malaysia including a new halotype of *Varroa jacobsoni*. *Apidologie* 33: 15–24.

Koeniger, N., Koeniger, G., and Delfinado-Baker, M. 1983. Observationson mites of the Asian honeybee species. *Apidologie* 14: 197–204.

Koeniger, N., Koeniger, G., Mardan, M., and Wongsiri, S. 1993. Possible effects of regular treatments of varroatosis on the host–parasite relationship between *A. mellifera* and *Varroa jacobsoni*. In: Connor, L.J., Rinderer, T., Sylvester, H.A., and Wongsiri, S. (Eds.), *Asian apiculture*. International Conference on Asian Honey Bees and Bee Mites, Bangkok 1992, 541–550.

Koeniger, N., and Muzafar, N. 1988. Lifespan of the parasitic honey bee mite *Tropilaelaps clareae* on *Apis cerana*, *A. dorsata* and *A. mellifera*. *Journal of Apicultural Research* 27: 207–212.

Paar, J., Oldroyd, B.P., Huettinger, E., and Kastberge, G. 2002. Driftingof workers in nest aggregation of the giant honey bee *Apis dorsata*. *Apidologie* 33: 553–561.

Peng, Y., Fang, Y., Xu, S., Ge, L., and Nasr, M.E. 1987. Response of foster Asian honey bee (*Apis cerana* Fabr.) colonies to the brood of European honey bee (Apis mellifera) infested with parasitic miteVarroa jacobsoni Qudemans. *Journal of Invertebrate Pathology* 49: 259–264.

Rath, W., and Delfinado-Baker, M. 1991. *Analysis of Tropilaelaps clareae populations collected from the debris of Apis dorsata and Apis mellifera*. Proc. Int. Symp. Recent Res. Bee Pathology, Ghent 1990, Belgium, 86–89.

Rath, W., and Dresher, W. 1990. Response of *Apis cerana* Fab. Towardsbrood infested with Varroa jacobsoni Qud. and infestation rate of colonies in Thailand. *Apidologie* 21: 311–321.

Rothenbuhler, W. 1964. Behaviour genetics of nest cleaning in honeybees: IV response of F1 and backcross generations todisease-killed brood. *American Zoologist* 4: 111–123.

Spivak, M., and Gilliam, M. 1998. Hygienic behaviour of honey bees and its application for control of brood diseases and *Varroa*. *Bee World* 124–134, 169–186.

Woodrow, A.W. 1941. Behavior of honeybees toward brood infected with American foulbrood. *American Bee Journal* 81: 363–366.

Woyke, J. 1984. Survival and prophylactic control of *Tropilaelaps clareae* infesting *Apis mellifera* colonies in Afghanistan. *Apidologie* 15: 421–434.

Woyke, J. 1985a. *'Tropilaelaps' clareae in Afghanistan and control methods applicable in tropical Asia*. Proc. 3rd Int. Conf. Apic. Trop. Climates, Nairobi, 163–166.

Woyke, J. 1985b. *Tropilaelaps clareae*, a serious pest of *Apis mellifera* in the tropics, but not dangerous for apiculture in temperate zones. *American Bee J.* 125: 497–499.

Woyke, J. 1996. *Different reaction of 'Apis dorsata' and 'Apis mellifera' to brood infestation by parasitic mites*. Proc 3rd AAA Conf. on Bee Res. and Beekeeping Dev, Hanoi, Vietnam, 172–175.

Woyke, J., Kruk, C., Wilde, J., and Wilde, M. 2004a. Periodic mass flights of the giant honey bee *Apis dorsata*. *Journal of Apicultural Research* 43: 180–185.

Woyke, J., Wilde, J., and Reddy, C.C. 2004b. Open-air-nesting honey bees *Apis dorsata* and *Apis laboriosa* diffr from the cavity-nesting *Apis* bees. *Journal of Invertebrate Pathology* 86: 1–65.

Woyke, J., Wilde, J., Reddy, C., and Nagaraja, N. 2005. Periodic mass flights of the giant honeybee *Apis dorsata* in successive days at two nesting sites in different environment conditions. *Journal of Apicultural Research* 44: 140–149.

Woyke, J., Wilde, J., and Wilde, M. 2001a. *Coexistence of Apis mellifera and Apis dorsata workers in the same colonies*. Proceedings of VII IBRA Conference on Tropical Bees and V Asian Apicultural Association Conference Chiang Mai, 115–120, March 19–25.

Woyke, J., Wilde, J., and Wilde, M. 2001b. *Swarming, migration and absconding of Apis dorsata colonies*. Proceedings of the VII International Conference on Tropical Bees and V Asian Apicult. Assoc. Conf. Chiang Mai, 183–188, March 19–25.

Woyke, J., Wilde, J., and Wilde, M. 2003. Periodic mass flights of *Apis laboriosa* in Nepal. *Apidologie* 34: 121–128.

Woyke, J. 1987. Length of stay of the parasitic mite*Tropilaelaps clareae* outside sealed honeybee brood cells as a basis for its effective control. *Journal of Apicultural Research* 26: 104–109.

VIDEOS

(8) (PDF) Open-air-nesting honey bees *Apis dorsata* and *Apis laboriosa* differ from the cavity-nesting *Apis mellifera* and *Apis cerana* in brood hygiene behaviour.
Available from: www.researchgate.net/publication/222137752_2004 Open-air nesting_honey_bees_Apis_dorsata_and_Apis_laboriosa_differ_from_the_cavity-nesting_Apis_mellifera_and_Apis_cerana_in_brood_hygiene_behaviour#fullTextFileContent [accessed 17.06 2021].

(2) (PDF) Hygienic behaviour of *Apis dorsata* differs from that of *Apis mellifera*. Available from: www.researchgate.net/publication/257132082_Hygienic_behaviour_of_Apis_dorsata_differs_from_that_of_Apis_mellifera [accessed 17.06 2021]

6 Genetic Diversity of *Apis dorsata* and *Apis laboriosa*

B. Fakrudin, J. Ugalat, T.N. Lakshmidevamma, C. Kumar, K.B. Rakesh and Ruchita Thimmarayappa

6.1 INTRODUCTION

Honeybees are the most important beneficial insects for humankind, primarily for producing honey, a natural food which is considered as liquid gold, and for playing a major role in crop pollination. Wax, royal jelly, bee venom and propolis are the secondary products from honeybees. They are an integral part of nature's effort to achieve a sustainable ecosystem. It is an important model organism for behavioral research as it is a colonial insect with complex social behavior and contains a vast genetic diversity and gene pools. The classification of *Apis* species has gained importance because of the economic, nutritional and ecological roles associated with these honeybee species. Honeybees belong to the order Hymenoptera and family Apidae. Currently, the genus *Apis* has ten commonly identified species *viz. A. andreniformis, A. binghami, A. cerana, A. dorsata, A. florea, A. koschevnikovi, A. laboriosa, A. mellifera, A. nuluensis* and *A. nigrocincta* (Arias and Sheppard 2005). The giant tropical honeybee *Apis dorsata* and the Himalayan giant honeybee *Apis laboriosa* differ in behavior and ecology from other honeybee species (Oldroyd and Wongsiri 2009).

6.2 *APIS DORSATA*

Apis dorsata, the giant tropical bee, is a honeybee of Southeast Asia, found mainly in forests of Nepal, Malaysia and Singapore (Robinson 2012). *Apis dorsata* was first described by Fabricius in 1793 and was later reintroduced by Maa (1953). Ruttner in 1988 reported about its homogenous appearance. This bee is wild in nature and as such cannot be domesticated. Social bees including *Apis dorsata* are known for their aggressive defense strategies and vicious behavior, particularly when disturbed (Suwannapong et al. 2012). Being open nesters, they thrive well under diversified ecosystems by extending pollination services to various plant species (Raghunandan and Basavarajappa 2014). Adult bees typically measure 17–20 mm in length (Wikipedia 2019). They form giant combs usually 3–4 feet in length and 2–3 feet wider in open places at a height of 1,500 m (Qamer et al. 2008; Hepburn and Radloff 2011).

DOI: 10.1201/9781003294078-6

However, they are also reported to build single wax combs less than 1.5 m in diameter (Robinson 2012). Nests are mainly built in exposed places far from the ground, e.g., tree limbs, under cliff overhangs and sometimes on buildings (Suwannapong et al. 2012). *Apis dorsata* is an important pollinator (Osamu and Tasen 2009; Inson and Malaipan 2011; Suwannapong et al. 2012; Hawkeswood and Sommung 2016) and honey producer (Suwannapong et al. 2012) which ranges through Southern Asia from Pakistan to Indonesia (Suwannapong et al. 2012; Robinson 2012). Colonies typically live in lofty communal nest sites of mostly 20–100 colonies at relatively high elevations (>1,000 m) during the dry season (Robinson 2012). As foraging decreases toward the end of the season, colonies abandon their combs and migrate to lower elevations, establishing new nests where there will be mass flowering in the monsoon season (Ahmad 1989; Venkatesh and Reddy 1989; Paar et al. 2000; Robinson 2012).

6.3 *APIS LABORIOSA*

The Himalayan giant honeybee *Apis laboriosa*, the largest individual bee species in the genus *Apis*, is a key pollinator and important honey producer in Himalayan regions (Joshi et al. 2004). *A. laboriosa* lives in mountainous areas from Nepal to southwestern China and has evolved adaptive behaviors to cope with harsh environments: nesting on inaccessible cliffs above 1,200 m, migrating seasonally and foraging at low temperatures (Sakagami et al. 1980; Batra 1996; Trung et al. 1996). This species is a member of the subgenus *Megapis*, along with *Apis dorsata*. *Apis laboriosa* was long considered to be a subspecies of *Apis dorsata* (Koeniger et al. 2011) until 1980, when Sakagami described *Apis laboriosa* as a separate species (Engel 1999; Sakagami et al. 1980), which was later supported by DNA (deoxyribonucleic acid) sequencing (Cao et al. 2012). *Apis laboriosa* builds a large open nest with a single comb, usually on rocky cliffsides, which gives an average yield of 55–132 lbs (25–60 kg) of honey per colony, per year (Thapa et al. 2018; Aryal et al. 2015), producing different types of honey depending on the season. Honey produced by *Apis laboriosa* has high moisture content and ferments quickly (Thapa et al. 2018).

6.4 ORIGIN AND GEOGRAPHICAL DISTRIBUTION

Apis dorsata is a giant tropical bee native to Asia. However, it has a smaller geographic distribution, mainly in the southern region of Asia. Their habitat is less diverse, as they inhabit cliffed areas and elevated urban areas. *Apis dorsata* is distributed from the Indian subcontinent to Southeast Asia (Otis 1996; Roubik et al. 1985), found throughout the southern countries of Asia, including Malaysia, Indonesia and the Philippines (Roubik et al. 1985), South China, Celebes and Timor, but not in Iran or the Arabian Peninsula. Among subspecies of *Apis dorsata*, the distribution patterns vary; *Apis dorsata dorsata* is only found in the Philippines. *Apis dorsata binghami* is noticed on Sulawesi, an island of Indonesia (Sakagami et al. 1980).

The Himalayan giant honeybee *A. laboriosa* is a spectacular but poorly understood species in large part because it usually nests on inaccessible cliff faces in the Himalayas (Cronin 1979; Sakagami et al. 1980; Roubik et al. 1985; Underwood 1990;

Joshi et al. 2004; Woyke et al. 2012); Gogoi et al. 2021). The first specimen was collected in the mountainous regions of western Yunnan and named by Frederick Smith (Moore et al. 1871), who noted several characteristics that distinguish *A. laboriosa* from lowland *A. dorsata*. This taxon was subsequently ignored until Maa (1953) undertook a reassessment of honeybee taxonomy, who stated its distribution as India (Sikkim; Assam); China (western Yunnan), probably also occurring in N. Burma and has more occurrence in Nepal (Sakagami et al. 1980). *A. laboriosa* is primarily found in the Hindu Kush Himalayan (HKH) region of southern Asia. It has a westernmost border of western Nepal and a southeast border that expands through the northern borders of Laos and Vietnam. It can be found as far north as the northern border of India into southern China. Distribution of *A. laboriosa* better defines its range (Otis 1996; Roubik et al. 1985; Gregory and Jack 2022) and extends it eastward to the mountains of northern Vietnam, Southward along the Arakan Mountains to west-central Myanmar, into the Shillong Hills of Meghalaya, India, and northwestward in Uttarakhand, India (Trung et al. 1996). As *Apis laboriosa* has a limited distribution, there is little evidence to suggest that this bee might become an invasive species in the United States (Gregory and Jack 2022)

6.5 SYMPATRIC OCCURRENCE OF *APIS LABORIOSA* WITH *APIS DORSATA*

Several researchers have indicated that the *A. laboriosa* is a subspecies of *A. dorsata*. *A. dorsata* forages and makes nests to a height of 1,000–3,000 m (Qamer et al. 2008; Hepburn and Radloff 2011) at relatively a lower elevation compared to *A. laboriosa*, which goes up to 4,000 m. However, in northeastern India, colonies occur during summer at sites as low as 850 m and some lower-elevation colonies maintain their nests throughout the winter. The three regions and sites in Arunachal Pradesh, India, where these colonies occur are (1) Western Arunachal: West Kameng District, Nag Mandir; (2) Central Arunachal: West Siang District, Tumbin and Siang District, Modi; and (3) Southeast Arunachal: Tirap District, Kala Pahar and Tutnyu. Nine locations in northern Vietnam observed the workers of *A. laboriosa* and *A. dorsata* foraging sympatrically; their co-occurrence supports the species status of *Apis laboriosa*. It has also been noted that these sister species co-occur in several locations in Asia (Sakagami et al. 1980; Roubik et al. 1985).

6.6 POLYMORPHISM AND ALLELIC FREQUENCY

The occurrence of two or more different forms or morphs in the population of a species is referred to as *polymorphism*; this can be assessed through differences in the morphological characteristics, biochemical compounds and molecular markers. *Apis* species are divided into three lineages: the cavity-nesting bees, *Apis mellifera*, *A. cerana*, *A. koschevnikovi*, *A. nigrocincta* and *A. nulensis*; open nesting dwarf bees, *A. florea* and *A. andreniformis*; and the giant bees, *A. dorsata* and *A. laboriosa*, grouped under subgenus *Megapis*. The nine honeybee species within the genus *Apis* share many similar morphological, behavioral and physiological traits (Arias and Sheppard 2005; Raffiudin and Crozier 2007).

The giant honeybee (*A. dorsata*) is distributed over vast geographic areas in Southeast Asia. On the basis of morphology, this species can be further classified into three subspecies: *A. dorsata dorsata*, *A. dorsata binghami*, and *A. dorsata breviligula*. *A. dorsata breviligula* is with a short tongue and medium forewing length, while *A. dorsata binghami* is with long tongue and long forewing (Gupta 2014). In addition, Sakagami et al. (1980) recognized the Himalayan giant honeybee as *A. laboriosa* that differs significantly from the giant honeybee *A. dorsata* of mainland Asia in many characteristics. The first detailed descriptions of the morphology, biology and geography of *A. laboriosa* by Sakagami et al. (1980) provided strong evidence to recognize it as a distinct species different from the lowland giant honeybee *A. dorsata*. Despite these many differences, Engel (1999) considered *A. laboriosa* to be a subspecies of *A. dorsata*; however, it deserves species status because of substantive differences in drone morphology and distinct morphometric differences between the two taxa collected sympatrically in northeastern India. Both these have a markable difference in performance and ecosystem (McEvoy and Underwood 1988; Kirchner et al. 1996; Otis 1996). The two taxa have also been distinguished on the basis of thoracic hair color, which was "tawny yellow" in laboriosa and "mostly dark" in dorsata. Additionally, the first two gastral tergites of laboriosa are black (gray in callow adults)—in dorsata of mainland Asia, they are orange-brown—and also genetic analysis indicated these two taxa have diverged sufficiently to consider *A. laboriosa* to be a distinct species (Arias and Sheppard 2005; Raffiudin and Crozier 2007; Lo et al. 2010; Chhakchhuak et al. 2016; Takahashi et al. 2018). A. *laboriosa* unequivocally differed from A. *dorsata* in 96 of 103 morphometric measurements and also structurally distinguished (Trung et al. 1996).

Morphological markers are limited in numbers to discriminate the species and subspecies of insects. By using various molecular marker techniques, the genetic variety and gene pool of insects may be studied better. Mitochondrial DNA (mtDNA), random amplified polymorphic DNA (RAPD), microsatellites, expressed sequence tags (ESTs) and amplified fragment length polymorphism (AFLP) markers are important molecular techniques which are being employed in insect ecological studies and have contributed significantly in understanding the genetic origin of insect variety (Behura 2006). Bees are mostly classified through their morphological characteristics, but their genetic variations are observed through mitochondrial DNA gene segments (Branchiccela et al. 2014; Ostroverkhova et al. 2015), single nucleotide polymorphism (Chapman et al. 2016) and allozymes (Smith and Glenn 1995) to differentiate the races between the species. The molecular analysis offers a compelling tactic to measure the extent of genetic difference between and among the honeybee species (Meemongkolkiat et al. 2019). Further, combination of morphological, molecular and physiological characteristics intensifies the classification. Morphological and molecular data on *Apis* has revealed the three subgenera: subgenus *Micrapis* (*A. andreniformis* and *A. florea*), subgenus *Megapis* (*A. dorsata*) and subgenus *Apis* (*A. mellifera*, *A. koschevnikovi*, *A. cerana*, *A. nigrocincta* and *A. nuluensis*) (Takahashi 2006).

Within the species, in honeybee and other social insects, polymorphism occurs because of castes. The phenomenon of existence of several morphological forms with separate functions in a species is also known as polymorphism. Honeybees are well-known as social and polymorphic insects. The true social organization (eusociality) of honeybees is well understood in the study of *Apis mellifera*. They live in colonies in hives and each bee colony includes several thousands of bees which consist of one queen, several hundred drones and tens of thousands of worker bees (50,000–80,000 or more). In honeybees and several other species, the signal to switch the developmental system to male or female involves a sex-determining locus—the "complementary sex determination" (*csd*) locus. The *csd* locus is highly polymorphic, with around a dozen different alleles, presumably at intermediate frequencies in populations, so that most diploid zygotes are heterozygous. Haploid zygotes are, of course, never heterozygotes, so that heterozygosity for different alleles can serve as a signal to control the developmental difference.

The honeybee genome sequence revealed that the *csd* gene is a fairly recent duplicate in these bees of another nearby gene, *fem*, located 12 kilobases away from *csd*. The very high DNA sequence polymorphism initially claimed for the honeybee *csd* gene is partly due to the inclusion of *fem* alleles, whose sequences are distinct from those of all *csd* alleles; synonymous site divergence is 17%, and non-synonymous site divergence is almost as high as 14.5%. Honeybees close relatives share the duplication but surprisingly, they do not evidently share any *csd* allele. The three bee species studied—*Apis mellifera*, *A. cerana* and *A. dorsata*—are estimated to have diverged in the last 10 million years—the synonymous site divergence estimates (based on *fem*, and another gene) are 10% and 13% for *A. mellifera* versus *A. cerana* and *A. dorsata*, respectively. However, *A. dorsata* is distinguished by a lack of body size variation between castes unlike in other honeybees—reproductive castes are larger than workers—but in *A. dorsata*, all castes are the same size. Three DNA microsatellite loci (A14, A76, A88) with a total of 27 alleles provided sufficient genetic variability to classify the workers into distinct subfamilies revealing the degree of polyandry of the queens (Moritz et al. 1995).

Comparative genomics analysis of *A. dorsata* and *A. laboriosa* has revealed that the genes in *A. laboriosa* genome have undergone stronger positive selection (2.5 times more genes) and more recent duplication/loss events (6.1 times more events) than those in the *A. dorsata* genome (Lin et al. 2021): this implies the potential molecular mechanisms underlying the high-altitude adaptation of *A. laboriosa*.

Potential exocrine compounds produced by both forms, *Apis dorsata* and *Apis laboriosa*, revealed that the cephalic and abdominal natural products of these two honeybees shared no common denominators. The sting shaft of workers of *A. dorsata* is the source of a large series of esters dominated by 1-acetoxy-2-decene. Other major constituents include isopentyl acetate, accompanied by isopentyl propionate, farnesyl acetate and several other esters. On the other hand, nothing but presumed structural lipids (e.g., ethyl palmitoleate) were identified from sting shafts of workers of *A. laboriosa*. By contrast, cephalic (including mandibular glands) extracts from workers of *A. laboriosa* contained γ-octanoic lactone, whereas comparable extracts of *A. dorsata* workers contained only structural lipids. The major qualitative differences in the chemical characteristics between *A. laboriosa* and *A. dorsata* are consistent with the designation of these two forms as distinct species.

In honeybees, many females mate with more than one male (polyandry) (Arnqvist and Nilsson 2000). The evolution of polyandry in social insects has led to increased genetic diversity (Hamilton 1987; Sherman et al. 1988; Schmid-Hempel 1997, 1998; Schmid-Hempel and Crozier 1999). Seasonal migration of 100–200 km is a common behavior of *A. dorsata*, and queens appear to return to their previously occupied sites (Koeniger and Koeniger 1991; Dyer and Seeley 1994). Home-site fidelity was reported in *A. dorsata* based on microsatellite analysis (A14, A76 and A88; Neumann et al. 2000). Moreover, based on inferred microsatellite genotypes (A14, A76, A88 and A107) of queens heading colonies at two sites in India, Paar et al. (2000) estimated that the probability of different queens having the same genotype is the product of the individual probabilities for each locus. *A. dorsata* aggregations are comprised of colonies that share more alleles than expected by chance (Paar et al. 2004). Although queens heading neighboring colonies are not close relatives, fixation indices show significant genetic differentiation among aggregation sites. However, there appears to be sufficient gene flow among aggregations to prevent high degrees of relatedness developing between colonies within aggregations (Paar et al. 2004). Geometric morphometric analysis of wing venation, together with other molecular studies, have indicated that the *A. dorsata* populations across Thailand are panmictic (Rattanawannee et al. 2012).

Genetic diversity of the population of diverse climatic zones and plant communities of *A. dorsata* populations from Yunnan, Guangxi and Hainan provinces of China—which are located hundreds of kilometers apart and at different altitudes—were assessed using the genotypes of derived workers' fathers' sampling. The microsatellite loci used were A14, A88, A76, A24 and B124. The total number of alleles detected per locus ranged from seven at locus A14 to 14 at locus A88, and average allelic richness and expected heterozygosity were 9.208 ± 2.101 and 0.745 ± 0.097, respectively. There was no significant genetic differentiation between parental queens and males in any of the populations by pairwise F_{ST} estimates ($p > 0.05$ in all cases). However, a significant difference was noticed across the populations from different regions in China. Whereas 15 single-locus DNA microsatellite markers have indicated high intracolonial relatedness within the studied solitary nests in Marang district, Malaysia (Sahebzadeh et al. 2012).

Genetic differentiation of *A. dorsata* in northern India using eight microsatellite loci indicated that *A. dorsata* aggregations are composed of colonies that share more alleles than would be expected by chance. Accordingly, a study conducted has indicated the low level of genetic variation both in and between *A. dorsata* populations of worker honeybee from Nankana and Narowal of Punjab, India and Pakistan by using RAPD marker technique (Qamer et al. 2021).

Molecular markers specific to the mitochondrial genes are commonly used in insects to analyze the extent of genetic diversity (Bouga et al. 2011) since the mitochondrial genome exhibits 10 times higher average mutation rates in comparison to the nuclear genome (Ballard and Whitlock 2004) owing to nucleotide imbalance (Song et al. 2003) and the insignificant effectual population size linked to active haploid heritage among eukaryotic taxa (Neiman and Taylor 2009). The polyandrous mating approach in honeybees and other eusocial insects shows to introduce vast substitution ratios in mitochondrial genes as compared to nuclear genes because of the maternal inheritance of mitochondrial DNA (Meusel and Moritz 1993). As a sampling strategy for population genetic studies of social insects like *A. dorsata*, mtDNA genes have the advantage over nuclear DNA markers that a mitotype of a single specimen represents the matriarchal line of each colony. However, PCR-RFLP of mitochondrial genes revealed limited genetic diversity and divergence of *A. dorsata* in Thailand. The genetic status of each geographic region is represented by the most common mitotype distributed across all geographic regions with comparable frequencies. This reflects a lack of genetic heterogeneity based on polymorphisms of mtDNA genes ($p > 0.05$). However, Insuan et al. (2007) recorded a relatively high level of genetic diversity in *A. dorsata* ($H_0 = 0.68–0.74$; average number of alleles per locus was 6.0–9.0) in Thailand, when analyzed with microsatellite (A14, A24 and A88) markers across the populations collected from three regions.

Since mtDNA is haploid and transmitted maternally, the effective population size estimated from mtDNA is generally smaller than that estimated from nuclear markers (Birky et al. 1998). This increases its sensitivity to genetic drift and founder effects compared to nuclear DNA markers (O'Connell et al. 1998). However, the mitochondrial amplicon sequence, mitochondrially encoded NADH dehydrogenase 2 (ND2), was 462 bp and 465 bp in *A. dorsata* and *A. laboriosa*, respectively, and the base frequencies statistics indicated the high AT content, mean, 86.3% for ND2 and 79.8% for COII (Cytochrome *c* oxidase II), in the study of understanding the genetic diversity among the species and at different locations of Yunnan, Guangxi and Hainan provinces, China. The mitochondrial, ND5 gene fragment in Asian and European honeybee species from Pakistan and other countries showed comparatively higher genetic diversity indices and variations than the COII gene.

Honeybees of five *Apis* species—*A. cerana*, *A. mellifera*, *A. dorsata*, *A. florea* and *A. andreniformis*—collected from 12 provinces throughout Thailand were shown the low genetic variation within species (haplotype diversity ranging from 0.212–0.394), while 19 polymorphic sites were detected between species for *cytb* and *28S rRNA*. Although the relative haplotype diversity was high, a low nucleotide divergence was found (0.7%) with migratory species. For *cytb*, the sequence divergence ranged from 0.24 to 3.88% within species and 7.35 to 13.07% between species. The divergence of *cytb* was higher than that of *28S rRNA*.

6.7 PHYLOGENETIC ANALYSIS

To comprehend evolutionary dynamics and adaptations to the local environment, *A. laboriosa* is usually compared with the closely related species *Apis dorsata*, a lowland giant honeybee that is widespread throughout tropical and subtropical Asia. These two species were grouped into a "giant honeybees" clade in phylogenetic studies (Willis et al. 1992; Engel and Schultz 1997; Arias and Sheppard 2005; Chhakchhuak et al. 2016).

The phylogeny of *apis* species from China has indicated the three major groups: giant honeybees (*A. dorsata* and *A. laboriosa*), cavity-nesting honeybees (*A. cerana*) and dwarf honeybees (*A. florea* and *A. andreniformis*) (Cao et al. 2012). The species status of *A. laboriosa* was supported in the study conducted by Cao et al. (2012). When comparing *A. dorsata* haplotypes from China with those from other Asian countries, Chinese *A. dorsata* specimens clustered with *A. d. dorsata* and separated from *A. d. breviligula* and *A. d. binghami* sequences, which means that Chinese *A.*

dorsata may belong to *A. d. dorsata*. Several clusters existing within *A. d. dorsata* revealed high divergence in the population, which also was suggested by Arias and Sheppard (2005). *A. d. binghami*, clustered with *A. d. breviligula* and *A. d. dorsata*, separated from *A. laboriosa*. Arias and Sheppard (2005) and Raffiudin and Crozier (2007) also shown the close genetic relationships among the *A. d. binghami*, *A. d. breviligula* and *A. d. dorsata*. Similarly, the phylogeography of *A. dorsata* in Thailand was observed from a bootstrapped neighbor-joining tree constructed from genetic distances between pairs of geographic regions.

Morphometry-based phylogenetic studies by Maa (1953) indicated the presence of four species under genus *Megapis—Megapisbreviligula*, *M. binghami*, *M. dorsata* and *M. laboriosa*—en Ruttner (1988), with his comprehensive work using morphological characteristics of *A. dorsata*, stated that there was very little evidence to support the idea of distinguishing *A. dorsata* into different subspecies. Sakagami et al. (1980) recognized the Himalayan giant honeybee as *A. laboriosa*, and the existence of a giant bee (*A. laboriosa*) was confirmed in Nepal.

Phylogeny of Chinese *Apis* species, as homogeneity of the two gene partitions, was accepted with the ILD test, and the ND2 and COII gene sequences were combined for phylogenetic analysis. Of the 1,061 characteristics, 283 were phylogenetically informative (26.67%). Thirty-nine haplotypes were found in the 52 specimens for the combined sequences. A Bayesian phylogenetic tree was constructed for the 39 haplotypes sequences (Cao et al. 2012). The monophyly of the species was confirmed. The species status of *A. laboriosa* also was supported. All *A. dorsata* samples clustered together and diverged from other Apis species despite their wide distributions in China. Maximum parsimony trees derived from the combined data set and the two genes separately were identical to the Bayesian tree, with slight variation in terminal branch resolution. Phylogenetic trees were constructed using the UPGMA (unweighted pair group method with arithmetic mean) and NJ (neighbor joining) algorithms for individual colonies. The Hainan Island population was divergent from the two China mainland populations (Yunnan and Guangxi) significantly.

Genetic relationships of *A. dorsata* between China and neighboring Asian countries has revealed the Dorsata specimens grouped with an *A. dorsata* sample from Palawan Island of Philippines and two *A. dorsata* samples from Sabah and Kuala Lumpur of Malaysia. Another two *A. dorsata* samples from Lambir (in Malaysian Borneo) and Cameron Highlands (in Peninsular Malaysia) clustered with an *A. dorsata* sample from Nepal. *A. dorsata* cluster from Thailand also was suggested. All *A. d. breviligula* samples from the Philippines grouped together and differentiated from other samples clearly (Cao et al. 2012). The differentiation of *A. d. binghami* from other species also was supported despite of lower bootstrap value. For *A. laboriosa*, Chinese samples grouped together and then clustered with one sample from Nepal, but separated from the other two Nepalese samples.

Phylogenetic relationships within Apis were studied using the two different genomic regions (ND2 mitochondrial gene and EF1-alpha intron). The phylogenetic analyses strongly supported the basic topology recoverable from morphometric analysis, grouping the honeybees into three major clusters: giant bees (*A. dorsata*, *A. binghami* and *A. laboriosa*), dwarf bees (*A. andreniformis* and *A. florea*) and cavity-nesting bees (*A. mellifera*, *A. cerana*, *A. koschevnikovi*, *A. nuluensis* and *A.*

nigrocincta). However, the clade of Asian cavity-nesting bees included paraphyletic taxa. Exemplars of *Apiscerana* collected from divergent portions of its range were less related to each other than were sympatric *A. cerana*, *A. nuluensis* and *A. nigrocincta* taxa. in nucleotide sequence divergence between allopatrically distributed western (*A. mellifera*) and eastern (*A. cerana*, *A. koschevnikovi*, *A. nigrocincta* and *A. nuluensis*) cavity-nesting species, around 18% for the mitochondrial gene and 10–15% for the nuclear intron suggested an earlier divergence for these groups than previously estimated from morphometric and behavioral studies (Arias and Sheppard 2005).

The genetic variation of mitochondrial DNA sequences in *Apis* species from China and other countries is indicated by Cao et al. (2012). Chinese *A. dorsata* grouped with one *A. dorsata* sample from Palawan Island of Philippines and two samples from Malaysia. Another two samples from Malaysia clustered with one Nepalese sample and the cluster from Thailand. These results support the hypothesis that glaciations and deglaciations during the Pleistocene could have greatly influenced the distribution and divergence of *A. dorsata* in China and Southeast Asia. In addition, the species status of *Apis dorsata breviligula* (Maa) and *Apis dorsata binghami* (Cockerell) and genetic variation may exist in *Apis laboriosa* (Smith) despite limited natural distribution.

The phylogenetic hierarchy of the genus *Apis* created on the COI gene presented three main groups. All sequences of each species fall into their respective groups—almost in accordance with Arias and Sheppard (2005), whose phylogenetic analysis reinforced the elementary topology recoverable as of morphometric study and assembling the honeybees hooked on three main groups of giant bees, dwarf bees and cavity-nesting bees. *A. mellifera* comprised three subgroups representing vast genetic interaction suggesting that mitochondrial genes represent high average genetic diversity. Phylogenetic relationship based on ND5 gene sequences (Arias and Sheppard [2005] and the U.S. National Center for Biotechnology Information [NCBI] database) has defined the more genetic interaction among all honeybee species. The phylogenetic hierarchy of the genus *Apis* built on ND5 gene presented six main groups. *A. cerana*, *A. dorsata* and one *A. cerana* from India fall into one subcluster of *A. mellifera*.

DNA sequences from three mitochondrial (*rrnL*, *cox2*, *nad2*) and one nuclear gene (*itpr*) from all nine known honeybee species (*Apis*); a tenth possible species, *Apis dorsata binghami*; and three out-group species (*Bombusterrestris, Melipona bicolor* and *Trigonafimbriata*) implied the Apis phylogenetic relationships using Bayesian analysis. The dwarf honeybees were confirmed as basal, and the giant and cavity-nesting species to be monophyletic. All nodes were strongly supported except that grouping *Apis cerana* with *A. nigrocincta*.

The NJ tree derived from pairwise genetic distances among the populations revealed three major clusters. The cluster 1 consisted of two populations from Maharashtra (280 and 300). The cluster 2 was subdivided into five subclusters: the subcluster C1 had populations derived from Karnataka and Andhra Pradesh; subcluster C2 consisted of populations from Jammu and Kashmir, along with the populations from Assam; the subcluster C3 consisted of populations sampled from Tamil Nadu; subcluster C4 had populations derived from Kerala; and the subcluster C5 consisted of rest of the populations sampled from Maharashtra. Careful observations revealed that the populations from Karnataka and Andhra Pradesh, though forming separate groups, they shared close genetic proximity. The populations from Jammu and Kashmir were intermixed with populations from Assam. The populations of Tamil Nadu and Kerala formed separate clusters. In Maharashtra, though a majority of the populations form a single group, three of the populations diverged at greater distance indicating the underlying genetic diversity (Fakrudin et al. 2013).

For a better global understanding of the relation of *A. dorsata* haplotypes identified from India, newly identified haplotype sequences, along with the available CoII sequences of *A. dorsata* from China and Japan, were compared to generate a global ML tree. The tree resulted in five different clades. The first one included *A. laboriosa* from Nepal, Indonesia and China; the second clade included *A. mellifera*; the third clade included *A. dorsatabreviligula*; the fourth clade included *A. dorsatadorsata* from China, the Philippines and Japan, along with certain locations from Jammu and Kashmir state in India; and the fifth clade included all other samples from India. A similar trend of clustering was reported in previous studies on mitochondrial DNA analysis, in which all Chinese *A. dorsata* populations clustered together with *A. d. dorsata* and separated from *A. d. breviligula* and *A. d. binghami* (Cao et al. 2012). Interestingly, the samples from Jammu and Kashmir, which fell under Hap1, were very much varied from other Indian samples but closely related to the China and Philippines ones in the NJ tree. The results indicated the presence of *A. d. dorsata* in sampled regions, whereas *A. laboriosa* was not observed.

6.8 POPULATION GENETIC STRUCTURE

Population structure can arise via processes such as genetic drift and restricted gene flow that cause heterogenous distribution of genetic variation within and among populations (Frankham 1995, 2010). The first step to quantify the effects of population structure is to choose an appropriate measure of population structure. A commonly used measure is Wright's F_{ST} (Wright 1943). Knowledge of genetic population structure can be of great use in the natural resource management and conservation programs for honeybees, especially considering their wide distribution and great variation (Cao et al. 2012).

Microsatellites are valuable tools for population structure studies due to their high levels of polymorphism, as well as being selectively neutral, displaying codominance and Mendelian inheritance (Goldstein 1999). Population structure analysis based on F_{ST} showed that *A. dorsata* populations were significantly differentiated at the nuclear microsatellite loci (overall $F_{ST} = 0.107 \pm 0.030$, $p = 0.000$). F_{ST} values in any pair of the *A. dorsata* populations (Yunnan, Guangxi and Hainan) were significantly different from zero. The Hainan Island population had higher F_{ST} scores compared with Yunnan (0.138) and Guangxi (0.150) in comparison with the F_{ST} between Yunnan and Guangxi (0.022). Allele frequency differences (genic differentiation) in the three pairs of populations were also highly significant. Similarly, using microsatellites, Paar et al. (2004) showed that there was a significant population differentiation between different geographical regions of *A. dorsata* in northern India.

Insuan et al. (2007) also found significant differentiation among *A. dorsata* populations from geographically different regions in Thailand using three microsatellite loci. Further, based on microsatellite markers, only significant genetic differentiation was observed among three sampling days in a drone congregation area of *A. dorsata* (Kraus et al. 2005). DNA microsatellites can also be used to estimate the extent of worker drifting between nests in *A. dorsata* aggregations (Paar et al. 2004). Significant genetic differentiation was found among *A. dorsata* populations from different regions in China (F_{ST} and AMOVA). The microsatellite analysis of 54 nests in three aggregations showed that no colonies were related as mother–daughter in India. Similar observations have been reported for different sets of markers by Sahebzadeh et al. (2012) in honeybees of Malaysia.

The genetic variation and migration rate among the two populations of giant honeybees *Apis dorsata* from two districts, i.e., Nankana and Narowal from Punjab, Pakistan, was carried out by RAPD-PCR method using ten oligonucleotide primers (Qamer et al. 2021). The whole hereditary variation among thirty populations presented GST = 0.516 and number of migrants $N_m = 0.814$. Gene flow based on N_m values from Nankana and Narowal and the monomorphic loci showed involvement of mutual gene groups between unlike populations (at the border of the city). The population samples from the main city showed monomorphic bands compared to the populations at the border of the city. A close level of polymorphism showed a low level of deviation in *A. dorsata* populations, proposing a somewhat homogeneousness in *A. dorsata* populations from Nankana and Narowal.

6.9 PRINCIPLE COMPONENT ANALYSIS

In addition to cluster analysis, the principal coordinate was used to confirm the results of cluster analysis. In the study *of A. dorsata* population for genetic diversity and population structure collected from different geographic regions of China (Yunnan, Guangxi and Hainan), the AMOVA revealed that differences between individuals within populations and within-individual differences were large (74% and 20 %, respectively; $p < 0.01$). Populations from different geographical regions were significantly differentiated. Likewise, a principal coordinate analysis (PCA) was carried out to show the distribution and relationships between the populations based on the bidimensional plot of the genetic distance in the rock honeybee *Apis dorsata f.* in Andhra Pradesh and Telangana, India (Raju and Naidu 2016).

6.10 CONCLUSION

A comprehensive review of evolutionary dynamics, distribution, phylogenetic relationship, morphometric variations and genetic diversity among the species of *Apis* would help in their management, utilization and conservation for the sustenance in ecosystems. *A. dorsata* was reported to be non-panmictic and the aggregations represented distinct breeding units with significant gene flow. Repeated interbreeding with invading Asian populations has resulted in the high genetic diversity of *A. dorsata*. Even though *A. dorsata* is a highly efficient pollinator of a range of outbreeding crops, it is difficult to manage the colony because of its violent nature and migratory behavior. Genetic divergence within *A. dorsata* and the genetic relationships between *A. dorsata* subspecies has been explained. Variations measured and clusters formed within *A. d. dorsata* revealed high divergence in the population and extensive sequence divergence was found between *A. d. dorsata* samples collected from different countries. Close genetic relationships among the *A. d. binghami, A. d. breviligula* and *A. d. dorsata* were indicated. Population structure analysis based on F_{ST} indicated the divergence of *A. dorsata* populations at the nuclear microsatellite loci. *A. laboriosa* is a significant and valid species based on biological and evolutionary species concept, which usually colonizes between the altitudes of 2,500 m and 4,000 m, whereas *A. dorsata* occurs below the altitude of 1,500 m. Morphometric measurements and DNA sequence data have been used to establish phylogenetic relationship of *Apis* species. The chemical profile of *A. laboriosa* beeswax differs unequivocally from that of *A. dorsata*. Molecular markers specific to the mitochondrial genes—owing to their high mutation rates in comparison to the nuclear genome—are commonly used to analyze the extent of genetic diversity. Phylogenetic relationships within *Apis* have also been studied using the different genomic regions. Differences in defense behavior and sequence divergence between *A. laboriosa* and *A. dorsata* was consistent with behavioral data and supports the species status of *A. laboriosa*. Further studies at morphometric and molecular level with larger sampling points in different geographical regions could give a valuable information on diversity and distribution of *Apis dorsata* and *Apis laboriosa*.

REFERENCES

Ahmad, R. 1989. A note on the migration of *Apisdorsata* in the Andaman and Nicobar Islands. *Bee World* 70: 62–65.

Arias, M.C., and Sheppard, W.S. 2005. Phylogenetic relationships of honey bees (Hymenoptera: Apinae: Apini) inferred from nuclear and mitochondrial DNA sequence data. *Mol. Phylogenet. Evol.* 37(1): 25–35.

Arnqvist, G., and Nilsson, T. 2000. The evolution of polyandry: Multiple mating and female fitness in insects. *Anim. Behav.* 60(2): 145–164.

Aryal, S., Thapa, R., and Jung, C. 2015. An overview of the beekeeping economy and its constraints in Nepal. *J. Apic. Res.* 30(3): 135–142.

Ballard, J.W.O., and Whitlock, M.C. 2004. The incomplete natural history of mitochondria. *Mol. Ecol.* 13: 729–744.

Batra, S.W.T. 1996. Biology of *Apislaboriosa* Smith, a pollinator of apples at high altitude in the Great Himalaya range of Garhwal, India, (Hymenoptera: Apidae). *J. Kans. Entomol. Soc.* 69(2): 177–181.

Behura, S.K. 2006. Molecular marker systems in insects: Current trends and future avenues. *Mol. Ecol.*15: 3087–3113.

Birky, C.W., and Walsh, J.B. 1998. Effects of linkage on rates of molecular evolution. *Proc. Natl. Acad. Sci. USA* 85: 6414–6418.

Bouga, M., Evangelou, V., Lykoudis, D., Cakmak, I., and Hatjina, F. 2011. Genetic structure of *Marchalinahellenica* (Hemiptera: Margarodidae) populations from Turkey: Preliminary mtDNA sequencing data. *Biochem. Genet.* 49: 683–694.

Branchiccela, B., Aguirre, C., Parra, G., Estay, P., Zunino, P., and Antunez, K. 2014. Genetic changes in *Apis mellifera* after 40 years of Africanization. *Apidologie* 45: 752–756.

Cao, L.F., Zheng, H.Q., Hu, C.Y., He, S.Y., Kuang, H.O., and Hu, F.L. 2012. Phylogeography of *Apis dorsata* (Hymenoptera: Apidae) from China and neighboring Asian areas. *Annals of the Entomological Society of America* 105(2): 298–304.

Chapman, N.C., Harpur, B.A., Lim, J., Rinderer, T.E., Allsopp, M.H., Zayed, A., and Oldroyd, B.P. 2016. Hybrid origins of Australian honeybees (*Apis mellifera*). *Apidologie* 47: 26–34.

Cronin, E.W. 1979. *The Arun: A natural history of the world's deepest valley.* Houghton Miffin & Co., Boston University Michigan USA.

Chhakchhuak, L., De Mandal, S., Gurusubramanian, G., Sudalaimuthu, N., Gopalakrishnan, C., Mugasimangalam, R.C., and Vanramliana, S.K.N. 2016. The complete mitochondrial genome of the Himalayan honey bee, *Apislaboriosa. Mitochondrial DNA Part A* 27(5): 3755–3756.

Dyer, F.C., and Seeley, T.D. 1994. Colony migration in the tropical honey bee *Apisdorsata* F. (Hymenoptera: Apidae). *Insectes Soc.* 41(2): 129–140.

Engel, M.S. 1999. The taxonomy of recent and fossil honey bees (Hymenoptera: Apidae; Apis). *J. Hymenopt. Res.* 8(2): 165–196.

Engel, M.S., and Schultz, T.R. 1997. Phylogeny and behavior in honey bees (Hymenoptera: Apidae). *Ann. Entomol. Soc. America* 90(1): 43–53.

Fakrudin, B., Vastrad, A.S., Basavanagoud, K., Babu, O., Nadu, T., Muthuraman, M., Mohankumar, S., Vijaykumar, K., Parthiban, T.P., and Preetha, B. 2013. *Morphometry and phylogeography of honey bees and stingless bees in India.* Monograph, 1. https://www.researchgate.net/publication/273384586_Monograph_On_Morphometry_and_Phylogeography_of_Honey_Bees_and_Stingless_Bees_In_India.

Frankham, R. 1995. Effective population size or adult population size in wildlife: A review. *Genet. Res.* 66: 95–107.

Frankham, R. 2010. Challenges and opportunities of genetic approaches to biological conservation. *Biol. Conserv.* 143(9): 1919–1927.

Gogoi, H., Taid, R., Boruah, B.B., Umbrey, C., and Ete, G. 2021. Response of foragers of the giant Himalayan honey bee, Apislaboriosa, to microclimates in agro-net and climate control greenhouses. *Journal of Apicultural Research* 1–8.

Goldstein, D.B. 1999. *Microsatellites: Evolution and applications.* Oxford University Press, Oxford, UK.

Gregory, M., and Jack, C. 2022. *Himalayan giant honey bee, cliff honey bee (suggested common names) Apislaboriosa Smith (Insecta: Hymenoptera: Apidae).* https://entnemdept.ufl.edu/Creatures/MISC/BEES/Himalayan_giant_honey_bee_cliff_honey_bee_Apis_laboriosa.htm.

Gupta, R.K. 2014. Taxonomy and distribution of different honeybee species. In: *Beekeeping for poverty alleviation and livelihood security.* Springer, Dordrecht, 63–103.

Hamilton, W.D. 1987. Kinship, recognition, disease, and intelligence: Constraints of social evolution. *Soc. Anim.* 11(1): 81–100.

Hawkeswood, T.J., and Sommung, B. 2016. Pollination of *Ruellia simplex* C. Wright (Acanthaceae) by the giant tropical bee, *Apisdorsata* (Hymenoptera: Apidae) in Bangkok, Thailand. *Calodema* 438: 1–5.

Hepburn, H.R., and Radloff, S.E., 2011. *Honeybees of Asia.* Springer Science & Business Media, Cham.

Inson, C., and Malaipan, S. 2011. Diversity of bees (Hymenoptera: Apoidea) as insect pollinators on physic nuts (Euphorbiaceae: *Jatropha curcas* L.) in Thailand. *Thai J. Agric. Sci.* 44: 263–269.

Insuan, S., Deowanish, S., Klinbunga, S., Sittipraneed, S., Sylvester, H.A., and Wongsiri, S. 2007. Genetic differentiation of the giant honey bee (*Apisdorsata*) in Thailand analyzed by mitochondrial genes and microsatellites. *Biochem. Genet.* 45(3–4): 345–361.

Joshi, S.R., Ahmad, F., and Gurung, M.B. 2004. Status of *Apislaboriosa* populations in Kaski district, Western Nepal. *J. Apic. Res.* 43(4): 176–180.

Kirchner, W.H., Dreller, C., Grasser, A., and Baidya, D. 1996. The silent dances of the Himalayan honeybee, *Apislaboriosa. Apidologie* 27: 331–339.

Koeniger, G., Koeniger, N., and Phiancharoen, M. 2011. Comparative reproductive biology of honeybees. *Honeybees of Asia* 159–206.

Koeniger, N., and Koeniger, G. 1991. An evolutionary approach to mating behaviour and drone copulatory organs in Apis. *Apidologie* 22: 581–590.

Kraus, F.B., Neumann, P., and Moritz, R.F.A. 2005. Genetic variance of mating frequency in the honeybee (*Apis mellifera* L.). *Insectes Soc.* 52(1): 1–5.

Lin, D., Lan, L., Zheng, T., Shi, P., Xu, J., and Li, J. 2021. Comparative genomics reveals recent adaptive evolution in Himalayan giant honeybee *Apislaboriosa. Genome Biology and Evolution* 13(10): 227.

Lo, N., Gloag, R.S., Anderson, D.L., and Oldroyd, B.P. 2010. A molecular phylogeny of the genus Apis suggests that the giant honey bee of the Philippines, *A. breviligula* Maa, and the plains honey bee of Southern India, *A. indica* Fabricius, are valid species. *Syst. Entomol.* 35(2): 226–233.

Maa, T.C. 1953. An inquiry into the systematics of the tribusApidini or honeybees (Hym.). *Treubia* 21: 525–640.

McEvoy, M.V., and Underwood, B.A. 1988. The drone and species status of the Himalayan honey bee, *Apislaboriosa* (Hymenoptera: Apidae). *J. Kans. Entomol. Soc.* 61(2): 246–249.

Meemongkolkiat, T., Rattanawannee, A., and Chanchao, C. 2019. Genetic diversity of Apis spp. in Thailand inferred from 28S rRNA nuclear and cytochrome b mitochondrial gene sequences. *Psyche* 11.

Meusel, M.S., and Moritz, R.F. 1993. Transfer of paternal mitochondrial DNA during fertilization of honeybee (*Apis mellifera* L.) eggs. *Curr. Genet.* 24: 539–543.

Moore, F., Walker, F., and Smith, F. 1871. Descriptions of some new insects collected by Dr. Anderson during the expedition to Yunan. *Proceedings of the Zoological Society of London* 244–249.

Moritz, R.F., Kryger, P., Koeniger, G., Koeniger, N., Estoup, A., and Tingek, S. 1995. High degree of polyandry in *Apisdorsata* queens detected by DNA microsatellite variability. *Behav. Ecol. Sociobiol.* 37(5): 357–363.

Neiman, M., and Taylor, D.R. 2009. The causes of mutation accumulation in mitochondrial genomes. *Proc. R. Soc. B: Biol. Sci.* 276: 1201–1209.

Neumann, P., Koeniger, N., Koeniger, G., Tingek, S., Krygerll, P., and Moritz, R.F.A. 2000. Homesite fidelity in migratory honeybees. *Nature* 406: 474–475.

O'Connell, M., Dillon, M.C., Wright, J.M., Bentzen, P., Merkouris, S., and Seeb, J. 1998. Genetic structuring among Alaskan Pacific herring populations identified using microsatellite variation. *J. Fish. Biol.* 53: 150–163.

Oldroyd, B.P., and Wongsiri, S. 2009. *Asian honey bees: Biology, conservation, and human interactions*. Harvard University Press, Cambridge.

Osamu, S., and Tasen, W. 2009. Bees of natural forest, teak plantations and agricultural fields in Thailand. *Esakia* 49: 7–13.

Ostroverkhova, N., Konusova, O., Kucher, A., Kireeva, T., Vorotov, A., and Belikh, E. 2015. Genetic diversity of the locus COI-COII of mitochondrial DNA in honeybee populations (*Apis mellifera* L.) from the Tomsk region. *Russian J. Genet.* 51: 80–90.

Otis, G.W. 1996. Distributions of recently recognized species of honeybees (Hymenoptera: Apidae: Apis) in Asia. *J. Kans. Entomol. Soc.* 69: 311–333.

Paar, J., Oldroyd, B.P., Huettinger, E., and Kastberger, G. 2004. Genetic structure of an *Apisdorsata* population: The significance of migration and colony aggregation. *J. Heredity* 95: 119–126.

Paar, J., Oldroyd, B.P., and Kastberger, G. 2000. Giant honeybees return to their nest sites. *Nature* 406(6795): 475–475.

Qamer, S., Ahmad, F., Latif, F., Ali, S.S., and Shakoori, A.R. 2008. Physicochemical analysis of *Apisdorsata* honey from terai forests, Nepal. *Pakistan J. Zool.* 40(1): 53–58.

Qamer, S., Al-Abbadi, A.A., Sajid, M., Asad, F., Khan, M.F., Khan, N.A., Sthanadar, A.A., Akhtar, M.N., Mahmoud, A.H., and Mohammed, O.B. 2021. Genetic analysis of honey bee, Apisdorsata populations using random amplified polymorphic DNA (RAPD) markers. *Journal of King Saud University-Science* 33(1): 101218.

Raffiudin, R., and Crozier, R.H. 2007. Phylogenetic analysis of honey bee behavioral evolution. *Mol. Phylogenet. Evol.* 43(2): 543–552.

Raghunandan, K.S., and Basavarajappa, S. 2014. Incidence of Galleria Mellonella infestation on Apisdorsata colonies at different regions of South-Western Karnataka. *Zoology* 3(2).

Raju, A.S., and Naidu, A.V.R. 2016. A study on morphometrics, floral rewards and floral calendar of the rock honey bee, *Apis dorsata F.* in Andhra Pradesh and Telangana, India. *Advances in Pollen Spore Research* 34: 41–125.

Rattanawannee, A., Chanchao, C., and Wongsiri, S. 2012. Geometric morphometric analysis of giant honeybee (ApisdorsataFabricius, 1793) populations in Thailand. *Journal of Asia-Pacific Entomology* 15(4): 611–618.

Robinson, W.S. 2012. Migrating giant honey bees (*Apisdorsata*) congregate annually at stopover site in Thailand. *PLoS One* 7(9): 1–9.

Roubik, D.W., Sakagami, S.F., and Kudo, I. 1985. A note on distribution and nesting of the Himalayan honey bee *Apislaboriosa* Smith (Hymenoptera: Apidae). *Journal of the Kansas Entomological Society* 746–749.

Ruttner, F. 1988. *Breeding techniques and selection for breeding of the honeybee.* Vol. 5. British Isles Bee Breeders' Association, Derby, 152.

Sahebzadeh, N., Mardan, M., Ali, A.M., Tan, S.G., Adam, N.A., and Lau, W.H. 2012. Genetic relatedness of low solitary nests of *Apisdorsata* from Marang, Terengganu, Malaysia. *PLoS One* 7(7): 1020.

Sakagami, S.F., Matsumura, T., and Ito, K. 1980. *Apislaboriosa* in Himalaya, the little-known world largest honeybee (Hymenoptera, Apidae). *Insectamatsumurana. New Series: J. Fac. Agri. Hokkaido Univ., Series Entomology* 19: 47–77.

Schmid-Hempel, P. 1997. Infection and colony variability in social insects. In: W.D Hamilton, J C Howard (Eds.), *Infection, polymorphism and evolution.* Springer, Cham, 43–51.

Schmid-Hempel, P. 1998. *Parasites in social insects.* Vol. 60. Princeton University Press, Princeton.

Schmid-Hempel, P., and Crozier, R.H. 1999. Ployandry versus polygyny versus parasites. *Philosophical Transactions of the Royal Society of London. Series B: Biological Sciences* 354(1382): 507–515.

Sherman, P.W., Seeley, T.D., and Reeve, H.K. 1988. Parasites, pathogens, and polyandry in social Hymenoptera. *American Nat.* 131(4): 602–610.

Smith, D.R., and Glenn, T.C. 1995. Allozyme polymorphisms in Spanish honeybees (*Apis mellifera iberica*). *J. Hered.* 86(1): 12–16.

Song, S., Wheeler, L.J., and Mathews, C.K. 2003. Deoxyribonucleotide pool imbalance stimulates deletions in HeLa cell mitochondrial DNA. *J. Biol. Chem.* 278: 43893–43896.

Suwannapong, G., Benbow, M.E., and Nieh, J.C. 2012. Biology of Thai honeybees: Natural history and threats. Chapter 1. *Bees: Biology, Threats and Colonies* 1–11.

Takahashi, J.I. 2006. Taxonomy and phylogeny of the genus Apis, 2: Ecology of the dwarf honeybee Apisflorea and A. andreniformis. *Honeybee Science-Tamagawa University (Japan)* 26(4).

Takahashi, J.I., Rai, J., Wakamiya, T., and Okuyama, H. 2018. Characterization of the complete mitochondrial genome of the giant black Himalayan honeybee (Apislaboriosa) from Nepal. *Conservation Genetics Resources* 10(1): 59–63.

Thapa, R., Aryal, S., and Jung, C. 2018. Beekeeping and honey hunting in Nepal: Current status and future perspectives. *Asian Beekeeping in the 21st Century* 111–127.

Trung, L.Q., Dung, P.X., and Ngan, T.X. 1996. A scientific note on the first report of *Apislaboriosa* F Smith, 1871 in Vietnam. *Apidologie* 27(6): 487–488.

Underwood, B.A. 1990. Time of drone flight of *Apislaboriosa* Smith in Nepal. *Apidologie* 21(6): 501–504.

Venkatesh, G., and Reddy, C. 1989. Rates of swarming and absconding in the giant honey bee, *Apisdorsata* F. *Ani. Sci.* 98(6): 425–430.

Wikipedia. 2019. *Apisdorsata*. https://en.wikipedia.org/wiki/Apis_dorsata.

Willis, L.G., Winston, M.L., and Honda, B.M. 1992. Phylogenetic relationships in the honey-bee (genus Apis) as determined by the sequence of the cytochrome oxidase II region of mitochondrial DNA. *Mol. Phylogenet. Evol.* 1(3): 169–178.

Woyke, J., Wilde, J., and Wilde, M. 2012. Swarming and migration of Apisdorsata and Apislaboriosa honey bees in India, Nepal and Bhutan. *Journal of Apicultural Science* 56(1): 81.

Wright, S. 1943. Isolation by distance. *Genetics* 28(2): 114.

7 Pests, Predators and Pathogens of *Apis dorsata* and *Apis laboriosa*

Dharam P. Abrol

7.1 INTRODUCTION

7.1.1 Species of Honeybees

Nine species of honeybees are currently recognized, which include giant honeybees *Apis dorsata* or rock bees and *Apis laboriosa* or Himalayan bees; dwarf honeybees *Apis florea* and *Apis andreniformes*; and cavity-nesting honeybees *Apis mellifera*, *Apis cerana*, *Apis koshevnikovi*, *Apis nigrocincta* and *Apis nuluensis*. Of these, *Apis mellifera* (European honeybee) is natively distributed across Europe, the Middle East, Asia and Africa. *Apis mellifera* in Europe is composed of up to ten subspecies based on morphometric and genetic data. The European honeybees are thought to have originated in Africa or the Middle East and evolved through four lineages, M (Western and Northern Europe), C (Eastern Europe), O (Eastern and Central Asia) and A (Africa) (Ruttner 1988; Whitfield et al. 2006). *Apis dorsata*, *Apis cerana* and *Apis florea* bees are natively found in South and Southeast Asia. *Apis koshenikovi* and *Apis nigrocincta* are found highly localized on the Indonesian archipelago (Hepburn and Radloff 2011). *Apis mellifera* in Europe and *Apis cerana* in Asia have been extensively used in commercial beekeeping.

7.2 IMPORTANCE OF BEES

Honeybees (*Apis mellifera* L.) play a vital role in terrestrial ecosystems by maintaining floral biodiversity and provide sustainable agriculture through pollination, thereby ensuring food security. Honeybees are extremely important economically and ecologically through their role as pollinators; the yields of 96% of animal-pollinated crops can be increased by the visits of honeybees (Potts et al. 2010). Honeybees are the most economically valuable pollinators of monoculture crops worldwide (Klein et al. 2007), and contribute billions of dollars annually to the economy of the United States alone (Southwick and Southwick 1992), including hundreds of millions of dollars in direct bee products such as honey and wax (Gallai et al. 2009). Other more esoteric benefits,

DOI: 10.1201/9781003294078-7

such as nitrate fertilization of plants with bee faeces (Mishra et al. 2013), are just beginning to be investigated.

Not only honeybees are known to pollinate crops for human consumption, but also bumblebees, solitary bees, stingless bees, wasps, flies, beetles, thrips, birds and possibly bats (Klein et al. 2007). There has been a suggestion that, through exploitative competition, introduced honeybees can have a detrimental impact on native pollinators and their relationships with the plants they pollinate (Aizen and Feinsinger 1994a; Kato et al. 1999; Do Carmo et al. 2004). However, the widespread nature of such effects is not proven (Goulson 2003; Paini 2004), and synergistic interactions between *Apis* and non-*Apis* pollinators have been demonstrated (Greenleaf and Kremen 2006).

Honeybees are attacked by various pests and pathogens including bacteria, fungus, viruses, microsporideans, mites, insects, pesticides and environmental changes. The interactions among these—possibly along with other unknown factors—are causing widespread decline in honeybee populations across the planet. colony collapse disorder (CCD) results in sudden disappearance of bees which is matter of great confusion among beekeepers throughout the world. Honeybee colonies contain several thousands of individuals, so occurrence of any disease-causing organism in the colony can result in immediate spread in entire colony and between the colonies. Easy modes of transport at local and global levels has further aggravated the problem of spread of diseases.

All living organisms are subjected to infestation or attack by natural enemies or predators, and honeybees of the genus *Apis* are no exception. Honeybee species both under managed, as well as natural, conditions are attacked by various pests such as Wax moths, hive beetles, ants, wasps, amphibians, reptiles, birds and mammals which affect bee hives, resulting in serious losses (Nagaraja and Rajagopal 2009). Such information on pests and predators of dwarf honeybee *A. florea* are not fully documented except some reports ants attacking colonies in Thailand (Seeley et al. (1982). *A. dorsata*, commonly known as "wild bee", nests in open arboreal conditions (Seeley et al. 1982), facing problems from predators at its nesting niche such as wasps (Ruttner 1988; Oldroyd and Wongsiri 2006; Kastberger et al. 2010, 2008), birds (Seeley et al. 1982; Kastberger and Sharma 2000) and mammals (Viraktamath et al. 2005; Basavarajappa et al. 2009). Because of rich nutrition available in their nests, *Apis dorsata* colonies attract a wide variety of pests and enemies (Hepburn and Radloff 2011). Some predatory animals feed on hive products, whereas others feed on brood at various regions of India and other parts of the world. Certain Vespidae family members (Muzaffar and Ahmed 1986; Chang et al. 1993; Ono et al. 1995; Srivastava et al. 1995). Also, some birds and other higher animals interfere with population buildup of *A. dorsata* colonies. Nagaraja and Rajagopal (2011) recorded 40 species of birds preying on honeybees in the field. Ambrose (1990), Thapa and Wongsiri (2003) reported drongo and swifts attacking *A. dorsata* colonies. Kastberger and Sharma (2000) and Thapa et al. (2000) observed counterattacks of bee-eaters on *A. dorsata* colonies in Assam and Nepal.

The Oriental honey buzzard, *Pernis ptylorhynchos* has been reported feeding on the brood and honey of *A. dorsata* in Nepal (Thapa and Wongsiri (2003).

Jadczak (1986) reported that several species of bears (such as *Helarctus malayanus*, *Melursus ursinus* and *Ursus thibetanus*) attacking *A. dorsata* colonies. Seeley et al. (1982), Besides, some species of martens (Thapa et al. 2000; Oldroyd and Wongsiri (2006) and monkeys have been reported to disturb *A. florea*, *A. cerana* and *A. dorsata* colonies (Seeley et al. 1982; Thapa et al. 2000; Basavarajappa 2012), Nagaraja and Rajagopal (2011). Moreover, vertebrate enemies such as rats, honey badgers, foxes, lizards and toads have been reported predating honeybee colonies (Caron 1978; Novogrodzki 1990). *Apis dorsata* colonies in general are attacked by various predators, parasites and pests (Abrol 2002; Nagaraja and Rajagopal 2011; Morse and Laigo 1969), which causes considerable damage. The predators' interferences to *A. dorsata* colonies have been studied by various researchers in India and other countries around the world (Seeley et al. 1982; Kastberger and Sharma 2000; Thapa and Wongsiri 2003; Gulati and Kaushik 2004; Viraktamath et al. 2005; Oldroyd and Wongsiri 2006; Kastberger et al. 2008, 2009; Abrol 2009a, 2009b; Nagaraja and Rajagopal 2009; Schmelzer and Kastberger 2009; Weihmann et al. 2012).

Raghunandan (2014) recorded distribution of various enemies of *A. dorsata* colonies in arid, semi-arid and Malnad regions of southwestern Karantaka. Of the total, 13 species belong to three groups: namely insects, birds and mammals. Among the insects, only the giant wasp (*Vespa cincta*) predated on *A. dorsata* colonies at different regions of southwestern Karnataka. The birds attacking *A. dorsata* colonies comprised of four orders, namely Coraciformes, Columbiformes, Passeriformes and Accipitriformes. The birds included *Merops orientalis*, *M. Persicus*, *Ocyceros birostris*, *Dicrurus adsimilis*, *Acridotheres fuscus*, *A. tristis*, *Muscicapa para* and *Corvus culminates*. Only two mammals namely *Funambulus palmarum* (Rodentia: Sciuridae) and *Pteropus giganteus* (Chiroptera: Pteropdidae) were observed showing predatory activity on the hive of *A. dorsata*.

Transfer of parasites like *Varroa destructor* or *Tropilaelaps* spp. from their original hosts to other honeybee species have posed an honest challenge for management of honeybee diseases. Viruses are spread by *Apis mellifera* beekeepers migrating or shipping bees to new areas and infecting and sometimes decimating *Apis cerana* colonies. In the developing world, beekeeping might be a declining industry; recent decades have seen a increase in colony losses in managed honeybee colonies, and feral honeybee colonies are declining at alarming rate because of pests and diseases.

7.3 PARASITES OF HONEYBEES

A number of parasites and predators have been identified causing severe damage to honeybee colonies. One infection by this agent may cause no harm to a colony; however, if exposed to a pesticide at the identical time, the colony might die as the mixture of pests, parasites and pesticides winds up in an inadvertent "meltdown".

7.4 PARASITIC MITES

The sudden emergence of bee mites has posed a superb threat to the beekeeping industry and a way forward for agricultural production. *Acarapis woodi*, *Varroa jacobsoni* and *Tropilaelaps clareae* are the foremost pests, but about 100 mostly harmless mite species are associated with honeybees (Sammataro et al. 2000). An important feature of social insects is division of labour, nest homeostasis and abundance of food available in the hive for development and reproduction.

7.5 *TROPILAELAPS CLAREAE*

Tropilaelaps clareae considered to be primary pest of *Apis dorsata*, has spread to other bee species *A. mellifera*, *A. laboriosa*, *A. cerana* and *A. florea*. *A. mellifera* may not be successful in Asian countries as it lacks a natural defence mechanism against harmolymph feeding mites compared with native bees like *A. dorsata*, *A. laboriosa* and *A. florea* infested with *T. clareae* and *T. koenigerum* (Koeniger et al. 1983). *Tropilaelaps* has been found to infest *A. mellifera* as an obligate brood parasite altogether in the Southeast Asian countries and is reported to cause serious damage to beekeeping with *A. mellifera* in tropical Asia (Hosamani et al. 2006). A single, alarming report of this mite in Kenya has not been repeated. Like *Varroa*, the female mites are dispersed by bees, but phoretic survival is of short because the piercing/grasping mouthparts it can feed only on soft tissues, like honeybee brood. Development requires about one week and therefore the adults—including the foundress mite—emerge with the adult bee and look for new hosts. Thanks to a shortened life cycle, *T. clareae* out-competes *Varroa* when both infest the identical colony of *A. mellifera*. Nevertheless, populations of both mites can survive within the identical apiary for 12 months, probably because their niches do not seem to be completely congruent. Irregular brood patterns, dead or malformed wingless bees at the hive's entrance and also the presence of fast-running, brownish mites on the combs are diagnostic clues for *T. clareae*.

Four species of *Tropilaelaps* are described within the mite family Laelapidae. Delfinado and Baker (1961) identified *Tropilaelaps clareae* from dead *A. mellifera* bees and field rats near beehives within the Philippines. Delfinado-Baker et al. (1985) collected parasitic mites *Tropilaelaps clareae* and *T. koenigerum* from brood nests of *Apis dorsata* and *A. laboriosa* in Nepal. *A. laboriosa* is recorded as a replacement host of both *Tropilaelaps* species, and *T. clareae* is assumed to be a widespread pest of *Apis cerana*, *A. dorsata*, *A. mellifera* and *A. laboriosa* in Asia. Twenty years later, *Tropilaelaps koenigerum* was observed parasitizing *A. dorsata* in the country (Delfinado-Baker and Baker 1982). Anderson and Morgan (2007) found both *Tropilaelaps mercedesae* and *Tropilaelaps thaii* parasitizing *A. dorsata* and *A. mellifera* in mainland Asia and *A. laboriosa* within the Himalayas, respectively. Almost rather like the reclassification of *V. jacobsoni* and *V. destructor* (Anderson and Trueman 2000), *T. mercedesae* was initially described as *T. clareae* (Anderson and Morgan 2007).

Tropilaelaps mites are believed to be indigenous parasites of the large honeybees, *A. dorsata*, *A. laboriosa*, and *A. breviligula* (Laigo and Morse 1968; Delfinado-Baker et al. 1985; Anderson and Morgan 2007). Since its discovery, all reports on *Tropilaelaps* are from the distribution range of the large honeybees and thus suggest coevolution (Anderson and Morgan 2007). *T. clareae* was first observed infesting

A. mellifera within the Philippines and was recently found parasitizing *A. breviligula* within the Philippines and Sulawesi Island in Indonesia (Anderson and Morgan 2007). Anderson and Morgan (2007) suggests that *T. mercedesae* parasitizes *A. dorsata* and *A. mellifera* throughout southern mainland Asia and Indonesia, but Sulawesi Island, moreover as *A. laboriosa* within the Himalayas. It is impressed that *T. mercedesae* as *T. clareae* classification (Delfinado-Baker 1982; Kapil and Aggarwal 1987; Delfinado-Baker et al. 1989; Wongsiri et al. 1989; Abrol and Putatunda 1995; Koeniger et al. 2002) should be revised in further studies. *T. koenigerum* could be a parasite of *A. dorsata* in mainland Asia and Indonesia apart from Sulwasesi and Borneo (Delfinado-Baker and Baker 1982; Anderson and Morgan 2007), and it is assumed to co-infest *A. dorsata* colonies in Borneo and Thailand with *T. mercedesae* (Delfinado-Baker et al. 1989; Koeniger et al. 2002). *T. koenigerum* was reported on *A. cerana* brood in India (Abrol and Putatunda 1995) and *T. mercedesae* in Thailand (Anderson and Morgan 2007). *T. koenigerum* has only been observed during *A. mellifera* debris in Thailand (Kavinseksan, personal communication). Conversely, *T. thaii* has only been observed parasitizing *A. laboriosa* in Vietnam (Anderson and Morgan 2007).

7.6 MANAGEMENT

Apis dorsata can defend against *Tropilaelaps* mites by their behavioual strategies. *A. dorsata* lives in the open, in single-comb nests hanging on cliffs, tree branches or eaves of human buildings. Migration in *A. dorsata* colonies is a regular feature, and brood rearing is stopped during this period. This a very advantageous strategy for bees to get rid of mites as *T. clareae* cannot survive on adult bees for more than three days because their chelicerae are not specialized for feeding on adult bees. When *T. clareae* populations are very high and beyond the grooming capacity, the colonies of *A. dorsata* migrate, creating broodless conditions to reduce the mite populations. As the mites cannot feed on adult bees and are unable to survive outside brood, a period of broodlessness serves as a non-chemical method to control mites. Strong colonies can defend against predators and pests. Use of non-chemical methods can be preferred to keep the population densities below their economic injury levels. Chemicals, if necessary, need be applied only when cultural, mechanical methods prove ineffective.

7.7 PREDATORS OF HONEYBEES

Asian honeybees are attacked by various pathogens, pests and predators (Oldroyd and Wongsiri 2006). Several species of birds prey upon honeybees which include Orange-rumped honey guide (*Indicator xanthonotus*) and the Malaysian honey guide (*I. archipelagicus*). Other birds preying on honeybees include the Oriental (*Pernis ptilorhyncus*) and barred (*P. celebensis*) honey buzzards. Besides, the bee-eaters (*Merops* spp.) and drongos (*Dicrurus*) are also seen attacking honeybee colonies.

Tingek et al. (2004) reported a Conopid fly *Physocephala parralleliventris Kröber* parasitizing *A. cerana, A. koschevnikovi* and *A. dorsata* in Borneo. This fly picks up the bees in flight and deposits a tiny larva on the integument which penetrates the bee's cuticle, consuming the bee from the inside. Spread of this fly bee could potentially be more dangerous. Emergence of small hive beetle *Aethina tumida* to Asian honeybees is yet another problem (Dietemann et al. 2009; Mostafa and Williams 2002; Neumann and Elzen 2004; Ellis and Hepburn 2006), though this pest lives saprophytically in a honeybee colony. It is matter of concern that that *A. tumida* may be parasites of some or all of the Asian cavity-nesting species. This is evident from the fact that *A. tumida* was able to complete its life cycle when introduced in the nests of Bombus impatiens (Stanghellini et al. 2000), indicating that this species could parasitize the Asian honeybee species.

Bears are most dangerous for honeybee colonies (Seeley et al. 1982; Thapa et al. 2000; Oldroyd and Wongsiri 2006). They can break the nest cavities with their strong claws. They feed on the combs which are filled with brood and honey. Martens also attack entire colonies, but have also been observed to prey on flying bees. Monkeys (*Macaca* sp.) have sometimes been observed robbing nests of dwarf honeybees *A. florea* or *A. cerana* (Seeley et al. 1982; Thapa et al. 2000). Other vertebrates occasionally attack rats, honey badgers, foxes, lizards and toads (Caron 1978; Novogrodzki 1990). However, major threats remain from human beings, who use all the nest components. Various species of birds such as drongos and swifts have been observed preying on flying honeybees (Ambrose 1978; Seeley 1994; Thapa and Wongsiri 2003). Honeybees have been reported to form a considerable part of the diet in bee-eaters as inferred from the crop content of *Merops orientalis* and *M. superciliosus* (Fry 1984). Birds are mostly interested in capturing individual bees and rarely attack colonies (Kastberger and Sharma 2000). The problem was more severe at low temperatures when colonies were mostly inactive and flocks of *M. orientalis* attacked *A. dorsata* colonies more intensively, and they can be picked easily from the ground (Thapa et al. 2000). Honeyguides (Indicatoridae) are specialized wax feeders, but mainly feed on abandoned nests (Oldroyd and Wongsiri 2006). Besides vertebrates, other arthropods impose considerable predation pressure (Seeley et al. 1982; Fuchs, personal observation), causing no serious harm to colonies. Wasps and ants are of major concern as they pose a severely reduced work force by attacking and robbing entire colonies. In particular, the larger vespine wasps of the genus *Vespa* play a major role as predators of Asian honeybees (Matsuura 1988); an array of hornets which attack Asian bees are listed in Oldroyd and Wongsiri (2006). Some giant wasps have the effectiveness to deplete entire colonies (Seeley et al. 1982; Ono et al. 1987; Kastberger et al. 2008). Some wasps such as *Vespa multimaculata* and *V. velutina* have a specialized strategy in which they hover in front of a colony entrance and catch returning foragers, which could deplete a colony by 20–30 (Sakagami and Akahira 1960).

Hornets not only damage individual bees but destroy entire colonies. A few *V. tropica* wasps can overpower an *A. florea* colony within three hours by successively seizing and killing nest workers and gradually depleting the hands (Seeley et al. 1982). By recruiting nestmates, *V. mandarinia* homets are able to mass-attack entire *A. cerana* colonies and, having killed the bees, raid pupae and larvae. So as to focus their attacks, they use pheromones to mark the besieged colonies (Ono et al. 1995). Ants are present almost everywhere and might feed on live and dead bees, larvae, pupae and honey (Akratanakul 1987). The weaver ants living on trees are widespread and more troublesome for bees. Specifically, *Oecophylla smaragdina* could be a major threat to colonies of dwarf honeybees, who share the identical habitat, but may additionally occasionally drive cavity-nesting *A. cerana* from their nests (Dyer 1991). These ants seize and pull down flying bees from a perch. Other species, like *Monomorium indicum*, *Diacamma rugosum* and *pheidologeton* sp., were observed disturbing *A. cerana* colonies, without causing serious damage. A few ant species, like *Dolichoderus bituberculatus*, attack *A. dorsata* colonies (Seeley et al. 1982). Aland and Mamlayya (2011) reported that a dipteran (identified as *Pyragra* sp.) was observed carrying a honeybee (*Apis dorsata*). The dipteran stabbed the bee with the assistance of strong proboscis and stout and spiny legs. The proboscis was inserted within the head region ventrolaterally so to inject saliva with neurotoxic and proteolytic enzymes which paralyzed and digested the insides of the prey. Then, the bee killer sucked the liquified meal through the proboscis. Within 20–25 minutes, the honeybee shrunk and died. One amongst the remarkable and astonishing behavioural observations noted was that after a hand snap, the two-winged insects had not left the honeybee. On the opposite hand, its loosened grip was tightened by the bee killer with the assistance of spiny legs. Sihag (2014) reported that *Apis dorsata* was free from any conspicuous viral, bacterial and fungal diseases and also failed to have any serious predators and enemies. Old colonies above 20% were infested by *Tropilaelaps clareae* and 100% by pyralid moth *Galleria mellonella*. The pyralid infestation could show even from the surface of the colonies when white pegenjaminructures of waxmoth cocoon appeared on the upper exposed honey part of the comb. However, during the whole study period, this honeybee was not found to be attacked by any predatory wasp or the bird on the nesting or the foraging site. No pollen or honey was found in deserted combs, indicating that none of the examined colony was found to indicate the presence of pollen or honey in its deserted comb—that each one the stored foods were completely exhausted before deserting the comb.

7.8 PESTS

Several vertebrates pest are encountered attacking honeybee colonies. They are especially more troublesome for *A. mellifera*. They include coraciiform bird birds like little green coraciiform bird (*Merops orientalis*), chestnut headed coraciiform bird (*Menjamindaenaulti*), the swifts (*Crypsiurus balasiensis*, *Chaetura* spp.), needletail (*Hirundapus cochinchinesis*), the woodpeckers (*Picus* spp.), the honeyguides (Indicatoridae), the black drongo (*Dicrurus macrocercus*), the ashy drongo (*Dicrurus leucophaeus*) and therefore the greater racket-tailed drongo (*Dicrurus paradiseus*) (Akratanakul 1990; Cervancia 1993; Wongsiri et al. 2005). The beekeepers either relocate the colonies or employ net trapping methods to limit the predation by birds. Invertebrates include wax moths which are known pests of honeybees throughout the globe. Two species of pyralids like greater wax moths pyralid and lesser wax moth *Achroia grisella*. Both species of moths lay eggs in cracks and crevices of bee hives (Akratanakul 1990). Bee moth larvae are especially more problematic for storage of unused combs (Pernal and Clay 2013). Besides physical damage of the store combs and damaging live honeybee colonies, they are also responsible for the spread of IAPV (Israeli acute paralysis virus) and BQCV (chronic bee paralysis virus) within the colonies (Triyasut et al. 2015); however, their replication was not confirmed. there is a requirement to further investigate possible roles of bee viruses in pyralid larvae. Wax moths cause heavy damage in *Apis cerana* colonies in Southeast Asia (Akratanakul 1990). Abandoned combs of *Apis dorsata* are seriously damaged. Chemical control methods are not commonly recommended. Various cultural methods like removing hive debris, managing strong colonies and burning infested combs are recommended. *G. mellonella* destroys natural colonies of *A. dorsata* and produces considerable losses to the beekeeping industry. it is a heavy destructive pest which causes considerable damage to both normal colonies and abandoned combs of *A. dorsata* (Williams 1976). Kapil and Sihag (1983), Basavarajappa (2012) have reported attacks of *G. mellonella* against the weak colonies or abandoned combs of *A. dorsata*. Swamy et al. (2009) have identified the larvae of *G. mellonella* within the comb with mass of silken wells and debris. Thus, its infestation brings considerable damage to the colony population and successively causes economic loss to the beekeeping industry. Moreover; several predators like birds and mammals also interfere with the traditional survival of *A. dorsata* population. Nagaraja and Rajagopal (2009) have recorded around 40 species of birds which are known to kill honeybees within the field. Seeley (1983), Ambrose (1990) and Thapa and Wongsiri (2003) have reported the interference of birds (e.g. drongo and swifts) to honeybee colonies. Kastberger and Sharma (2000) and Thapa et al. (2000) have reported the counter attack of bee-eaters (*Merops orientalis* and *M. persicus*) on *A. dorsata* colonies in Assam and Nepal. Thapa and Wongsiri (2003) have observed the Oriental hawk (*Pernis ptylorhynchos*) feeding on the brood and honey of *A. dorsata* in Nepal. Jadczak (1986) has reported attacks by bears (*Helarctus malayanus*, sloth bear and *Ursus thibetanus*) on *A. dorsata* in the United States. Seeley et al. (1982), Thapa et al. (2000), Oldroyd and Wongsiri (2006) have recorded the attack of martens (*Martes flavigula* and *M. foina*) to the colony, together with flying bees. Moreover, Seeley et al. (1982), Thapa et al. (2000), Basavarajappa (2012) and Nagaraja and Rajagopal (2009) has reported monkeys (*Macaca radiata*) occasionally disturbing *A. florea*, *A. cerana* and *A. dorsata* colonies. Predation by other vertebrates like rats, honey badgers, foxes, lizards and toads on honeybees has been reported several investigators (Caron 1978; Novogrodzki 1990). In addition, current human activities also threaten the welfare of the *A. dorsata* population considerably (Basavarajappa 2012). Various human activities—together with honey hunting, beeswax and pollen harvest from the hives—become a prominent threat to the colonies of *A. dorsata* within the wild. Reports on the possible impact of pests and predators—including man—on the survival of *A. dorsata* population are scanty. Various predators, parasites and pests cause damage to *A. dorsata* colonies at different parts of India and elsewhere (Morse and Laigo 1969; Abrol 2002; Nagaraja and Rajagopal 2009). Although *A. dorsata* colonies under arboreal conditions within the wild are free for frequent hunting, various pests, predators, enemies and human activities interfere often with their populations. Information on these interferences amidst agri-horticultural ecosystems and in residential areas of south-western Karnataka is scanty.

7.9 SMALL HIVE BEETLE (*AETHINA TUMIDA*)

This beetle has been emerging as a heavy pest of honeybees in Asia since its discovery in the Philippines (Brion 2015). It is an opportunistic scavenger (Neumann and Elzen 2004) most serious in mild climates. However, at higher temperature, their lives are shortened (de Guzman and Frake 2007) and availability of food stores throughout the year promotes fecundity.

7.10 WASPS

Wasps are serious predators of honeybees in Asia (Matsuura 1988). They aim for honeybees at the colony entrances (Cervancia 1993). Sometimes they have been found to predate *A. cerana* drones at drone congregation areas (Koeniger et al. 1994). Honeybees can defend their colonies by shaking movements (Koeniger et al. 1996; Kastberger et al. 1998; Tan et al. 2010; Khongphinitbunjong et al. 2012), and *Apis cerana* colonies can kill the wasps by heat and suffocation by balling reaction (Ono et al. 1987; Tan et al. 2005). Other methos include killing wasps by flapping, by toxic baiting and or by reducing the hive entrance (Cervancia 1993).

7.11 ANTS

Several species of ants attack honeybees, including weaver ants (*Oecophylla smaragdina*), black ants (*Monomorium* spp.), fire ants (*Solenopsis* spp.), and *Formica* spp. (Akratanakul 1990).

7.12 BEE LICE (*BRAULA COECA*)

B. coeca wingless flies do not seem to be considered serious pests of honeybees (Pernal and Clay 2013). However, larvae consume wax, pollen and honey, and they damage combs with tunnelling. The adults eat nectar and pollen. Treatments kind of like parasitic mites are found to be effective (Kulincevic et al. 1991)

7.13 BACTERIA

Probiotic bacteria have involved the context of human nutrition in improving digestive functions; likewise in systems against infectious diseases worldwide (Pineiro and Stanton 2007). Lactobacilli inhabits nutrition rich environments, like humans, animals, food and plants (Tannock 2004; Vasquez and Olofsson 2009). The gastrointestinal tract of worker honeybees contains multiple carboxylic acid bacteria (LAB), like *Lactobacillus*, *Bacillus*, *Corynebacterium*, *Clostridium*, *Streptococcus*, *Escherichia*, *Achromobacter*, *Citrobacter*, *Pseudomonas*, *Enterobacter*, *Flavobacterium*, *Klebsiella*, *Proteus* and *Bifidobacterium* (Rada et al. 1997; Snowdon and Cliver 1996; Gilliam 1997; Jeyaprakash et al. 2003; Mohr and Tebbe 2006; Lyapunov et al. 2008; Yoshiyama and Kimura 2009; Killer et al. 2010). Three *Bifidobacterium* species are isolated from the honeybee channel, of which two species—*Bifidobacterium asteroides* and *Bifidobacterium coryneforme*—were found within the *A. mellifera* intestine; however, *Bifidobacterium* indicum was identified in Malaysia and Philippines (Scardovi and Trovatelli 1969). On the other hand, honey and bee products have been utilized in human health and nutrition for hundreds of years. So far, two phylotypes of lactobacilli are identified within the honey stomachs and in fresh honey of the European honeybee and bumblebee (Olofsson and Vasquez 2008; Vasquez et al. 2009; Tajabadi et al. 2011). Tajabadi et al. (2013) isolated and identified 17 *Lactobacillus* strains from a full honeycomb of the honeybee *Apis dorsata* using selective media. *Melissococcus pluton*, the causative organism of European foulbrood (EFB) of honeybees, has previously been reported from honeybee and *A. cerana* (Bailey 1983). Sakagami et al. (1980) reported this bacterium from *Apis laboriosa*. Allen et al. (1990) collected diseased larvae from *A. laboriosa* colonies at two cliff face sites within the same river valley within the Gandaki zone of Nepal which were examined for bacterial and viral pathogens. Of the ten colonies at one site (altitude 4,000 feet), nine were harvested and every one was diseased, some so severely as to have little healthy brood. Combs had the dense white colonies up to 1 mm in diameter, resembling *M. pluton* from *A. mellifera* (Bailey and Collins 1982). *A. laboriosa* suffers from a disease analogous to EFB which the mortality of brood in colonies was caused by *M. pluton*. The fastidiousness, serological reactions and other characteristics of the isolate show that it is closely associated with the *M. pluton* isolates from *A. mellifera* and *A. cerana*, and clearly separated from all other known bacterial species. The studies on the opposite native Asian honeybee species, particularly *A. dorsata*, an in-depth relative of *A. laboriosa*, are required to present further insight into the relationships between the various isolates. *M. plutonius* is most closely associated with the *Enterococci* (Cai and Collins 1994). Historically, little genetic variation has been seen within *M. plutonius*, with only very minor differences in RFLP (restriction fragment length polymorphism) patterns (Djordjevic et al. 1999), culture characteristics and serology (Allen and Ball 1993) being reported among isolates from diverse sources. Very recently, a subtype of *M. plutonius* was isolated in Japan. This was designated "atypical" *M. plutonius*, as its less stringent culture characteristics, higher virulence in vitro and more diverse PFGE (pulsed field gel electrophoresis) patterns distinguished it from the sort strain and every one previously identified "typical" *M. plutonius* (Arai et al. 2012; Takamatsu et al. 2013).

7.14 IMPACT OF CLIMATE CHANGE ON HONEYBEE DISEASES

Climate change has a predominant impact on honeybees at different levels. It can have an instant influence on honeybee behaviour and physiology. Climate can change the quality of the floral resources, impacting the harvest by colonies and their development. The distribution range of honeybees can change with new pests, parasites and pathogens. Beekeepers have also to vary management practices. They prefer moving their hives to new foraging areas and importing new races to test in new environments. Different diseases are present in numerous parts of the world. Some known pathogens distributed worldwide include: *Varroa destructor* within the case of honeybee and *Apis cerana*; bacteria that cause American and European foulbrood; *Nosema apis* and *N. cerana*; and various viruses affecting honeybees. These pathogens tend to possess different haplotypes of varying virulence. Other pathogens or haplotypes have more limited distribution ranges, like *Tropilaelaps*, which so far has been found only in Asia (Ruttner 1988). Global climate change ends up in the movement of honeybee colonies, and different races may inherit contact resulting in transfer of diseases as observed just in the case of *Varroa destructor* and *Apis mellifera*. Within the space of a few decades last century, two extremely homogeneous haplotypes of this honeybee parasite were sufficient to invade virtually the full honeybee distribution range (Solignac et al. 2005). How will the pathogen/bee interaction evolve? Pathogen species infesting different honeybee races or species are going to be brought into contact with new hosts. The recent discovery of *Nosema cerana* (Higes et al. 2006) and also the IAPV (Cox-Foster et al. 2007) among honeybees could be a crucial example that has been spread by people generally. Global temperature change could modify the interactions among these different pathogens. *Tropilaelaps* could be a noteworthy case in point. The social nature of the honeybee, which is characterized by trophallaxis and shut contact within a caste structured colony which can carry with it tens of thousands of individuals (Winston 1987), can exacerbate the spread and effects of pathogens. To combat pathogens, however, honeybees have evolved a diffusion of inherited behavioural traits. Perhaps the best studied of these traits is the active ejection of diseased individuals from the colony (Rothenbuhler 1964). Nevertheless, in recent years, because the honeybee has increased its geographical range as a result of modem commercial beekeeping practices, it has encountered pathogens against which it has only some active defences. Two such pathogens do not seem to be microorganisms but species of ectoparasitic mites. At present, these mites are considered the foremost serious threats to modern beekeeping (Bradbear 1988; Fries 1993; Paxton 1992)

7.15 PATHOGENS AND THEIR TRANSMISSION IN SOCIAL INSECTS

Many efforts have been made to understand disease transmission in solitary insects but similar efforts still need to be made in social insects (Onstad and Carruthers 1990). Two aspects of the biology of social insects are of interest in regards to the transmission and population dynamics of pathogens and to the pathology process. First, colonies of social insects are often capable of compensating for individual worker mortality, thus reducing the effects of pathogens. Second, transmission of pathogens also occurs when infected individuals enter an uninfected colony (Royce and Rossignol 1990). Some studies have attempted to examine the interaction between pathogen populations in social insect colonies and their effects on colony populations, mainly because of the difficulties associated with handling. Certainly, several questions could be addressed regarding processes in host–parasite interactions, both regulatory (e.g. density-dependent constraints on population growth within individual hosts) and destabilizing (e.g. parasite reproduction within a host that directly increases parasite population sizes) (Anderson and May 1979). Probably one of the best systems for examining these questions is the relationship between honeybees and parasitic mites, as suggested by Royce and Rossignol (1990). Mites are large enough to handle, they can be extremely pathogenic to honeybees and honeybee colonies themselves are amenable to experimental manipulation. However, before any modelling can be easily attempted, information regarding the biology of the host (honeybee) and parasite (mite) is necessary, as well as the hygienic behaviour of *A. dorsata* and *A. laboriosa*. Concerning *A. dorsata*, Woyke (1984, 1985, 1996) suggested that the brood cells killed by *Tropilaelaps clareae* were not opened by worker bees, which absconded as a strategy to eliminate mites.

Both the open-nesting and cavity-nesting bees show hygienic behaviour, but it differs significantly from series to species. *A. cerana* opens cells infested by *V. destructor* but does not remove drone brood infested with *V. destructor* mite (Koeniger et al. 1983; Rath and Dresher 1990), but on the other hand plugs the cappings of the dead drone brood having pore (Boecking 1999). Hard-cap structure of drone cells in case of *A. cerana* prevents the hygienic behaviour of nurse bees. Woyke (1984, 1996) found that *A. dorsata* workers of the migrated colonies do not open sealed cells with brood killed by *T. clareae*, indicating a mechanism of mite elimination. It seems that brood hygiene behaviour of the open-air nesting bees *A. dorsata* and *Apis laboriosa* differs from the cavity-nesting *A. mellifera* and *A. cerana*. They do not open undamaged cells with dead brood. This behaviour seems to be more efficient than the cavity-nesting honeybees. It is useful for migrating *A. dorsata* and *A. laboriosa* to temporarily disuse a part of the comb cells for reducing the rise of mite population.

Giant honeybees (Apis dorsata and *A. laboriosa*) are the oldest honeybee species after the dwarf honeybees, having evolved about 5–10 million years ago (Ruttner 1988; Oldroyd and Wongsiri 2006). Unlike the more recent cave-dwelling honeybees (e.g., *A. cerana*, *A. mellifera*), giant honeybees build their nests in the open, suspending them with overhead supports such as tree branches, rocks or buildings (Paar et al. 2004). In the open, they are directly exposed to a variety of predators, particularly to birds and wasps (Roepke 1930; Seeley et al. 1982; Kastberger and Sharma 2000). This predatory pressure apparently gave rise to the evolution of a series of defence strategies (Oldroyd and Wongsiri 2006; Butler 1962; Morse and Laigo 1969; Kastberger and Sharma 2000; Kastberger et al. 1998; Breed et al. 2004).

Defence behaviour of honeybees generally involves physical contact with aggressors. A prominent example is their capacity to recruit stinging guards (Breed et al. 2004) and to mobilize a whole army of defenders. In giant honeybees, mass mobilization of stinging guards may occur within seconds, thereby attaining the status of the most dangerous stinging insects in the world (Morse and Laigo 1969; Seeley et al. 1982; Kastberger and Sharma 2000; Koeniger and Fuchs 1975). Against wasps, honeybees have developed specific defence behaviours such as heat balling of wasps that come into direct contact with the honeybee nest. Heat balling has been reported for *A. cerana*, *A. mellifera* (Ono et al. 1995; Ken et al. 2005) and *A. dorsata* (Kastberger et al. 2001). The bees by their thoracic muscles can raise temperature above 45°C, which is lethal for wasps. Giant honeybees (*Apis dorsata*) have evolved a variety of defence strategies against predatory wasps, such as "shimmering", whereby hundreds of bees flip their abdomens upwards. Shimmering repels predatory hornets, forcing them to hunt free-flying bees rather than foraging bees directly from the honeybee nest.

7.16 VIRUSES

7.16.1 Viruses in *Apis dorsata*

Roberts and Anderson (2014) reported a novel serotype of sacbrood virus (SBV) infecting *Apis mellifera* in New Guinea which has earlier been found infecting *A. mellifera* and *Apis cerana*. They also reported SBV from *Apis dorsata* colonies for first time in Indonesia. Viruses of honeybees are nearly ubiquitously distributed throughout the planet, with 18 species isolated (Bailey and Ball 1991; Allen and Ball 1996; Chen and Siede 2007). Among them, seven varieties—BQCV, DWV, KBV, SBV, ABPV, CBPV and IAPV—were most prevalent (Chen and Siede 2007). In Asia, these seven viruses are reported to infect different honeybee species throughout the region (Sanpa and Chantawannakul 2009; Ai et al. 2012; Forsgren et al. 2015). Supporting their genomic structures, SBV and DWV are classified under Iflaviridae, whereas BQCV, ABPV, KBV, and IAPV belong to the Dicistroviridae.

Of the viruses, SBV is that the first virus discovered in a very. mellifera (White 1913). Historical records from Asia showed that sacbrood disease was first detected from *Apis cerana* in Thailand in 1976 (Bailey et al. 1982). It absolutely was named Thailand sacbrood virus (TSBV, also referred to as Chinese sacbrood virus) since it absolutely was serologically associated with SBV but is physiochemically distinct (Bailey et al. 1982). TSBV caused deaths of more than 90 percent of managed *A. cerana* populations in Kashmir (Abrol and Bhat 1990) and was also found during *A. dorsata* and *A. florea* in India (Allen and Ball 1996). In *A. cerana*, it had been observed in India, Thailand, Republic of Korea, Japan, Nepal, China and Vietnam (Shah and Shah 1988; Allen and Ball 1996; Choi et al. 2008; Kojima et al. 2011; Grabensteiner et al. 2007; Ai et al. 2012; Forsgren et al. 2015). SBV infection of *A. mellifera* colonies was detected in Thailand, Japan, China, and Vietnam (Sanpa and Chantawannakul 2009; Yoo et al. 2009; Kojima et al. 2011; Ai et al. 2012; Forsgren et al. 2015).

DWV was more abundant than SBV in four honeybee hosts. In *A. mellifera*, it absolutely was detected in Sri Lanka, Nepal, Thailand, Japan, China and Vietnam (Berényi et al. 2007; Sanpa and Chantawannakul 2009; Kojima et al. 2011; Ai et al. 2012; Forsgren et al. 2015). DWV infections have been reported of *A. cerana* in China, Japan and Vietnam (Kojima et al. 2011; Ai et al. 2012; Li et al. 2012; Forsgren et al. 2015) and of *A. florea* and *A. dorsata* in China (Zhang et al. 2012). The levels of DWV virus was detected in *A. mellifera* colonies infested with *V. destructor* and *T. mercedesae* mites in Thailand (Chantawannakul et al. 2006; Khongphinitbunjong et al. 2015).

7.17 DICISTROVIRIDAE

BQCV was isolated from queen larvae and pupae of *A. mellifera* which were dead within their cells (Bailey and Woods 1977). The virus is most common in *Apis cerana* (Sanpa and Chantawannakul 2009; Ai et al. 2012; Yang et al. 2013; Reddy et al. 2013; Kojima et al. 2011; Forsgren et al. 2015); China, Vietnam, Thailand, and Japan (Li et al. 2012; Choe et al. 2012; Yang et al. 2013; Forsgren et al. 2015; Mookhploy et al. 2015); and in *A. florea* and *A. dorsata* from China and Thailand (Zhang et al.

2012; Mookhploy et al. 2015). Genetic variations of BQCV strains across four species of honeybees are studied. The capsid coding region and the phylogenetic analysis revealed that BQCV isolates from northern Thailand, China, and Japan displayed an in-depth relationship within Asia (Mookhploy et al. 2015). Similar findings were reported with the BQCV strains in Korea and Japan, where the viral isolates from the identical country or continent showed high levels of similarity (Kojima et al. 2011; Noh et al. 2013). The role of BQCV in honeybee mortality is not yet fully understood, especially in association with other parasites (Bailey et al. 1983). Zhang et al. (2012) found that the host-range breadth of pathogens is an important factor with serious consequences. Phylogenetic analyses indicate that these viruses have moved from *A. mellifera* to wild bees. This information has important consequences for management of viruses.

7.18 FUNGI

Two species, namely *N. apis* and *N. ceranae*, have been reported from honeybees in Asia (Klee et al. 2007). *Nosema apis* was first described from *A. mellifera* by Zander (1909) from European countries. *N. ceranae* was first detected from *A. cerana* colonies in China (Fries et al. 1996; Botías et al. 2012). *N. ceranae* has been reported to parasitize all honeybee species such as A. mellifera, A. cerana, A. florea, *A. dorsata*, and A. koschevnikovi (e.g., Klee et al. 2007; Chaimanee et al. 2010; Botías et al. 2012). It was found that *N. ceranae* isolated from *A. mellifera* was more infective than the isolate obtained from *A. cerana* and *A. mellifera* (Chaimanee et al. 2013). It was interesting to find that *A. cerana* workers infected by *N. ceranae* had lower bacterial populations of *Bifidobacterium* and Pasteurellaceae that produce several antibiotic compounds important to host defence against parasites (Li et al. 2012). *N. ceranae* has been reported to enhance the spread of other bee diseases, such as Chalkbrood (Hedtke et al. 2011). Both parasites can be managed by the fungicide fumagillin, with varying degrees of success (Akratanakul 1990; Williams et al. 2008, 2011; Huang et al. 2013).

7.19 *ASCOSPHAERA APIS*

The fungus *A. apis*, which causes chalkbrood disease in honeybees, was first described within the primary twentieth century in *A. mellifera* in Europe (Maassen 1913) and now is found throughout the planet. In Asia, it has been reported in China, Israel, Japan, Korea, the Philippines, Russia, Thailand and Turkey (Oldroyd and Wongsiri 2006; Aronstein and Murray 2010). In Turkey, it had been believed to have been introduced from contaminated beeswax which was imported to the country (Tutkun et al. 1993). Within the Philippines and Thailand, beekeepers often reported higher incidence of mycosis during the wet season (Cervancia 1993; Chantawannakul and Puchanichanthranon 2005). The infection spreads through the ingestion of the fungus by honeybee larvae (Heath 1982; Aronstein and Murray 2010). The fungus can produce lytic enzymes (protease and beta-N-acetylglucosaminidase) which could be a very important factor that assists fungal invasion within the bee larvae; the enzymatic patterns from Thai isolates were rather a bit like the previous report of Spanish isolates (Theantana and Chantawannakul 2008). In *A. mellifera*, chalkbrood disease may end in lowered colony productivity, but it rarely finally ends up in colony death (Heath 1982; Aronstein and Murray 2010). It is also not considered to be a serious disease of honeybees in Asia, although it has been reported more widespread than stonebrood disease which is caused by fungi from the genus (Akratanakul 1990; Ra et al. 2012).

7.20 *PAENIBACILLUS LARVAE*

P. larvae, which causes American foulbrood (AFB), is distributed throughout Asia (Genersch 2010) The disease was reported from India on *A. cerana* (Singh 1961) and *A. mellifera* in Taiwan (Yen and Chyn 1971). Bee larvae become infested when spores are ingested.

7.21 *MELISSOCOCCUS PLUTONIUS*

M. plutonius may well be a ubiquitously distributed bacterium of honeybees which has been detected throughout Asia, including *A. mellifera* (Akratanakul 1990), *A. cerana* (Diwan et al. 1971; Bailey 1974; Zhou et al. 2000; Rana et al. 2012), *A. laboriosa* (Allen et al. 1990), and *A. florea* (Saraithong et al. 2015a). *M. plutonius* primarily affects honeybee larvae, causing a disease called European foulbrood (EFB) characterized by dead, twisted, brood within their cells. A phenotypically and geneIically distinct *M. plutonius* strain has been isolated from *A. cerana* japonica; it is frequently found and more virulent than the quality strain (Arai et al. 2014; Takamatsu et al. 2014)

7.22 CONCLUSIONS AND FUTURE PERSPECTIVES

Native honeybees are exposed to a greater risk of pathogens and parasites from interspecies transmission routes due to introduction of exotic honeybee *Apis mellifera*. Similarly, *Apis mellifera* may be infested with pests and parasites from native bees. The examples are *V. destructor* and *Nosema ceranae* (Fries 2010; Rosenkranz et al. 2010). The native bees have strong defence mechanisms against parasites compared to *A. mellifera* (Khongphinitbunjong et al. 2012; Chaimanee et al. 2010). Transmission of diseases also occurs due to migration of bees (Kavinseksan et al. 2003). Due to their unique bacterial community, Asian honeybees inhibit growth of bacterial infections. The gut bacteria has been reported to differ in *A. mellifera*, *A. cerana*, *A. dorsata* and *A. florea* (Disayathanoowat et al. 2011; Saraithong et al. 2015a, 2015b). Some of the bacterial isolates from midgut have been reported to inhibit growth of American foul brood (Disayathanoowat et al. 2011, 2012). Honeybee hives also have actinomycetes which are main microbial groups producing natural antibiotics (Promnuan et al. 2009, 2011). In addition, propolis and bee bread also help in protecting against pests and diseases (Simone-Finstrom and Spivak 2012; Simone et al. 2009). Loss of foraging sources, habitat alteration and pesticide exposure are main reasons for the decline of bee populations (Stanley et al. 2015). The protection and conservation of pollinators needs the utmost priority. Complete information on pests and parasites is still lacking in some regions. Furthermore, there is a need of research on biomedical properties of honey and propolis to promote beekeeping in Asia (Wanjai et al. 2012; Sanpa et al. 2015).

REFERENCES

Abrol, D.P. 2002. *Pest, predators and pathogens of honeybee, Apis mellifera L. in Jammu, India*. Proceedings 6th AAA Int. Conf. & World Apiexpo, Bangalore, India, 112.

Abrol, D.P. 2009a. *Honeybee diseases and their management*. Kalyani Publishers, New Delhi, 70–102.

Abrol, D.P. 2009b. *Bees and beekeeping*. Kalyani Publishers, New Delhi, 210–256.

Abrol, D.P., and Bhat, A.A. 1990. Studies on Thai sac brood virus affecting indigenous honey bee *Apis cerana indica* Fab colonies—prospects and future strategies. *J. Anim. Morphol. Physiol.* 37: 101–108.

Abrol, D.P., and Putatunda, B.N. 1995. Discovery of an ectoparasitic mite *Tropilaelaps koenigerum* Delifinado-Baker & Baker on *Apis dorsata, A. meliffera* L. and *Apis cerana* F. in Jammu and Kashmir, India. *Current Sci* 68: 90.

Ai, H., Yan, X., and Han, R. 2012. Occurrence and prevalence of seven bee viruses in *Apis mellifera* and *Apis cerana* apiaries in China. *J. Invertebr. Pathol* 109: 160–164.

Aizen, M.A., and Feinsinger, P. 1994a. Habitat fragmentation, native insect pollinators, and feral honey bees in Argentine 'chaco serrano'. *Ecological Applications* 4(2): 378–392.

Akratanakul, P. 1987. FAO agricultural services bulletin 6814 beekeeping with oriental honeybees. In: *Beekeeping in Asla*. http://mvw.fao.orq/docreplx0083e/xO083e00.HTM.

Akratanakul, P. 1990. *Beekeeping in Asia*. Food and Agriculture Organisation of the United Nations (FAO) Agricultural Services Bulletin 68/4. Rome, Italy.

Aland, S.R., and Mamlayya, A.B. 2011. A note on predation of honey bee, *Apis dorsata* by a robber fly, *Pyragra* sp. *Bionotes* 13(3): 123.

Allen, M.F., and Ball, B.V. 1993. The cultural characteristics and serological relationships of isolates of *Melissococcus pluton*. *Journal of Apicultural Research* 32: 80–88.

Allen, M.F., and Ball, B.V. 1996. The incidence and world distribution of honey bee viruses. *Bee World* 77: 141–162.

Allen, M.F., Ball, B.V., and Underwood, B.A. 1990. An Isolate of Melissococcus pluton from *Apis laboriosa*. *Journal of Invertebrate Pathology* 55: 439–440.

Ambrose, J.T. 1978. Birds. In: Morse, R.A. (Ed.), *Honey bee pests, predators, and diseases*. Cornell University Press, Ithaca.

Ambrose, J.T. 1990. Birds. In: Morse, R.A., and Nowogrodzki, R. (Eds.), *Honey bee pests, predators and diseases*. Cornell University Press, Ithaca, NY, 243–260.

Anderson, D.L., and Morgan, M.J. 2007. Genetic and morphological variation of bee-parasitic *Tropilaelaps* mites (Acari: Laelapidae): New and re-defined species. *Exp. Appl. Acarol.* 43: 1–24.

Anderson, D.L., and Trueman, J. 2000. *Varroa jacobsoni* (Acari: Varroidae) is more than one species. *Experimental and Applied Acarology* 24: 165–189.

Anderson, R.M., and May, R.M. 1979. Population biology of infectious diseases: Part 1. *Nature* 280: 361–367.

Arai, R., Miyoshi-Akiyama, T., Okumura, K., Morinaga, Y., Wu, M., Sugimura, Y., Yoshiyama, M., Okura, M., Kirakae, T., and Takamatsu, D. 2014. Development of duplex PCR assay for detection and differentiation of typical and atypical *Melissococcus plutonius* strains. *J. Vet. Med. Sci.* 76: 491–498.

Arai, R., Tominaga, K., Wu, M., Okura, M., Ito, K., Okamura, N., Onishi, H., Osaki, M., Sugimura, Y., Yoshiyama, M., and Takamatsu, D. 2012. Diversity of *Melissococcus plutonius* from honey bee larvae in Japan and experimental reproduction of European foulbrood with cultured atypical isolates. *PLoS One* 7: e33708.

Aronstein, K.A., and Murray, K.D. 2010. Chalkbrood disease in honey bees. *J. Invertebr. Pathol.* 103: S20–S29.

Bailey, L. 1983. Melissococcus pluton, the cause of European foulbrood of honey bees (Apis spp.). *J. Appl. Bacteriol.* 55: 65–69.

Bailey, L. 1974. An unusual type of *Streptococcus pluton* from the Eastern hive bee. *J. Invertebr. Pathol.* 23: 246–247.

Bailey, L., and Ball, B.V. 1991. *Honey bee pathology.* Academic Press Ltd., London, 193.

Bailey, L., Ball, B.V., Carpenter, J.M., and Woods, R.D. 1982. A strain of sacbrood virus from *Apis cerana. J. Invertebr. Pathol.* 39: 264–265.

Bailey, L., Ball, B.V., and Perry, J.N. 1983. Association of viruses with two protozoal pathogens of the honey bee. *An. Appl. Biol.* 103: 13–20.

Bailey, L., and Collins, M.D. 1982. Reclassification of 'Streptococcus pluton' (White) in a new genus Melissococcus, as Melissococcus pluton nom. rev.; comb. nov. *Journal of Applied Bacteriology* 53(2): 215–217. https://doi.org/10.1111/j.1365-2672.1982.tb04679.x.

Bailey, L., and Woods, R.D. 1977. Two more small RNA viruses from honey bees and further observations on sacbrood and acute bee-paralysis viruses. *J. Gen. Virol.* 37(1): 175–182.

Basavarajappa, S. 2012. *Study on the biological constraints of rock bee, Apis dorsata F. in southern Karnataka.* UGC Major Research Project Final Report, New Delhi, 10–180.

Basavarajappa, S., Raghunandan, K.S., and Hegde, S.N. 2009. *Pest of rockbee, Apis dorsata F. and their natural control under wild conditions.* Proc. Nat. level UGC sponsored Conf. on New Horizons in IPM. KLE Society, Haveri, Karnataka, 38–43.

Berényi, O., Bakonyi, T., Derakhshifar, I., Koglberger, H., Topolska, G., Ritter, W., Pechhacker, H., and Nowotny, N. 2007. Phylogenetic analysis of deformed wing virus genotypes from diverse geographic origins indicates recent global distribution of the virus. *Appl. Environ. Microbiol.* 73: 3605–3611.

Boecking, O. 1999. Sealing up and non-removal of diseased and Varroa jacobsoni infested drone brood cells is part of the hygienic behaviour in Apis cerana. *J. Apic. Res.* 38: 159–168.

Botías, C., Anderson, D.L., Meana, A., Garrido-Bailón, E., Martín-Hernández, R., Higes, M. 2012. Further evidence of an oriental origin for *Nosema ceranae* (Microsporidia: Nosematidae). *J. Invertebr. Pathol.* 110(1): 108–113.

Bradbear, N. 1988. World distribution of major honeybee diseases and pests. *Bee World* 69: 15–39.

Breed, M.D., Guzmán-Novoa, E., and Hunt, G.J. 2004. Defensive behavior of honey bees: Organisation, genetics, and comparison with other bees. *Ann Rev Entomic.* 49: 271–298.

Brion, A.C.B. 2015. Small hive beetle poses threat to bee industry. *The Philippine Star*, February 22.

Butler, C.G. 1962. *The world of the honeybee.* Collins, London.

Cai, J., and Collins, M.D. 1994. Evidence for a close phylogenetic relationship between Melissococcus pluton, the causative agent of European foulbrood disease, and the genus enterococcus. *International Journal of Systematic Bacteriology* 44: 365–367.

Caron, D.M. 1978. Other insects. In: Morse, R.A. (Ed.), *Honey bee pests, predators, and diseases.* Cornell University Press, London, 158–196.

Cervancia, C.R. 1993. *Philippines beekeeping status of research and development.* Proc. Beenet Asia: Workshop on Priorities in R&D on Beekeeping in Tropical Asia. Kuala Lumpur, Malaysia, 49–63.

Chaimanee, V., Pettis, J.S., Chen, Y., Evans, J.D., Khongphinitbunjong, K., and Chantawannakul, P. 2013. Susceptibility of four different honey bee species to *Nosema ceranae. Veterinary Parasitology* 193(1–3): 260–265.

Chaimanee, V., Warrit, N., and Chantawannakul, P. 2010. Infections of *Nosema ceranae* in four different honeybee species. *J. Invertebr Pathol.* 105: 207–210.
Chang, Y.D., Lee, M.Y., Yim, Y.H., and Youn, Y.N. 1993. Species and visiting patterns of wasps (Hymenoptera: Vespidae) in an apiary. *Korean J. Apic. Res.* 8: 22–28.
Chantawannakul, P., and Puchanichanthranon, T. 2005. Inhibitory effects of some medicinal plant extracts on the growth of *Ascosphaera apis*. *Acta Hort.* 678: 183–186.
Chantawannakul, P., Ward, L., Boonham, N., and Brown, M. 2006. A scientific note on the detection of honeybee viruses using real-time PCR (TaqMan) in *Varroa* mites collected from a Thai honeybee (*Apis mellifera*) apiary. *J. Invertebr. Pathol.* 91: 69–73.
Chen, Y.P., and Siede, R. 2007. Honey bee viruses. In: Karl Maramorosch, A.J.S., and Frederick, A.M. (Eds.), *Advances in virus research*. Vol. 70. Academic Press, San Diego, CA, 33–80.
Choe, S.E., Nguyen, L.T.K., Noh, J.H., Koh, H.B., Jean, Y.H., Kweon, C.H., and Kang, S.W. 2012. Prevalence and distribution of six bee viruses in Korean *Apis cerana* populations. *J. Invertebr. Pathol.* 109(3): 330–333.
Choi, Y.S., Lee, M.L., Lee, M.Y., Lee, K.G. 2008. Occurrence of seven honey bee viruses and research of disease occurrence in Korean apiaries. *Korean J. Apic.* 23(2): 153–159.
Cox-Foster, D.L., Conlan, S., Holmes, E.C., Palacios, G., Evans, J.D., Moran, N.A., Quan, P.L., Briese, T., Hornig, M., Geiser, D.M., Martinson, V., van Engelsdorp, D., Kalkstein, A.L., Drysdale, A., Hui, J., Zhai, J.H., Cui, L.W., Hutchison, S.K., Simons, J.F., Egholm, M., Pettis, J.S., and Lipkin, W.I. 2007. A metagenomic survey of microbes in honey bee colony collapse disorder. *Science* 318(5848): 283–287.
de Guzman, L.I., and Frake, A.M. 2007. Temperature affects *Aethina tumida* (Coleoptera: Nitidulidae) development. *Journal of Apicultural Research* 46(2): 88–93.
Delfinado, M.D., and Baker, E.W. 1961. *Tropilaelaps*, a new genus of mites from the Philippines (Laelaptidae[s.lat]: Acarina). *Fieldiana Zool* 44: 53–56.
Delfinado-Baker, M. 1982. New records for *Tropilaelaps clareae* from colonies of *Apis cerana indica*. *Am. Bee J.* 122: 382.
Delfinado-Baker, M., and Baker, E.W. 1982. A new species of *Tropilaelaps* parasitic on honey bees. *Am. Bee J.* 122: 416–417.
Delfinado-Baker, M., Baker, E.W., and Phoon, A.C.G. 1989. Mites (Acari) associated with bees (Apidae) in Asia, with description of a new species. *Am. Bee J.* 129(609–610): 612–613.
Delfinado-Baker, M., Underwood, B.A., and Baker, E.W. 1985. The occurrence of *Tropilaelaps* mites in brood nests of *Apis dorsata* and *A. laboriosa* in Nepal, with descriptions of nymphal stages. *Am. Bee J.* 125: 703–706.
Dietemann, V., Pirk, C.W.W., and Crewe, R. 2009. Is there a need for conservation of honeybees in Africa? *Apidologie* 40(3): 285–295. https://doi.org/10.1051/apido/2009013.
Disayathanoowat, T., Young, J.P.W., Helgason, T., and Chantawannakul, P. 2011. T-RFLP analysis of bacterial communities in the midguts of *Apis mellifera* and *Apis cerana* honey bees in Thailand. *FEMS Microbiol Ecol.* 79(2): 273–281.
Diwan, V.V., Kshirsagar, K.K., Ramama Rao, A.V., Raghunath, D., Bhambure, C.S., and Godbole, S.H. 1971. Occurrence of a new bacterial disease of Indian honey bee *Apis indica* F. *Curr. Sci.* 40: 196–197.
Djordjevic, S.P., Smith, L.A., Forbes, W.A., and Hornitzky, M.A.Z. 1999. Geographically diverse Australian isolates of Melissococcus pluton exhibit minimal genotypic diversity by restriction endonuclease analysis. *FEMS Microbiol. Lett.* 173: 311–318.
Do Carmo, R.M., Franceschinelli, E.V., and de Silveira, F.A. 2004. Introduced honeybees (*Apis mellifera*) reduce pollination success without affecting the floral resource taken by native pollinators. *Biotropica* 36: 371–376. https://doi.org/10.1111/j.1744-7429.2004.tb00329.x.

Dyer, F.C. 1991. Bees acquire route-based memories but not cognitive maps in a familiar landscape. *Animal Behaviour* 41(2): 239–246. https://doi.org/10.1016/S0003-3472(05)80475-0.

Ellis, J.D., and Hepburn, H.R. 2006. An ecological digest of the small hive beetle Aethina tumida, a symbiont in honey bee colonies Apis mellifera. *Insectes Sociaux* 53: 8–19.

Forsgren, E., Wei, S., Guiling, D., Zhiguang, L., Tran, T.V., Tang, P.T., Truong, T.A., Dinh, T.Q., and Fries, I. 2015. Preliminary observations on possible pathogen spill-over from *Apis mellifera* to *Apis cerana*. *Apidologie* 46: 265–275.

Fries, I. 1993. *Nosema apis*—a parasite in the honey bee colony. *Bee World* 74(1): 5–19.

Fries, I. 2010. *Nosema ceranae* in European honey bees (*Apis mellifera*). *J. Invertebr. Pathol.* 103: S73–S79.

Fries, I., Feng, F., Da Silva, A., Slemenda, S.B., and Pieniazek, N.J. 1996. *Nosema ceranae* n. sp. (Microspora, Nosematidae), morphological and molecular characterization of a microsporidian parasite of the Asian honey bee *Apis cerana* (Hymenoptera, Apidae). *Eur. J. Protistol.* 32(3): 356.

Fry, C.H. 1984. *The bee eaters*. 1st edn. Poyser, Calton. ISBN 10: 0856610372. ISBN 13: 9780856610370.

Gallai, N., Vaissière, B.E., Potts, S.G., Salles, J., Kareiva, P., Daily, G., Ricketts, T., Tallis, H., and Polasky, S. 2009. Assessing the monetary value of global crop pollination services: The theory and practice of ecosystem service valuation 9 in conservation. *Ecological Economics* 68: 810–821.

Genersch, E. 2010. American Foulbrood in honeybees and its causative agent, *Paenibacillus larvae*. *J. Invertebr. Pathol.* 103: S10–S19.

Gilliam, M. 1997. Identification and roles of non-pathogenic microflora associated with honeybees. *FEMS Microbiology Letters* 155: 1–10. http://dx.doi.org/10.1111/j.1574-6968.1997.tb12678.x.

Goulson, D. 2003. Effects of introduced bees on native ecosystems. *Annu. Rev. Ecol. Evol. System.* 34: 1–26.

Grabensteiner, E., Bakonyi, T., Ritter, W., Pechhacker, H., and Nowotny, N. 2007. Development of a multiplex RT-PCR for the simultaneous detection of three viruses of the honeybee (*Apis mellifera* L.): Acute bee paralysis virus, Black queen cell virus and Sacbrood virus. *J. Invertebr. Pathol.* 94(3): 222–225.

Greenleaf, S.S., and Kremen, C. 2006. Wild bees enhance honeybees' pollination of hybrid sunflower. *Proc. Natl. Acad. Sci. U.S.A.* 103: 13890–13895. https://doi.org/10.1073/pnas.0600929103.

Gulati, R., and Kaushik, H.D. 2004. Enemies of honeybees and thei– management - a review. *Agricultural Reviews* 25: 189–200.

Heath, L.A.F. 1982. Development of chalk brood in a honey bee colony; Chalkbrood pathogens: A review. *Bee World* 63(3): 119–135.

Hedtke, K., Jensen, P.M., Bruun, A., and Genersch, E. 2011. Evidence for emerging parasites and pathogens influencing outbreaks of stress-related diseases like chalkbrood. *J. Invertebr. Pathol.* 108: 167–173.

Hepburn, R., and Radloff, S.E. 2011. *Honeybees of Asia*. Springer, Berlin, 669.

Higes, M., Martin, R., and Meana, A. 2006. – Nosema ceranae, a new microsporidian parasite in honeybees in Europe. *J. Invertebr. Pathol.* 92(2): 93–95.

Hosamani, R.K., Gulati, R., and Sharma, S.K. 2006. Bioecology and management of honeybee mite, *Tropilaelaps clareae* Delfinado and Baker—a review. *Agric. Rev.* 27(3): 191–199. http://dx.doi.org/10.1016/j.jip.2003.08.007.

Huang, W.F., Solter, L.F., Yau, P.M., and Imai, B.S. 2013. *Nosema ceranae* escapes fumagillin control in honey bees. *PLOS Pathogens* 9(3): e1003185.

Jadczak, A.M. 1986. *Honeybee diseases and pests*. Maine Department of Agriculture, Food and Rural Resources, Augusta, 24.

Jeyaprakash, A., Hoy, M.A., and Allsopp, M.H. 2003. Bacterial diversity in worker adults of Apis mellifera capensis and Apenjamindra scutellata (Insecta: Hymenoptera) assessed using 16S rRNA sequences. *Journal of Invertebrate Pathology* 84: 96–103.

Kapil, R.P., and Aggarwal, K. 1987. Some observations on the concurrent parasitization of *Apis florea* by *Tropilaelaps clareae* and *Euvarroa sinhai*. *Exp. Appl. Acarol.* 3: 267–269.

Kapil, R.P., and Sihag, R.C. 1983. Wax moth and its control. *Indian Bee J.* 45: 47–49.

Kastberger, G., Raspotnig, G., Biswas, S., and Winder, O. 1998. Evidence of Nasonov scenting in colony defence of the Giant honeybee Apis dorsata. *Ethology* 104: 27–37.

Kastberger, G., and Sharma, D.K. 2000. The predator-prey interaction between blue-bearded bee eaters (*Nyctyonis athertoni*) and Giant honeybees (*Apis dorsata*). *Apidologie* 31: 727–736.

Kastberger, G., Schmelzer, E., and Kranner, I. 2008. Social waves in giant honeybees repel hornets. *PLoS One* 3(9): 1–15.

Kastberger, G., Weihmann, F., and Hoetzl, T. 2010. Complex social waves of giant honeybees provoked by a dummy wasp support the special-agent hypothesis. *Comm. Int. Biol.* 3(2): 179–180.

Kastberger, G., Winder, O., and Steindl, K. 2001. Defence strategies in the Giant honeybee *Apis dorsata*. *Proceedings of the Deutsche Zoologische Gesellschaft, Osnabrück* 94(1): 7.

Kastberger, G., Thenius, R., Stabentheiner, A., and Hepburn, R. 2009. Aggressive and docile colony defence patterns in Apis mellifera. A retreater–releaser concept. *J. Insect Behav.* 22: 65–85.

Kato, M., Shibata, A., Yasui, T., and Nagamasu, H. 1999. Impact of introduced honeybees, Apis mellifera, upon native bee communities in the Bonin (Ogasawara) Islands. *Researches on Population Ecology* 41: 217–228.

Kavinseksan, B., Wongsiri, S., de Guzman, L.I., and Rinderer, T.E. 2003. Absence of *Tropilaelaps* infestation from recent swarms of *Apis dorsata* in Thailand. *J. Apic. Res.* 42: 49–50.

Ken, T., Hepburn, H.R., Radloff, S.E., Yusheng, Y., Yiqiu, L., Danyin, Z., and Neumann, P. 2005. Heat-balling wasps by honeybees. *Naturwissenschaften* 92: 492–495.

Khongphinitbunjong, K., de Guzman, L.I., Burgett, M., Rinderer, T., and Chantawannakul, P. 2012. Behavioral responses underpinning resistance and susceptibility of honeybees to *Tropilaelaps mercedesae*. *Apidologie* 43: 590–599.

Khongphinitbunjong, K., de Guzman, L.I., Tarver, M.R., Rinderer, T.E., and Chantawannakul, P. 2015. Interactions of *Tropilaelaps mercedesae*, honey bee viruses and immune responses in *Apis mellifera*. *J. Apic Res.* https://doi.org/10.1080/00218839.2015.1041311.

Killer, J., Kopec, J., Mrazek, J., Rada, V., Dubna, S., and Marounek, M. 2010. Bifidobacteria in the digestive tract of bumblebees. *Anaerobe* 16: 165–170. http://dx.doi.org/10.1016/j.anaerobe.2009.07.007.

Klee, J., Besana, A.M., Genersch, E., Gisder, S., Nanetti, A. et al. 2007. Widespread dispenjamindaenicrosporadian *Nosema ceranae* and emergent pathogen of western honey bee, *Apis mellifera*. *J. Invertebr. Pathol.* 96: 1–10.

Klein, A.M., Vaissière, B.E., Cane, J.H., Steffan-Dewenter, I., Cunningham, S.A., Kremen, C. et al. 2007. Importance of pollinators in changing landscapes for world crops. *Proceedings of the Royal Society B: Biological Sciences* 274: 303–313.

Koeniger, G., Koeniger, N., Anderson, D.L., Lekprayoon, C., and Tingek, S. 2002. Mites from debris and sealed brood cells of *Apis dorsata* colonies in Sabah (Borneo) Malaysia, including a new haplotype of *Varroa jacobsoni*. *Apidologie* 33: 15–24.

Koeniger, N., and Fuchs, S. 1975. Zur Kolonieverteidigung der asiatischen Honigbienen. *Tierpsychologie* 37: 99–106.

Koeniger, N., Koeniger, G., and Delfinado-Baker, M. 1983. Observations on enja of the asian honeybee species (*Apis cerana, Apis dorsata, Apis florea*). *Apidologie* 14: 197–204.

Koeniger, N., Koeniger, G., Gries, M., Tingek, S., and Kelitu, A. 1996. Observations on colony defense of *Apis nuluensis* and predatory behaviour of the hornet, *Vespa multimaculata* Pérez, 1910. *Apidologie* 27: 341–352.

Koeniger, N., Koeniger, G., and Mardan, M. 1994. Mimicking a honeybee queen? *Vespa affinis indosinensis* Pérez 1910 hunts drones of *Apis cerana* F. *Ethology* 98: 149–153.

Kojima, Y., Toki, T., Morimoto, T., Yoshiyama, M., Kimura, K., and Kadowaki, T. 2011. Infestation of Japanese native honey bees by tracheal mite and virus from non-native European honey bees in Japan. *Microbiol. Ecol.* 62: 895–906.

Kulincevic, J.M., Rinderer, T.E., and Mladjan, V.J. 1991. Effects of fluvalinate and amitraz on bee lice (Braula-Coeca Nitzsch) in honey bee (*Apis mellifera* L.) colonies in Yugoslavia. *Apidologie* 22: 43–47.

Laigo, F.M., and Morse, R.A. 1968. The mite *Tropilaelaps clareae* in *Apis dorsata* colonies in the Philippines. *Bee World* 49: 116–118.

Li, J., Qin, H., Wu, J., Sadd, B.M., Wang, X., Evans, J.D., and Peng, W. 2012. The prevalence of parasites and pathogens in Asian honey bees *Apis cerana* in China. *PLoS One* 7(11): e47955.1–12.

Lyapunov, Y.E., Kuzaev, R.Z., Khismatullin, R.G., and Bezgodova, O.A. 2008. Intestinal Enterobacteria of the hibernating Apis mellifera mellifera L. bees. *Microbiology* 77: 373–379. http://dx.doi.org/10.1134/S0026261708030181.

Maassen, A. 1913. Weitere Mitteilungen uber der seuchenhaften Brutkrankheiten der Bienen [Further communication on the epidemic brood disease of bees]. *Mitteilungen aus der Kaiserlichen Biologischen Anstalt fur Land- und Forstwirtschaft* 14: 48–58.

Matsuura, M. 1988. Ecological studies on vespine wasps (Hymenoptera: Vespidae) attacking honeybee colonies. *Appl. Entomol. Zool.* 23: 428–440.

Mohr, K.I., and Tebbe, C.C. 2006. Diversity and phylotype consistency of bacteria in the guts of three bee species (Apoidea) at an oilseed rape field. *Environmental Microbiology* 8: 258–72. http://dx.doi.org/10.1111/j.1462-2920.2005.00893.x.

Mookhploy, W., Kimura, K., Disayathanoowat, T., Yoshiyama, M., Hondo, K., and Chantawannakul, P. 2015. Capsid gene divergence of Black queen cell virus isolates in Thailand and Japan honey bee species. *J. Econ. Entomol.* 1–5. https://doi.org/10.1093/jee/tov102.

Morse, R.A., and Laigo, F.M. 1969. Apis dorsata in the Phillippines. *Phillip. Assoc. Entomol.* 96.

Mishra, A., Afik, O., Cabrera, M.L., Delaplane, K.S., and Mowrer, J.E. 2013. Inorganic nitrogen derived from foraging honey bees could have adaptive benefits for the plants they visit. *PLoS One* 8(7): e70591. https://doi.org/10.1371/journal.pone.0070591.

Mostafa, A.M., and Williams, R.N. 2002. New record of the small hive beetle in Egypt and notes on its distribution and control. *Bee World* 83(3): 99–108.

Muzaffar, N., and Ahmed, R. 1986. Studies on hornets attacking honeybees in Pakistan. *Pak. J. Agric. Res.* 7: 59–63.

Nagaraja, N., and Rajagopal, D. 2009. *Honeybees-diseases, parasites, pests, predators and their management.* MJP Publishers, Chennai, 123–161.

Nagaraja, N., and Rajagopal, D. 2011. *Honeybees – diseases, parasites, pests, predators and their management.* MJP Publishers, Chennai, 74–90.

Neumann, P., and Elzen, P.J. 2004. The biology of the small hive beetle (*Aethina tumida*, Coleoptera: Nitidulidae): Gaps in our knowledge of an invasive species. *Apidologie* 35: 229–247.

Noh, J.H., Reddy, K.E., Choe, S.E., Yoo, M.S., Doan, H.T.T. et al. 2013. Phylogenetic analysis of black queen cell virus genotypes in South Korea. *Virus Genes* 46(2): 362–368.

Novogrodzki, R. 1990. Amphibians and reptiles. In: Morse, R.A. (Ed.), *Honeybee pests, predators and diseases.* Cornell University Press, Ithaca and London, 227–256.

Oldroyd, B., and Wongsiri, S. 2006. *Asian honey bees: Biology, conservation, and human interactions.* Harvard University Press, London.

Olofsson, T.C., and Vasquez, A. 2008. Detection and identification of a novel lactic acid bacterial flora within the honey stomach of the honeybee apis mellifera. *Current Microbiology* 57: 356–363. http://dx.doi.org/10.1007/s00284-008-9202-0.

Ono, M., Igarashi, T., Ohno, E., and Sasaki, M. 1995. Unusual thermal defense by a honeybee against mass attack by hornets. *Nature* 377: 334–336.

Ono, M., Okada, I., and Sasaki, M. 1987. Heat production by balling in the Japanese honeybee, *Apis cerana japonica* as a defensive behavior against the hornet, *Vespa simillima xanthoptera*. *Experientia* 43: 1031–1032.

Onstad, D.W., and Carruthers, R.I. 1990. Epizootiological models of insect diseases. *Annual Review of Entomology* 35: 399–419.

Paar, J., Oldroyd, B.P., Huettinger, E., Kastberger, G., and Schaeffer, S. 2004. Genetic Structure of an *Apis dorsata* Population: The significance of migration and colony aggregation. *Journal of Heredity* 95(2): 119–126.

Paini, D. 2004. Impact of the introduced honey bee (Apis mellifera) (Hymenoptera: Apidae) on native bees: A review. *Austral Ecology* 29(4): 399–407.

Paxton, R. 1992. The bee mite marches on: Varroajacobsoni found in the UK. *Bee World* 73: 94–99.

Pernal, S.F., and Clay, H. 2013. *Honey bee diseases and pests*. 3rd edn. Canadian Association of Professional Apiculturists, Beaverlodge, AB, 68.

Pineiro, M., and Stanton, C. 2007. Probiotic bacteria: Legislative frame work requirements to evidence basis. *Journal of Nutrition* 137: 850–853.

Potts, S.G., Biesmeijer, J.C., Kremen, C., Neumann, P., Schweiger, O., and Kunin, W.E. 2010. Global pollinator declines: Trends, impacts and drivers. *Trends Ecol. Evol.* 25: 345–353. https://doi.org/10.1016/j.tree.2010.01.007.

Promnuan, Y., Kudo, T., and Chantawannakul, P. 2009. Actinomycetes isolated from beehives in Thailand. *World J. Microb. Biot.* 25(9): 1685–1689.

Promnuan, Y., Kudo, T., Ohkuma, M., and Chantawannakul, P. 2011. *Actinomadura apis* sp. nov., isolated from a honey bee (*Apis mellifera*) hive, and the reclassification of *Actinomadura cremea* subsp. *rifamycini* Gauze et al. 1987 as *Actinomadura rifamycini* (Gauze et al. 1987) sp. nov., comb. nov. *Int. J. Syst. Evol. Microbiol* 61(9): 2271–2277.

Ra, D.K., Jeong, C., Lee, J.H., Lee, Y.M., Kim, K.H., Han, T.H., and Lee, S.M. 2012. Prevalence of honeybee diseases in Incheon area in 2011. *Korean J. Vet. Serv* 35: 111–117.

Rada, V.M.A., Chova, M., Huk, J., Marrounek, M., and Duskova, D. 1997. Microflora in the honeybee digestive tract, counts, characteristics and sensitivity to veterinary drugs. *Apidologie* 28: 357–365. http://dx.doi.org/10.1051/apido:19970603.

Raghunandan, K.S. 2014. *Bioecology of apis dorsata in few areas of South Western Karnatka*. Phd thesis, Department of Studies in Zoology University of Mysore Manasagangotri Mysore, Karnataka.

Rana, B.S., Rao, K.M., Chakravarty, S.K., and Katna, S. 2012. Characterization of *Melisococcus plutonius* causing European foulbrood disease in *Apis cerana* F. *J. Apic. Res.* 51: 306–311.

Rath, W., and Dresher, W. 1990. Response of Apis cerana Fab. Towards brood infested with Varroa jacobsoni Oud. and infestation rate of colonies in Thailand. *Apidologie* 21: 311–321.

Reddy, K.E., Noh, J.H., Kim, Y.-H., Yoo, M.S., Doan, H.T.T. et al. 2013. Analysis of the nonstructural and structural polyprotein regions, and complete genome sequences of Israel acute paralysis viruses identified from honeybees (*Apis mellifera*) in Korea. *Virology* 444(1–2): 211–217.

Roberts, J.M.K., and Anderson, D.L. 2014. A novel strain of sacbrood virus of interest to world apiculture. *Journal of Invertebrate Pathology* 118: 71–74.

Roepke, W. 1930. *Beobachtungen an indischen Honigbienen, insbesondere an 'Apis dorsata'*. H. Veenman & Zonen, Wageningen.

Rosenkranz, P., Aumeier, P., and Ziegelmann, B. 2010. Biology and control of *Varroa destructor*. *J. Invertebr. Pathol.* 103: S96–S119.

Rothenbuhler, W. 1964. Behaviour genetics of nest cleaning in honeybees: IV response of F1 and backcross generations to disease-killed brood. *Am. Zool.* 4: 111–123.

Royce, L.A., and Rossignol, P.A. 1990. Epidemiology of honey bee parasites. *Parasitology Today* 6: 348–353.

Ruttner, F. 1988. *Biogeography and taxonomy of honey bees*. Springer Verlag, Berlin, 284.

Sakagami, S.F., and Akahira, Y. 1960. Studies on the Japanese honeybee, Apis ceruna carzna Fabricius. VIII: Two opposing adaptations in the post-stinging behaviour of honeybees. *Evolution* 142940.

Sakagami, S.F., Matsumura, T., and Ito, K. 1980. *Apis laboriosa* in Himalaya, the little known world's largest honey bee (Hymenoptera, Apidae). *Insecta Matsumurana* 19: 47–78.

Sammataro, D., Gerson, U., and Needham, G. 2000. Parasitic mites of honey bees: Life history, implications, and impact. *Ann Rev Ento* 45: 519–548.

Sanpa, S., and Chantawannakul, P. 2009. Survey of six bee viruses using RT-PCR in Northern Thailand. *J. Invertebr. Pathol.* 100: 116–119.

Sanpa, S., Popova, M., Bankova, V., Tunkasiri, T., Eitssayeam, S., and Chantawannakul, P. 2015. Antibacterial compounds from propolis of *Tetragonula laeviceps* and *Tetrigona melanoleuca* (Hymenoptera: Apidae) from Thailand. *PLoS One* 10(5): e0126886.

Saraithong, P., Li, Y., Saenphet, K., Chen, Z., and Chantawannakul, P. 2015a. Bacterial community structure in *Apis florea* larvae analyzed by denaturing gradient gel electrophoresis and 16S rRNA gene sequencing. *Insect Sci.* 22: 606–618. https://doi.org/10.1111/1744-7917.12155.

Saraithong, P., Li, Y., Saenphet, K., Chen, Z., and Chantawannakul, P. 2015b. Midgut bacterial communities in the giant Asian honeybee (*Apis dorsata*) across four developmental stages: A comparative study. *Insect Sci.* https://doi.org/10.1111/1744-7917.12271.

Scardovi, V., and Trovatelli, L.D. 1969. New species of bifid bacteria from Apis mellifica L. and *Apis indica* F. A contribution to the taxonomy and biochemistry of the genus Bifidobacterium. *Zentralbl Bakteriol Parasitenkd Infektionskr Hyg* 123: 64–88.

Schmelzer, E., and Kastberger, G. 2009. 'Special agents' trigger social waves in giant honeybees (Apis dorsata). *Naturwissenschaften* 96: 1431–1441.

Seeley, T.D. 1983. The Ecology of temperate and tropical honeybee societies. *Am. Scientist.* 71(3): 264–272.

Seeley, T.D. 1994. Honey bee foragers as sensory units of their colonies. *Behav Ecol Sociobiol* 34: 51–62.

Seeley, T.D., Seeley, R.H., and Aratanakul, P. 1982. Colony defence strategies of the honeybees in Thailand. *Ecological Monographs* 52: 43–63.

Shah, F.A., and Shah, T.A. 1988. Thai sacbrood disease of *Apis cerana*. *Indian Bee J.* 50: 110–112.

Sihag, R.C. 2014. Phenology of migration and decline in colony numbers and crop hosts of giant honeybee (*Apis dorsata* F.) in semiarid environment of Northwest India. *Journal of Insects* 9, Article ID 639467. http://dx.doi.org/10.1155/2014/639467/.

Simone, M., Evans, J.D., and Spivak, M. 2009. Resin collection and social immunity in honey bees. *Evolution* 63: 3016–3022.

Simone-Finstrom, M.D., and Spivak, M. 2012. Increased resin collection after parasite challenge: A case of self-medication in honey bees? *PLoS One* 7(3): e34601. https://doi.org/10.1371/journal.pone.0034601.

Singh, S. 1961. Appearance of American foulbrood disease in Indian honey bee (*Apis indica* Fabr.). *Indian Bee J.* 23(7–9): 46–50.

Snowdon, J.A., and Cliver, D.O. 1996. Microorganism in honey. *International Journal of Food Microbiology* 31: 1–26.

Solignac, M., Cornuet, J.M., Vautrin, D., Le Conte, Y., Anderson, D., Evans, J., Cros-Arteil, S., and Navajas, M. 2005. The invasive Korea and Japan types of *Varroa destructor*, ectoparasitic mites of the Western honeybee (*Apis mellifera*), are two partly isolated clones. *Proc. Roy. Soc. Lond., B, Biol. Sci.* 272(1561): 411–419.

Southwick, E., and Southwick, L.J. 1992. Estimating the economic value of honey bees (Hymenoptera: Apidae) as agricultural pollinators in the United States. *Journal of Economic Entomology* 85: 621–633.

Srivastava, S., Kumar, A., Kashyap, N.P., Chandel, Y.S., and Raj, D. 1995. A new device to protect honeybees from the predatory wasps at the hive entrances of *Apis mellifera* colonies. *Uttar Pradesh J. Zool.* 15: 165–171.

Stanghellini, M.S., Ambrose, J.T., and Hopkins, D.I. 2000. Bumble bee colonies as potential alternative hosts for the small hive beetle (Aethina tumida Murray). *Am. Bee J.* 140: 71–75.

Stanley, J., Sah, K., Jain, S.K., Bhatt, J.C., and Sushil, S.N. 2015. Evaluation of pesticide toxicity at their field recommended doses to honeybees, *Apis cerana* and *A. mellifera* through laboratory, semi-field and field studies. *Chemosphere* 119: 668–674.

Swamy, B.C., Hosmani, V., and Nagaraja, M.V. 2009. Influence of different species of honeybee combs on the life stages and biological parameters of greater wax moth, *Galleria mellonella. Karnataka J. Agric. Sci.* 22: 670–671.

Tajabadi, N., Mardan, M., Abdul Manap, M.Y., and Mustafa, S. 2013. Molecular identification of Lactobacillus spp. Isolated from the honey comb of the honey bee (*Apis dorsata*) by 16S rRNA gene sequencing. *Journal of Apicultural Research* 52(5): 235–241.

Tajabadi, N., Mardan, M., Abdul Manap, M.Y., Shuhaimi, M., Meimandipour, A., and Nateghi, L. 2011. Detection and identification of Lactobacillus bacteria found in the honey stomach of the Giant honeybee Apis dorsata. *Apidologie* 42: 642–649. http://dx.doi.org/10.1007/s13592-011-0069-x.

Takamatsu, D., Arai, R., Miyoshi-Akiyama, T., Okumura, K., Okura, M., Kirikae, T. et al. 2013. Identification of mutations involved in the requirement of potassium for growth of typical melissococcus plutonius strains. *Applied and Environmental Microbiology* 79: 3882–3886.

Takamatsu, D., Morinishi, K., Arai, R., Sakamoto, A., Okura, M., Osaki, M. 2014. Typing of *Melissococcus plutonius* isolated from European and Japanese honey bees suggests spread of sequence types across borders and between different *Apis* species. *Vet. Microbiol.* 171: 221–226.

Tan, K., Hepburn, H.R., Radloff, S.E., Yu, Y., Liu, Y., Zhou, D., and Neumann, P. 2005. Heat-balling wasps by honeybees. *Naturwissenschaften* 92: 492–495.

Tan, K., Li, H., Yang, M.X., Hepburn, H.R., and Radloff, S.E. 2010. Wasp hawking induces endothermic heat production in guard bees. *J. Insect Sci.* 10: 1–6.

Tannock, G.W. 2004. A special fondness for lactobacilli. *Applied and Environmental Microbiology* 70: 3189–3194. http://dx.doi.org/10.1128/AEM.70.6.3189-3194.2004.

Thapa, R., and Wongsiri, S. 2003. Flying predators of the giant honeybees, *Apis dorsata* and *Apis laboriosa* in Nepal. *Am. Bee J.* 143: 540–542.

Thapa, R., Wongsiri, S., Oldroyd, B.P., and Prawn, S. 2000. *Migration of Apis dorsata in Northern Thailand.* Proc. 4th AAA, Kathmandu, Nepal, 39–43.

Theantana, T., and Chantawannakul, P. 2008. Protease and beta-N acetylglucosaminidase of honey bee chalkbrood pathogen *Ascosphaera apis*. *J. Apic. Res.* 47(1): 68–76.

Tingek, S., Koeniger, G., Koeniger, N., and Gries, M. 2004. *Phylocephala pallaliventris (Krober) (Conopidae), a parasitic fly of three Apis species in northern Borneo.* Proceedings of 7th Asian Apicultural Association Conference, Los Banos, Philippines, 364–365.

Triyasut, P., Mookhploy, W., Kimura, K., Yoshiyama, M., Khongphinitbunjong, K., and Chantawannakul, P. 2015. First detection of honey bee viruses in wax moth. *Chiang Mai J. Sci.*, in press.

Tutkun, E., Maden, S., Inci, A., and Yilmarz, B. 1993. General situation of chalkbrood disease in honey bees in Turkey. *Turk. Entomol. Derg.* 17(2): 65–68.

Vasquez, A., and Olofsson, T.C. 2009. The lactic acid bacteria involved in the production of bee pollen and bee bread. *Journal of Apicultural Research* 48: 189–195. http://dx.doi.org/10.3896/IBRA.1.48.3.07.

Vasquez, A., Olofsson, T.C., and Sammataro, D. 2009. A scientific note on the lactic acid bacterial flora in honeybees in the USA- A comparison with bees from Sweden. *Apidologie* 40: 26–28. http://dx.doi.org/10.1051/apido:2008063.

Viraktamath, S., Basalingappa, S., and Lingappa, S. 2005. Biology and seasonal incidence of the braconid wasp, *Apanteles Galleriae* and its parasitization of greater wax moth, *Galleria mellonella*. *Ind. Bee J.* 67: 182–187.

Wanjai, C., Sringarm, K., Santasup, C., Pak-Uthai, S., and Chantawannakul, P. 2012. Physicochemical and microbiological properties of longan, bitter bush, sunflower and litchi honeys produced by *Apis mellifera* in Northern Thailand. *J. Apic. Res.* 51: 36–44.

White, G.F. 1913. *Sacbrood, a disease of bees.* Circular No. 169, US Department of Agriculture, Bureau of Entomology, Washington, DC, 1–5.

Weihmann, F., Hoetzl, T., and Kastberger, G. 2012. Training for defense? From stochastic traits to synchrony in giant honeybees (*Apis dorsata*). *Insects.* 3: 833–856.

Whitfield, C.W., Behura, S.K., Berlocher, S.H., Clark, A.G., Johnston, J.S., Sheppard, W.S. et al. 2006. Thrice out of Africa: Ancient and recent expansions of the honey bee. *Apis Mellif. Sci.* 314: 642–645. https://doi.org/10.1126/science.1132772.

Williams, G.R., Sampson, M.A., Shutler, D., and Rogers, R.E.L. 2008. Does fumagillin control the recently detected invasive parasite *Nosema ceranae* in western honey bees (*Apis mellifera*)? *J. Invertebr. Pathol.* 99: 342–344.

Williams, G.R., Shutler, D., Little, C.M., Burgher-MacLellan, K.L., and Rogers, R.E.L. 2011. The microsporidian *Nosema ceranae*, the antibioticlmagilin-B (R), and western honey bee (*Apis mellifera*) colony strength. *Apidologie* 42: 15–22.

Williams, J.L. 1976. Status of the greater wax moths, *Galleria mellonella* in the United States Beekeeping industry. *Am. Bee. J.* 166(11): 524–526.

Winston, M. 1987. *The biology of the honey bee.* Harvard University Press, Cambridge.

Wongsiri, S., Tangkanasing, P., and Sylvester, H.A. 1989. *The resistance behavior of Apis cerana against Tropilaelaps clareae*. Proceedings First Asia-Pacific Conf. of Enotmol. Chiang Mai, Thailand, pp. 828–836.

Wongsiri, S., Thapa, R., Chantawannakul, P., Chaiyawong, T., Thirakhupt, K., and Meckvichai, W. 2005. Bee eating birds and honey bees predation in Thailand. *Am. Bee. J.* 145(5): 419–422.

Woyke, J. 1984. Survival and prophylactic control of Tropilaelaps clareae infesting Apis mellifera colonies in Afghanistan. *Apidologie* 15: 421–434.

Woyke, J. 1985. *Tropilaelaps clareae in Afghanistan, and control methods applicable in tropical Asia*. Proceedings 3rd International Conference Apic. Trop. Climates, Nairobi, 163–166.

Woyke, J. 1996. *Different reaction of Apis dorsata and Apis mellifera to brood infestation by parasitic mites*. Proceedings 3rd AAA Conference on Bee Res. and Beekeeping Dev., Hanoi, Vietnam, 172–175.

Yang, B., Peng, G., Li, T., and Kadowaki, T. 2013. Molecular and phylogenetic characterization of honey bee viruses, Nosema microsporidia, protozoan parasites, and parasitic mites in China. *Ecol. Evol.* 3(2): 298–311.

Yen, D.F., and Chyn, L.C. 1971. Studies on a bacterial disease of honey bee in Taiwan. *Plant Protection Bulletin* 13: 12–17.

Yoo, M.S., and Yoon, B.S. 2009. Incidence on honey bee disease in Korea 2009. *Korean J. Apicul* 24: 273–278.

Yoshiyama, M., and Kimura, K. 2009. Bacteria in the gut of Japanese honeybee, Apis cerana japonica and their antagonistic effect against Paenibacillus larvae, the causal agent of American foulbrood. *Journal of Invertebrate Pathology* 102: 91–96. http://dx.doi.org/10.1016/j.jip.2009.07.005.

Zander, E. 1909. Tierische Parasiten als Krankenheitserreger bei der Biene. *Muchener Bienenzeitung* 31: 196–204.

Zhang, X., He, S.Y., Evans, J.D., Pettis, J.S., Yin, G.F., and Chen, Y.P. 2012. New evidence that deformed wing virus and black queen cell virus are multi-host pathogens. *J. Invertebr. Pathol.* 109(1): 156–159.

Zhou, T., Feng, F., and Dong, B. 2000. Study on the pathogen of European foulbrood in the Chinese honey bee (*Apis cerana cerana* F.). *Acta Entomol. Sinica* 43: 104–108.

8 Dance Communication of Giant Honeybees

Patrick L. Kohl, Benjamin Rutschmann and Axel Brockmann

8.1 INTRODUCTION

Recruitment communication for food or other resources is widespread among social insects (Leonhardt et al. 2016). Even when only considering the three groups of eusocial bees (bumblebees, stingless bees, and honeybees), a large diversity of communication mechanisms can be found. In bumblebees (tribe Bombini), recruitment is limited to a general food alert, which motivates other workers to leave the nest and search for the advertised resource (Dornhaus and Chittka 2001), while workers of stingless bees (tribe Meliponini) and honeybees (tribe Apini) can recruit nestmates to specific locations in the field. In several species of Meliponini, foragers guide nestmates by applying scent marks to vegetation along their flight paths, and others might use aerial pheromone trails or direct guidance for spatial recruitment (Grüter 2020). Honeybees, in turn, evolved the ability to communicate the distance and direction of foraging sites in a symbolic manner using a nest-based mechanism, the waggle dance (von Frisch 1967). This behaviour, which was originally discovered in the western honeybee (*Apis mellifera*), has often been referred to as the "dance language" (von Frisch 1967; Dyer 2002).

Upon return from a rewarding foraging trip, a worker honeybee may soon engage in a highly stereotyped figure-of-eight locomotor pattern consisting of circuits alternatingly headed to the left and the right. Each circuit consists of a waggle phase (or "waggle run"; the middle part of the eight) in which the bee walks straight and waggles her abdomen, and a return phase in which she turns back to her starting position (see Figure 8.4). Foragers may perform several dances after a foraging trip and a single dance typically consists of several waggle and return phases (Seeley 1995). The body orientation of the dancer during the waggle phases and the duration of the waggle phases contain the information on the direction and distance of the foraging site (von Frisch 1967). For all honeybee species, the reference direction for orientation in the field is the solar azimuth, i.e., the sun's position along the local horizon. The bees' dance angle with respect to a common reference in the nest indicates the direction of the foraging site in relation to the position of the sun (von Frisch 1967; Dyer 2002). The waggle duration, in turn, is a continuously increasing function of foraging distance, with longer waggle durations being associated with greater distances (von Frisch 1967; Dyer 2002). Dancers are typically followed by one or several other bees. These dance followers can use the spatial information of the dance and may

DOI: 10.1201/9781003294078-8

decide to fly to the advertised location (Riley et al. 2005; Menzel et al. 2011). Besides the figure-eight locomotor pattern, depending on the honeybee species, dancers produce wing vibrations, make visually conspicuous abdomen movements, and/or release pheromones (Wenner 1962; Thom et al. 2007; Towne 1985; Esch 1961). It is largely unknown whether these channels are also used for spatial communication (Michelsen et al. 1992; Landgraf et al. 2018). Although they can contain information on distance and/or direction, they probably primarily serve as a food alert or may attract other bees to the dancer (Esch 1961). Spatial communication using the waggle dance allows honeybees to exploit rewarding resources as a group even when these are located far away from the nest (several kilometres) (Ratnieks and Shackleton 2015). Since the probability of dancing and the number of waggle phase signals produced per dance increase as a function of the quality of a resource, and since the likelihood of recruitment to a patch depends on the number of waggle dance signals produced, dance communication can lead to optimal spatial allocation of foragers among resource locations in a self-organized manner (Seeley 2012). Apart from the context of resource collection (e.g., nectar, pollen, water, plant resin), dances are used to communicate the locations of potential nest sites (Seeley 2010; Makinson et al. 2014, 2011; Oldroyd et al. 2008) and are involved in the coordination of long-distance colony migrations (Hepburn 2011; Dyer and Seeley 1994; Koeniger and Koeniger 1980; Robinson 2021).

All extant honeybee species use the waggle dance (Dyer 2002; Lindauer 1956). We can expect that the fundamentals of dance communication are conserved across the genus because they most likely evolved once in a common ancestor (I'Anson Price and Grüter 2015; Raffiudin and Crozier 2007; George et al. 2020). However, the few comparative studies conducted so far have revealed functionally important interspecific variation that is likely attributable to differing ecological pressures (Dyer 1991, 2002; Kohl et al. 2020; Lindauer 1956; Towne 1985). It is therefore likely that future work on Asian honeybees will provide further insights into the mechanisms and evolution of the dance language (Nieh 2011).

The bees whose dances certainly still hold a range of secrets are the giant honeybees (subgenus *Megapis*). As we will outline in the following, investigating the dance language of giant honeybees will likely provide new insights into how honeybees use behaviour in the different contexts of foraging, nest site selections and seasonal migrations, why some honeybee species produce sounds while dancing, and how bees orient themselves when foraging during low light conditions. Furthermore, there is a large potential to use the dance language as a tool to understand the ecology of giant honeybees: by decoding waggle dances, human observers can infer the bees' spatio-temporal foraging patterns and potentially identify their migration routes—information urgently needed for the conservation of these enigmatic and important pollinators.

8.2 THE "DANCE FLOOR" OF GIANT HONEYBEES

Colonies of giant honeybees build single-comb nests attached to tree branches or overhangs at rocks or buildings in the open. A large fraction of a colonies' workforce

FIGURE 8.1 (a) *Apis dorsata* colony nesting at a window recess. The single comb nest is protected by a living curtain of worker bees. Foragers depart, arrive, and dance at the "mouth" of the colony (here: in the lower right quarter of the curtain) (Photo credit: Bharath Kumar). (b) *Apis dorsata* pollen forager advertising her foraging site with a waggle dance (centre). She is surrounded by at least nine dance followers.

is hanging in a protective curtain around the comb. During hours of foraging, workers depart and arrive at a circumscribed region of the bee curtain, the so-called mouth of the colony (Duangphakdee et al. 2011). The mouth can be regarded as the colonies' traffic hub and information centre: returning foragers unload their honey crop to food processing bees, provide taste samples to other foragers, and importantly, perform waggle dances (Figure 8.1). Compared to the cavity-nesting honeybee species including *Apis mellifera*, which dance on a smooth beeswax comb inside a dark cavity, the substrate of the "dance floor" of giant honeybees is rather irregular. One might think that dancing on top of other bees makes it harder to keep waggle dance movements continuous and precise. In fact, a comparative study showed that when dancing for nest sites on the surface of a swarm cluster (which is a collective structure of bee bodies comparable to a bee curtain around a nest), *A. mellifera* workers dance with less directional precision than when dancing for foraging sites on beeswax combs (Beekman et al. 2015). However, the same study showed that the dances of *A. dorsata* are not less precise than those of other honeybee species, suggesting that giant honeybees are well capable of precisely aligning their bodies despite dancing on top of other bees (Beekman et al. 2015). Unfortunately, it is unclear whether the dance activity of giant honeybees is limited to the surface of the colony (Oldroyd and Wongsiri 2006). Returning foragers often switch from the outer to the inner side of the bee curtain between foraging flights so they might also perform dances behind the curtain, out of our sight. Whether giant honeybees also dance on their beeswax comb might be revealed using endoscopic infrared cameras carefully submersed into the bee curtain or using sound/vibration sensors placed on the comb.

8.3 THE WAGGLE DANCE OF GIANT HONEYBEES

The waggle dance of giant honeybees resembles that of the Western honeybee, meaning that a person familiar with the behaviour of *A. mellifera* will immediately recognize waggle dances of giant honeybees as such. However, there is variation in several details, some of which are likely functionally important. Giant honeybees waggle their abdomen both laterally and dorsoventrally (i.e., left and right and up and down), meaning that their dances are visually more conspicuous that those of the cavity-nesting species *A. mellifera* and *A. cerana*, which only make lateral (i.e., left to right) abdomen movements. The exaggerated abdomen waggling might be an adaptation to dancing in the open and might serve to visually draw attention to the dancer on the bee curtain (Towne 1985). Evidence for this function comes from the comparison with the dwarf honeybees, which also nest and dance in the open and which also dance with visually conspicuous waggle run postures (Towne 1985). The cavity-nesting species *A. mellifera* and *A. cerana* accompany their waggle movements with wing buzzes and thus use the acoustic/vibrational channel during dance communication (Esch 1961; Wenner 1962; Towne 1985). This makes sense in the dark hive, as it allows dances to be located by potential dance followers even without direct contact to the dancer. Curiously, among the giant honeybees, dances with and without sound production occur. Kirchner and Dreller (1993) showed that *A. dorsata* dancers emit sounds during the waggle phase of their dances in a similar way as *A. mellifera* dancers do: the probability of sound production is positively associated with the quality of the food reward and the length of the sound pulses correlates with the duration of the waggle phase. Therefore, their dance sounds contain information on resource quality and distance. The same team of researchers, however, found no such sound production in the Himalayan giant honeybee, *A. laboriosa* (Kirchner et al. 1996). The functional explanation is that colonies of *A. dorsata* but not of *A. laboriosa* regularly forage and dance during twilight hours and even at night, which is when using the acoustic channel makes as much sense as when dancing in a dark hive during the day (Dyer 2002; Kirchner et al. 1996; Young, Kodabalagi et al. 2021). The apparent diversity in the use of dance sounds that exists among these closely related species makes the group of giant honeybees an interesting subject for the study of the role of sound production in dance communication. For example, it would be interesting to test whether, the quality of the food source being equal, the probability of sound production in *A. dorsata* is higher when dancing at night than during the day. Furthermore, it would be possible to test whether the use of dance sounds differs between geographic subspecies of *A. dorsata*, and if so, whether this correlates with differences in their day versus night-time activity schedules.

8.3.1 Direction Communication

Flight direction is encoded by the orientation of the dancer during the waggle phase. The dances of *A. dorsata* contain clear direction information even when the advertised food source is as close as 3 metres to the hive, i.e., most dances for natural resource locations (which are normally not directly at the nest site) contain clear waggle phases during which the bees are heading into a certain direction (Towne and Gould 1988; Gardner et al. 2009). But how do giant honeybees orient their dances when dancing on the bee curtain, i.e., how do they encode the sun-related flight direction of the horizontal plane onto their vertical dance plane? Since the pioneering study by Lindauer (1956) on the dance language of Asian honeybee species, it has been assumed and often re-cited that giant honeybees orientate their dances in the same way as *A. mellifera* does (I'Anson Price and Grüter 2015; Dyer 1991). However, this has never been verified using experimental data. The cavity-nesting species like *A. mellifera*, when dancing on a vertical wax comb in the dark cavity, use gravity as the reference direction (von Frisch 1967): dances with waggle runs pointing exactly upwards indicate foraging sites in the direction of the current solar azimuth, dances oriented downwards indicate locations in the direction opposite to the sun, and dances oriented orthogonally to the plumb line to the right indicate locations 90° from the sun's position, and so forth. Given that the solar azimuth moves along the local horizon, bees dancing with respect to gravity need to gradually adjust their dance angles over the day when advertising one and the same location. The dance floor of giant honeybees is also a vertical surface, so they could also use gravity as a reference when dancing. However, the important difference compared to the dance floor of *A. mellifera* is that it is always in the open, not in the dark (Lindauer 1956). Interestingly, when *A. mellifera* workers face a similar situation and dance on a vertical surface with the view of the sky (which naturally occurs on the swarm cluster) they do *not* take gravity as the (sole) reference for dance orientation. In the case when they can directly see the sun, they orientate their dances with respect to its azimuth as it appears on the horizon of their current field of view on the vertical dance plane (i.e., they behave as if the whole world had been flipped by 90 degrees). When the sun is not directly visible to the dancers, they do an interesting compromise and dance with angles between the expected "gravity dance angle" and the expected "sun dance angle" (Dyer 1985b; Edrich 1977; von Frisch 1962). Even though this behaviour appears overly complicated compared to the neat way of orienting with respect to the vertical alone, it might exactly be what giant honeybees do, too. To test the possibility that the sun influences their dance orientation, researchers need to provide dance orientation data collected in controlled feeder experiments in which the exact position of the nest (including the orientation of the dance floor), the feeder, and the sun are known. The realized dance angles then need to be statistically compared not only to the expected gravity dance angle but also to the expected sun dance angle, which requires knowing the sun's apparent azimuth on the vertical plane of the dance floor (von Frisch 1962).

8.3.2 Distance Communication

Distance is encoded in the duration of the waggle phase, with longer waggle runs signalling greater distances (von Frisch 1967). The general shape of the distance/duration curve is probably similar among all honeybee species in that it follows a linear increase for the closer range of distances and flattens at longer distances (Kohl and Rutschmann 2021). However, different honeybee populations differ in the slopes of their distance functions (Bosch 1957; Lindauer 1956). The explanation for these so-called "distance dialects" is that the distance curves are evolutionary adaptations to the typical foraging range of the bees ("adaptive tuning hypothesis") (Dyer 1991; Kohl et al. 2020). Under the assumption that there is an upper limit to the duration of the waggle phase, the bees face a trade-off between the precision of distance communication and the maximum foraging distance they can communicate. Steeper distance/duration curves have the advantage that they allow for a more precise distance communication since a given range of waggle durations corresponds to a smaller range of distances. However, they reach the maximum waggle duration more quickly, so the communicated foraging range is restricted (Dyer 2002). In line with the prediction, honeybee (sub)species with smaller foraging ranges have been found to have steeper distance/duration functions (Kohl et al. 2020).

The distance/waggle duration function of giant honeybees has so far only been reported for one population (Kohl et al. 2020). It required training forager bees to an artificial feeder, moving the feeder stepwise to various distances from the colony, filming the corresponding waggle dances back at the nest using a camcorder, and making computer-assisted measurements of waggle durations from videos. The available data from Bangalore, India, show that *A. dorsata* workers have a distance/duration curve with a lower slope than the sympatric dwarf honeybees, *A. florea*, and Asiatic hive bees, *A. cerana* (Figure 8.2). It corresponds to a relatively larger foraging range of the giant honeybees compared to the other two species. At 100-metre feeder distance, *A. dorsata* foragers dance with a waggle duration of about 0.6 seconds, and every other 100 metres, this duration increases by another 0.2 seconds. When compared to the distance curve of temperate-adapted *A. mellifera*, the giant honeybees communicate with a relatively steep distance/duration function (Figure 8.2). It would be interesting to know whether *A. dorsata* populations from other regions communicate with similar distance codes as the bees from Bangalore. Dialect difference between closely related populations or species are expected to arise when these

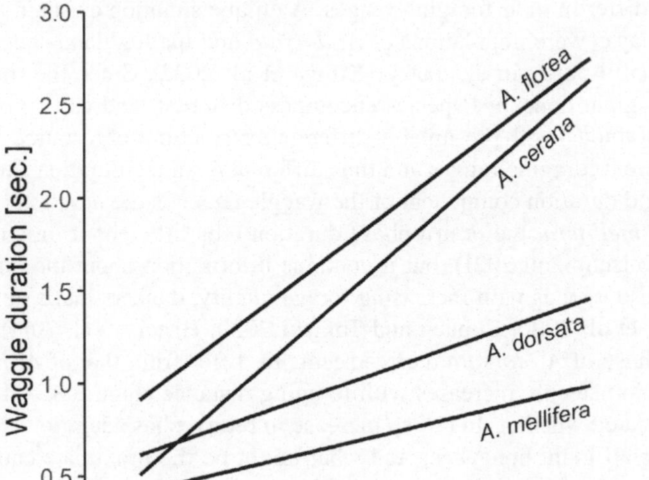

FIGURE 8.2 Distance dialects of *A. cerana*, *A. florea* and *A. dorsata* from Bangalore, India (data from Kohl et al. 2020), and of *A. mellifera* from Germany (data from Kohl and Rutschmann 2021). *Apis dorsata* has a relatively low distance dialect slope compared to the other Asian honeybee species.

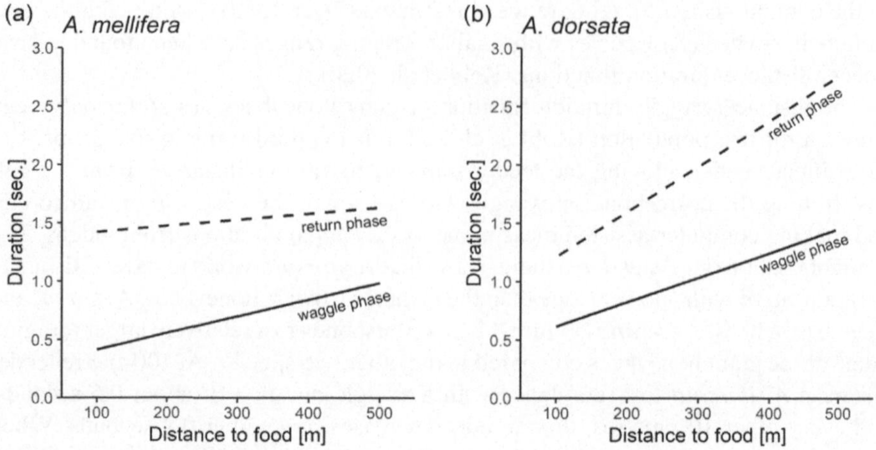

FIGURE 8.3 Distance-related increase in the duration of the waggle phase and the return phase in dances of (a) *A. mellifera* (data from Kohl and Rutschmann [2021]) and (b) *A. dorsata* (data from [Kohl et al. 2020]). In *Apis dorsata*, return phase duration increases much more rapidly.

populations differ in their foraging ranges. A unique situation exists in the foothills of the Himalayas were populations of *A. dorsata* and the highland species *A. laboriosa* seasonally occur in sympatry (Kitnya et al. 2022). Since the main habitats of these two giant honeybee species encompass different landscape types (lowland versus mountainous) with presumably different distributions of resources, they might require different foraging ranges and thus different distance-duration functions.

The second duration component of the waggle dance is the duration of the return phase. In *A. mellifera*, the return phase duration tells little about foraging distance (Kohl and Rutschmann 2021), but it contains information about the quality of the advertised resource, as with increasing forage quality, dancers make quicker return runs (Seeley et al. 2000; Łopuch and Tofilski 2020; Hrncir et al. 2011). Curiously, the return phase of *A. dorsata* dances seems to differ from that of *A. mellifera*, in that its duration steeply increases with foraging distance (Figure 8.3). It still needs to be investigated whether this steep increase in return phase duration with distance is typical for all giant honeybees and what might be the proximate cause. Perhaps giant honeybee dancers walk greater stretches during the waggle phase than *A. mellifera* so that they have a longer way back to their starting point (Tautz et al. 1996). Furthermore, it needs to be investigated whether giant honeybees tune the duration of the return phase according to resource quality, as *A. mellifera* does (Hrncir et al. 2011; Seeley et al. 2000; Łopuch and Tofilski 2020).

8.4 DANCING AT LOW LIGHT CONDITIONS

While most honeybee species—and bees in general—are diurnal animals, the giant honeybee *A. dorsata* is one of the few species that regularly forages during twilight hours and moonlit nights (Dyer 1985a; Young, Kodabalagi et al. 2021). The proportion of a colony's foraging trips that take place after sunset and before sunrise has been estimated to reach 30%, meaning that twilight and nocturnal foraging activity plays an important role for their overall energy intake (Young, Kohl et al. 2021). Although the resource use of *A. dorsata* during dim light foraging still requires scientific study (for a study on nocturnal foraging in carpenter bees, see Somanathan et al. [2020]), the potential benefits certainly involve the escape from direct competition with other flower visitors and the exploitation of resources that are exclusively available or more abundant at night. What is more challenging to understand is *how* do the bees orient themselves at low light levels? During the day, honeybees determine their flight direction visually using the sun-compass and local landmarks. They also use their eyes to estimate the distance travelled: they take as a proxy the amount of image motion passed by the retina during flight ("optical flow") (Dyer 2002; Srinivasan 2000; Esch and Burns 1996; Esch et al. 2001). Given that *A. dorsata* foragers can evidently forage when the sun is completely down, which orientation mechanisms do they use when the light is fading (Spitschan et al. 2016)? Their waggle dances might provide answers, since they are performed as frequently at night as one would expect, given their foraging activity (Dyer 1985a; Young, Kodabalagi et al. 2021). When the sun is below the horizon, giant honeybees might use the azimuth of the sun where it was last seen on the local horizon, an extrapolated position of the sun given the rate of movement it showed in the hours before sunset, or the actual position of the sun (however they might know it) as the direction reference (Edrich 1981). The possibility that the bees use the moon as an alternative reference can be safely excluded because the course of the moon does not affect the orientation of night dances (Dyer 1985a) and bees also forage when there is no moon visible on the night sky (Young, Kohl et al. 2021). Apart from direction orientation, distance estimation might be hampered at night because lower contrasts lead to lower rates of perceived image motion during flight. To investigate these questions, *A. dorsata* workers need to be trained to a feeder which is moved to locations at different distances from the nest. This will reveal how they orient their dances during the day and how distance is encoded in waggle duration. Then, the same feeder experiment needs to be repeated at night to reveal the reference for night-time dance orientation and whether the bees perceive distance differently at night.

8.5 NEST SITE AND MIGRATION DANCES

During seasons of colony growth, honeybees mainly use the waggle dance for the communication of food (e.g., nectar, pollen) and water locations. When colonies have reached their peak size and reproduce by fission, however, the bees leaving the old nest as a swarm also use their dance to communicate about new nesting sites (Seeley 2010; Lindauer 1955). The process of nest-site selection has been investigated in detail in *A. mellifera* (Seeley 2010), and experiments with *A. florea* and *A. dorsata* have suggested that the mechanisms are similar in the open-nesting species (Makinson et al. 2011, 2014). After separation of the swarm from the old colony and the formation of a cluster in the vicinity, a subset of the swarms' workers, the scouts, explore the surroundings for suitable nesting sites and indicate these locations back at the cluster using waggle dances. Other scouts follow the dances and examine the advertised sites themselves. Only if the communicated locations are considered suitable will they also dance for them. Since the bees stop promoting their discovered sites after a while, only information on good locations is passed on, and the swarm finally reaches a consensus on a single site. Then, the scout bees initiate the mass take-off and guide the swarm to the new home (Seeley 2010). Apart from the process of swarm dispersal, honeybees in tropical and subtropical regions regularly leave their nests without colony fission as a response to unfavourable conditions at the nest site or its surroundings (Hepburn 2011; Oldroyd and Wongsiri 2006). Depending on the cause and the organization of these movements, different types of colony relocations can be distinguished (Table 8.1). Particularly notable are the seasonal migrations of the giant honeybees. Two times a year, in predictable time windows, most—if not all—colonies leave their nests to start a journey to sites in a different region (Woyke et al. 2012; Koeniger and Koeniger 1980). The migrations can span up to 200 km and typically go from the lowlands to highlands and vice-versa. They are completed in several shorter moves, in between which the colonies rest for several days without building new comb (so-called bivouacs) (Koeniger and Koeniger 1980; Robinson 2021). During these breaks, the bees resume foraging to fill their crops with nectar which is needed to fuel the journey. While the evolutionary benefits

TABLE 8.1
Types of Colony Relocations in Giant Honeybees

Name	Cause	Navigational Information Communicated by Dancers	Acquisition of Navigational Information by Dancers	Organization	Place of Dance Performance
Nest relocation ("Prepared absconding")	Local decline of environmental conditions, e.g., reduced food availability, prolonged disturbance	Several discrete locations	Experience	Consensus building	Colony mouth
Swarming	Colony reproduction	Several discrete locations	Experience	Consensus building	Colony mouth
Seasonal migration	Seasonal change in resource availability	A single direction; variation of distance information	Innate or experience (unknown)	Informed elite*	Whole colony

References: Koeniger and Koeniger 1980; Dyer and Seeley 1994; Hepburn 2011; Neumann et al. 2000; Paar et al. 2000; Robinson 2021

* *Informed elite:* If older bees guide the colony during migrations, they could be called "informed elite".

of colony migrations clearly lie in tracking seasonal changes in flower availability associated with rainy and dry seasons (Dyer and Seeley 1994), it is a mystery how colonies coordinate these long-distance migrations, especially because they manage to return to their original nesting sites after several month have passed (Oldroyd and Wongsiri 2006; Neumann et al. 2000; Paar et al. 2000). In contrast to nest site selection, whereby the bees first search for and then dance for multiple discrete sites among which to choose, colony migrations are apparently preceded by dances of bees that did not recently leave the colony, and which all indicate the same direction (Dyer and Seeley 1994; Koeniger and Koeniger 1980). The so-called migration dances can contain very long waggle phases indicating distances beyond the typical foraging range of the bees (Dyer and Seeley 1994). Furthermore, migration dances are not performed in a circumscribed area of the colony but all over the bee cluster, with dancing bees crossing large stretches until reaching the edge of the colony (Figure 8.4) (Duangphakdee et al. 2011; Dyer and Seeley 1994; Koeniger and Koeniger 1980; Robinson 2021) The observations that migration dancers are performed even when there is no flight activity in the colony, and that dances of different bees indicating the same direction can vary greatly in the distances signalled, suggest that the bees might indicate innate migration directions without having been at any discrete destination before (Dyer and Seeley 1994). However, Robinson's (2021) observations suggest the alternative hypothesis that migrations might be led by bees that lived long enough to memorize the migration route from a preceding journey. He reported on observations of migration dances of bivouacking colonies at a stopover site along a migration route in Thailand (Robinson 2021). The resting colonies foraged and danced normally for nectar sources—but prior to the

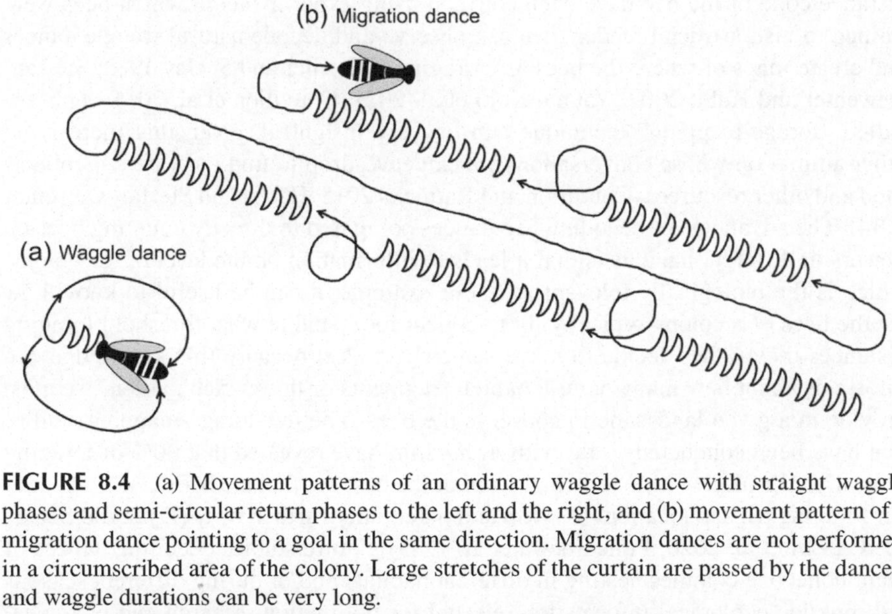

FIGURE 8.4 (a) Movement patterns of an ordinary waggle dance with straight waggle phases and semi-circular return phases to the left and the right, and (b) movement pattern of a migration dance pointing to a goal in the same direction. Migration dances are not performed in a circumscribed area of the colony. Large stretches of the curtain are passed by the dancer, and waggle durations can be very long.

next move, distinct migration dances were performed by apparently darker-coloured, older bees (Robinson 2021). Interestingly, contrary to prior observations, Robinson did not observe large variation in waggle durations between the migration dances of different bees, and waggle durations correlated with the distance travelled by the colonies, with dances for closer sites taking the form of ordinary waggle dances. It therefore seems that the important difference in the organization of nest site selection and migration is not the form of the dance but whether the bees use a "democratic" process to decide where to go (nest site dances) or whether a minority of bees (an "informed elite") signals a discrete destination or a direction of travel that is apparently predetermined (migration dances). What is urgently needed to better understand how the bees use the dance for nest-site selection and colony migration are longitudinal natural history observations. For example, a continuous record of the dances performed by individual colonies from the arrival until the departure from their nesting site would certainly provide valuable insights. Such records might tell us how dances for forage, dances for nest sites in preparation for swarming and migration dances are spatially and temporally organized in a giant honeybee colony. Furthermore, to test the possibility that migrations are led by old bees that memorize the destination from a preceding migration, migration dancers and their dance followers could be marked prior to take-off and their colonies be resurveyed for marked bees upon return (Oldroyd and Wongsiri 2006).

8.6 THE DANCE LANGUAGE AS A WINDOW INTO THE ECOLOGY OF GIANT HONEYBEES

Knowledge about the information content of the waggle dance allows us to answer questions about the ecology of the bees. When the orientation mechanism and the distance code of the bee have been clarified using experiments in which bees were trained to visit artificial feeders, we can observe and decode natural waggle dances and create maps of where the bees are foraging (Visscher and Seeley 1982; Steffan-Dewenter and Kuhn 2003; Young, Kohl et al. 2021; Couvillon et al. 2014). This so-called "forage-mapping" technique can be very insightful, given that there is no other animal on whose conversations we can eavesdrop to find out where it collects food and other resources (Couvillon and Ratnieks 2015; Härtel and Steffan-Dewenter 2014). The advantage of decoding bee dances compared to directly counting bees on flowers in different habitats is that it leads to information on the level of the colony, which is the biologically relevant unit. For example, it can be useful to know how far the bees of a colony typically fly to collect food, and at what threshold foraging distances they rather decide to move somewhere else, because this knowledge can inform us about how many natural habitat fragments or flower-rich gardens we must provide in a given landscape to conserve the bees. The few forage-mapping studies that have been conducted so far with *A. dorsata* have revealed that 90% of foraging typically take place within less than 1 km of the nest but that the bees also make longer foraging trips to locations 3–4 km and potentially further away (Dyer and Seeley 1991; Kohl et al. 2020; Punchihewa et al. 1985). Future studies decoding dances of giant honeybee colonies nesting in different habitats and/or during different seasons will provide ecological information relevant for pollination research and biological conservation.

8.7 HOW TO STUDY THE DANCES OF GIANT HONEYBEES

As outlined in this chapter, many aspects of dance language and the foraging behaviour of giant honeybees are still to be clarified (see Table 8.2). The nests of giant honeybees are easy to spot, and no special observation hives are required since their

TABLE 8.2
Major Questions regarding the Dance Communication and Foraging Behaviour of Giant Honeybees and Possible Experimental Approaches

Questions	Experiments
(A) Sensory and Behavioural Mechanisms	
Do giant honeybees, besides dancing on top of the bee curtain, dance on their beeswax comb?	Monitoring of dance behaviours behind the bee curtain using endoscopic cameras or sound/vibration sensors attached to the comb.
Do giant honeybees directly orient their dances with respect to the sun's position when dancing on the curtain in the open?	Comparison of the actual dance angle for a given feeder location with the calculated "gravity dance angle" and "sun dance angle"
Does foraging at low light conditions compromise the visual estimation of flight distance and thus waggle run duration?	Comparison of waggle durations for given feeder distances at different light conditions during the day and night
How do giant honeybees use the sun-compass after sunset, and how do they orientate their dances during the night?	Comparison of dance orientations for a given feeder location during the day and night
Which aspects of the dance communication in *A. dorsata* are modulated by the reward quality of the food source?	Comparing various aspects of waggle dances performed by bees foraging at feeders with different sugar concentrations
Do bees that perform migration dances know the destination due to prior experience?	Comparison of migratory dances between natural colonies and artificially established colonies composed of naïve workers without migration experience.
(B) Ecological Variation and Evolutionary Adaptations	
Questions	**Experiments**
Do giant honeybee species differ in their foraging ranges and thus show different distance/waggle duration curves (distance dialects)?	Forage mapping (dance decoding) of colonies to estimate foraging ranges, and feeder experiments to determine distance-calibration curves
Are dance sounds in *A. dorsata* and cavity-nesting honeybees an adaptation to perform dance communication in low light conditions?	Comparing dance sound production of bees foraging at a feeder with the same food reward during different light conditions
Does the walking pattern during the waggle run differ between open- and cavity-nesting honeybee species and affect the duration of the return run?	High-speed camera recordings to reconstruct the walking pattern of foragers from different species
(C) Landscape Ecology/Conservation Biology	
Questions	**Experiments**
How does natural and anthropogenic variation in landscape composition affect the spatio-temporal foraging patterns of giant honeybee colonies?	Landscape mapping and dance decoding of colonies in different landscapes

dances are performed in the open. The difficulty is that colonies often nest at inaccessible locations high above the ground. Therefore, a good spotting scope and a tripod will always be helpful tools for the giant honeybee researcher (Dyer and Seeley 1991). However, when bees need to be trained to visit an artificial feeder, when a video camera needs to be placed exactly perpendicular to the plane of the comb, or when individual paint marks on the bees' bodies need to be distinguished, one typically needs to physically get close to a colony. While climbing up a bee tree or building an observation platform are possible solutions (Kastberger and Sharma 2000), it is much more convenient to rely on colonies that already nest at accessible sites. The former situation may regularly be found in regions where giant honeybees occur despite the lack of rocks or large trees for nesting, as is the case in the peat swamp forest in southern Thailand. There, *A. dorsata* colonies attracted to traditional rafters may be used for observations at ground level (Chuttong et al. 2019). In cities, giant honeybee colonies often nest at buildings and may be accessible via staircases, balconies, or rooftops despite nesting at great heights (Kastberger et al. 1996). In Bangalore, we were able to use custom-built extension arms to obtain close-up video recordings of *A. dorsata* colonies nesting at buildings of the National Centre for Biological Sciences and the University of Agricultural Sciences (Figure 8.5a) (Kohl et al. 2020). Unfortunately, not every building is suited as a temporal research station. Especially at residential buildings, studying colonies from close distance for several days or weeks might not be possible. Here, moving an entire *A. dorsata* colony to a different location is the ideal solution (Sharma and Gupta 2014). For the transfer of a giant honeybee nest, one needs smoke, a clip as a support of the comb, a large knife, and a light-proof cloth bag fitting the colony (Fred C. Dyer, personal communication). After temporally driving the bees off the comb with smoke, a flexible bamboo

FIGURE 8.5 (a) Video-recording of dances of an *A. dorsata* colony from the roof of a building using a camera extension arm. (b) To transport an *A. dorsata* colony, the bees need to be temporally driven off the comb with smoke so that a clip can be attached to the comb. (c) After the bees have settled again, the nest can be cut off the substrate and be moved. (d) *A. dorsata* colony in an observation box (same colony as in b and c).

stick of 5–7 cm diameter that has been cut into two halves lengthwise can be used as a clip for the beeswax comb (Sharma and Gupta 2014) (Figure 8.5b). The ideal position of the clip is the area directly beneath the honey stores that has been used for brood rearing. Fresh combs on which no single brood cycle has been completed are not suited for transport because they are not rigid enough to be supported by the clip; furthermore, nests with combs that are bent and do not form a plane surface cannot easily be clipped. After the bamboo clip has been fixed to the comb, one needs to give the bees enough time to get back and settle—a hasty removal will likely result in leaving the queen behind. After the bees have re-established the bee curtain around the comb (Figure 8.5c), two people need to cover the colony with the cloth bag from below and grab the protruding handles of the bamboo clip. A third person carefully cuts the comb off its original substrate with a long knife. After covering the top of the comb with lobes of the cloth bag and closing all openings with adhesive tape, the colony can be transported and placed at a prepared stand at the desired location (Figure 8.5d). An alternative technique described by Mahindre (2000), which requires more preparation but is more elegant, is the use of "attraction planks", wooden boards smeared with beeswax, which are fixed beneath tree branches or overhangs at buildings known to be used by giant honeybees for nesting. Once colonies have settled on the planks, built comb, and laid eggs, the boards can be moved to a more convenient location without needing to drive the bees off their nest or to cut the comb off its substrate. While such nest relocation techniques offer great potential for giant honeybee research, it should only be done in company with an experienced beekeeper and following proper preparation, including the supply of full body protection to the participants, the procurement of anti-allergy emergency kits, the information of adjacent inhabitants, and the mastering of smoke.

For many research questions, one will need to make giant honeybee foragers visit an artificial feeder. To this end, simply placing a sugar feeder near a nest will rarely be successful, either because the bees show no intrinsic interest or because the feeder is discovered by bees not coming from the observation colony. We succeeded in training *A. dorsata* workers in Bangalore by presenting a piece of old wax comb (collected from another, abandoned *A. dorsata* nest) filled with scented 2 M sucrose solution directly to the bees at the mouth of the colony (Kohl et al. 2020). The colony was nesting in about 15 m height at a window recess and was accessible via the staircase of the building. We kept a safety distance by using a 2-metre plastic pole as a handle for the piece of comb and moved very slowly to avoid the risk of eliciting an attack. The bees on the curtain immediately showed interest in the bait and crawled onto it. We then pulled back the pole and put the piece of comb with the bees into a small box. With the box, we went down and outside the building, where we released the bees at about 25 m distance from the nest. Some of them memorized the new place of the feeder and came back—our feeding station was established! Another way of making bees visit a feeder was described by Rathore and Wells (1995). They sprayed the colony with a mist of scented sugar solution using a water sprayer which made the bees search the surroundings for the same resource which was provided at a feeder. When the study colony is not directly accessible, an indirect way to make bees of that colony visit an artificial feeder is to adopt the method of bee-lining and to catch several foragers on flowers in the immediate surroundings using a small box (Seeley 2016). The caught bees are presented a highly concentrated and scented sucrose solution inside the box and left undisturbed for a couple of minutes. After release, some the bees will memorize the position of the box and return to the artificial food source. These bees can then be individually paint-marked, and if they appear on the curtain of the observation colony, the trial was successful. The problem with this method is that it is never sure whether the bees that were initially caught belong to the colony of interest. Therefore, it might only be useful in cases when the colony is nesting singly so that there is a high chance that foragers on flowers in the immediate surroundings come from that single colony. These descriptions may serve as an inspiration for others to start investigating the dance language of giant honeybees. The scientific literature still lacks reliable protocols for mastering these bees and there are many prejudices concerning their aggressiveness. Therefore, publishing the methods of successful endeavours will be invaluable for the bee research community and promote the study of giant honeybees—an activity urgently needed since a deeper understanding of the biology of giant honeybees is the basis for the conservation of these important pollinators.

ACKNOWLEDGEMENTS

Patrick L. Kohl and Benjamin Rutschmann are grateful to the National Centre for Biological Sciences and the GKVK University of Agricultural Sciences in Bangalore for hosting them and for supporting the study of giant honeybees on their campuses. We thank Johann Kaderschabeck for crafting observation equipment for *Apis dorsata* and Bharath Krishnan for contributing the photograph of an *A. dorsata* colony. Research visits of Patrick L. Kohl and Benjamin Rutschmann to Bangalore were supported by the Bavarian-Indian Centre. Axel Brockmann's research is supported by NCBS-TIFR institutional funds (No. 12P4167) and the Department of Atomic Energy, Government of India (No. 12-R&D-TFR-5.04–0800 and 12-R&D-TFR-5.04–0900).

REFERENCES

Beekman, M., Makinson, J.C., Couvillon, M.J., Preece, K., and Schaerf, T.M. 2015. Honeybee linguistics—a comparative analysis of the waggle dance among species of *Apis*. *Frontiers in Ecology and Evolution* 3:11.

Bosch, R. 1957. RassenmäßigeUnterschiedebei den Tänzen der Honigbiene (Apismellifica L.) *Zeitschrift für Vergleichende Physiologie* 49: 289–320.

Chuttong, B., Somana, W., and Burgett, M. 2019. Giant honey bee (*Apis dorsata* F.) rafter beekeeping in Southern Thailand. *Bee World* 96: 66–68.

Couvillon, M.J., and Ratnieks, F.L.W. 2015. Environmental consultancy: Dancing bee bioindicators to evaluate landscape 'health'. *Frontiers in Ecology and Evolution* 3:44.

Couvillon, M.J., Schürch, R., and Ratnieks, F.L.W. 2014. Dancing bees communicate a foraging preference for rural lands in high-level agri-environment schemes. *Current Biology* 24: 1212–1215.

Dornhaus, A., and Chittka, L. 2001. Food alert in bumblebees (*Bombus terrestris*): Possible mechanisms and evolutionary implications. *Behavioral Ecology and Sociobiology* 50: 570–76.

Duangphakdee, O., Hepburn, H.R., and Tautz, J. 2011. The dance language. In: Hepburn, H.R., and Radloff, S.E. (Eds.), *Honeybees of Asia*. Springer, Berlin, Heidalberg 313–332.

Dyer, F.C. 1985a. Nocturnal orientation by the Asian honey bee *Apis dorsata. Animal Behaviour* 33: 769–774.

Dyer, F.C. 1985b. Mechanisms of dance orientation in the Asian honey bee *Apis florea* L. *Journal of Comparative Physiology A* 157: 183–198.

Dyer, F.C. 1991. Comparative studies of dance communication: Analysis of phylogeny and function. In: Smith, D.R. (Ed.), *Diversity in the genus Apis*. Westview, Boulder, CO, 177–198.

Dyer, F.C. 2002. The biology of the dance language. *Annu. Rev. Entomol.* 47: 917–949.

Dyer, F.C., and Seeley, T.D. 1991. Dance dialects and foraging range in three Asian honey bee species. *Behav. Ecol. Sociobiol.* 28: 227–233.

Dyer, F.C., and Seeley, T.D. 1994. Colony migration in the tropical honey bee *Apis dorsata* F. (Hymenoptera: Apidae). *Insectes Sociaux* 140: 129–140.

Edrich, W. 1977. Interaction of light and gravity in the orientation of the waggle dance of honey bees. *Animal Behaviour* 25: 342–363.

Edrich, W. 1981. Night-time sun-compass behaviour of honey bees at the equator. *Physiological Entomology* 6: 7–13.

Esch, H. 1961. Über Die Schallerzeugung Beim Werbetanz Der Honigbiene. *Zeitschrift Für Vergleichende Physiologie* 45: 1–11.

Esch, H., and Burns, J. 1996. Distance estimation by foraging honeybees. *The Journal of Experimental Biology* 199: 155–162.

Esch, H., Zhang, S., Srinivasan, M.V., and Tautz, J. 2001. Honeybee dances communicate distances measured by optic flow. *Nature* 411: 581–583.

Gardner, K.E., Seeley, T.D., and Calderone, N.W. 2009. Do honeybees have two discrete dances to advertise food sources? *Animal Behaviour* 75: 1291–1300.

George, E.A., Pimplikar, S., Thulasi, N., and Brockmann, A. 2020. Similarities in dance follower behaviour across honey bee species suggest a conserved mechanism of dance communication. *Animal Behaviour* 169: 139–155.

Grüter, C. 2020. Recruitment and communication in foraging. In: *Stingless bees: Their behaviour, ecology and evolution*. Springer, Cham, 341–372.

Härtel, S., and Steffan-Dewenter, I. 2014. Ecology: Honey bee foraging in human-modified landscapes. *Current Biology* 24: 524–526.

Hepburn, H.R. 2011. Absconding, migration and swarming. In: Hepburn, H.R., and Radloff, S.E. (Eds.), *Honeybees of Asia*. Springer, Cham, 133–158.

Hrncir, M., Maia-Silva, C., Mc Cabe, S.I., and Farina, W.M. 2011. The recruiter's excitement—features of thoracic vibrations during the honey bee's waggle dance related to food source profitability. *Journal of Experimental Biology* 214: 4055–4064.

I'Anson Price, R., and Grüter, C. 2015. Why, when and where did honey bee dance communication evolve? *Frontiers in Ecology and Evolution* 3: 125.

Kastberger, G., and Sharma, D.K. 2000. The predator-prey interaction between blue-bearded bee eaters (*Nyctyornis athertoni* Jardine and Selby 1830) and giant honeybees (*Apis dorsata* Fabricius 1798). *Apidologie* 31: 727–736.

Kastberger, G., Winder, O., Hoetzl, T., and Raspotnig, G. 1996. Behavioural features of a periodic form of massed flight activity in the giant honeybee *Apis dorsata. Apidologie* 27: 381–395.

Kirchner, W.H., and Dreller, C. 1993. Acoustical signals in the dance language of the giant honeybee, *Apis dorsata. Behavioral Ecology and Sociobiology* 33: 67–72.

Kirchner, W.H., Dreller, C., Grasser, A., and Baidya, D. 1996. The silent dances of the himalayan honeybee, *Apis laboriosa. Apidologie* 27: 331–339.

Kitnya, N., Otis, G.W., Chakravorty, J., Smith, D.R., and Brockmann, A. 2022. Apis Laboriosa confirmed by morphometric and genetic analyses of giant honey bees (Hymenoptera, Apidae) from sites of sympatry in Arunachal Pradesh, North East India. *Apidologie* 53: 1–17.

Koeniger, N., and Koeniger, G. 1980. Observations and experiments on migration and dance communication of *Apis dorsata* in Sri Lanka. *Journal of Apicultural Research* 19: 21–34.

Kohl, P.L., and Rutschmann, B. 2021. Honey bees communicate distance via non-linear waggle duration functions. *PeerJ* 9: e11187.

Kohl, P.L., Thulasi, N., Rutschmann, B., George, E.A., Steffan-Dewenter, I., and Brockmann, A. 2020. Adaptive evolution of honeybee dance dialects. *Proceedings of the Royal Society B* 287: 20200190.

Landgraf, T., Bierbach, D., Kirbach, A., Cusing, R., Oertel, M., Lehmann, K., Greggers, U., Menzel, R., and Rojas, R. 2018. Dancing honey bee robot elicits dance-following and recruits foragers. *ArXiv* 1803.07126.

Leonhardt, S.D., Menzel, F., Nehring, V., and Schmitt, T. 2016. Ecology and evolution of communication in social insects. *Cell* 164: 1277–1287.

Lindauer, M. 1955. Schwarmbienen Auf Wohnungssuche. *Zeitschrift für vergleichende Physiologie* 37: 263–324.

Lindauer, M. 1956. Über Die Verständigung Bei Indischen Bienen. *Zeitschrift für vergleichende Physiologie* 38: 521–557.

Łopuch, S., and Tofilski, A. 2020. Impact of the quality of food sources on the wing beating of honey bee dancers. *Apidologie* 51: 631–641.

Mahindre, D.B. 2000. Developments in the management of *Apis dorsata* colonies. *Bee World* 81: 155–163.

Makinson, J.C., Oldroyd, B.P., Schaerf, T.M., Wattanachaiyingcharoen, W., and Beekman, M. 2011. Moving home: Nest-site selection in the red dwarf honeybee (*Apis florea*). *Behavioral Ecology and Sociobiology* 65: 945–589.

Makinson, J.C., Schaerf, T.M., Rattanawannee, A., Oldroyd, B.P., and Beekman, M. 2014. Consensus building in giant Asian honeybee, *Apis dorsata*, swarms on the move. *Animal Behaviour* 93: 191–199.

Menzel, R., Kirbach, A., Haass, W.D., Fischer, B., Fuchs, J., Koblofsky, M., Lehmann, K. et al. 2011. A common frame of reference for learned and communicated vectors in honeybee navigation. *Current Biology* 21: 645–650.

Michelsen, A., Andersen, B.B., Storm, J., Kirchner, W.H., and Lindaner, M. 1992. How honeybees perceive communication dances, studied by means of a mechanical model. *Behav. Ecol. Sociobiol.* 30: 143–150.

Neumann, P., Koeniger, N., Koeniger, G., Tingek, S., Kryger, P., and Moritz, R.F. 2000. Home-site fidelity in migratory honeybees. *Nature* 406: 474–475.

Nieh, J.C. 2011. The evolution of honey bee communication: Learning from Asian species. *Formosan Entomologist* 31: 1–14.

Oldroyd, B.P., Gloag, R.S., Even, N., Wattanachaiyingcharoen, W., and Beekman, M. 2008. Nest site selection in the open-nesting honeybee *Apis florea*. *Behavioral Ecology and Sociobiology* 62: 1643–1653.

Oldroyd, B.P., and Wongsiri, S. 2006. *Asian honey bees: Biology, conservation, and human interactions*. Harvard University Press, Cambridge.

Paar, J., Oldroyd, B.P., and Kastberger, G. 2000. Giant honeybees return to their nest sites. *Nature* 406: 475.

Punchihewa, R.W.K., Koeniger, N., Kevan, P.G., and Gadawski, R.M. 1985. Observations on the dance communication and natural foraging ranges of *Apis cerana*, *Apis dorsata* and *Apis florea* in Sri Lanka. *Journal of Apicultural Research* 24: 168–175.

Raffiudin, R., and Crozier, R.H. 2007. Phylogenetic analysis of honeybee behavioural evolution. *Molecular Phylogenetics and Evolution* 43: 543–552.

Rathore, R.R.S., and Wells, H. 1995. Training Asian rockbees (*Apis dorsata*) to forage at specific locations. *Indian Bee Journal* 57: 8–9.

Ratnieks, F.L.W., and Shackleton, K. 2015. Does the waggle dance help honey bees to forage at greater distances than expected for their body size? *Frontiers in Ecology and Evolution* 3: 31.

Riley, J.R., Greggers, U., Smith, A.D., Reynolds, D.R., and Menzel, R. 2005. The flight paths of honeybees recruited by the waggle dance. *Nature* 435: 205–207.

Robinson, W.S. 2021. Surfing the sweet wave: Migrating giant honey bees (Hymenoptera: Apidae: *Apis dorsata*) display spatial and temporal fidelity to annual stopover site in Thailand. *Journal of Insect Science* 21: 1–12.

Rudolf, B. 1957. Rassenmäßige Unterschiede Bei Den Tänzen Der Honigbiene (*Apis mellifica* L.). *Zeitschrift Für Vergleichende Physiologie* 49: 289–320.

Seeley, T.D. 1995. *The wisdom of the hive: The social physiology of honey bee colonies.* Princeton University Press, Princeton.

Seeley, T.D. 2010. *Honeybee democracy.* Princeton University Press, Princeton.

Seeley, T.D. 2012. Progress in understanding how the waggle dance improves the foraging efficiency of honey bee colonies. In: Giovanni Galizia, C., Eisenhardt, D., and Giurfa, M. (Eds.), *Honeybee neurobiology and behavior: A tribute to Randolf Menzel.* Springer, Cham, 77–78.

Seeley, T.D. 2016. *Following the wild bees: The craft and science of bee hunting.* Princeton University Press, Princeton, NJ.

Seeley, T.D., Mikheyev, A.S., and Pagano, G.J. 2000. Dancing bees tune both duration and rate of waggle-run production in relation to nectar-source profitability. *Journal of Comparative Physiology—A Sensory, Neural, and Behavioral Physiology* 186: 813–819.

Sharma, D., and Gupta, R.K. 2014. Management of Asian honeybees. In: Gupta, R.K., Reybroeck, W., Van Veen, J.W., and Gupta, A. (Eds.), *Beekeeping for poverty alleviation and livelihood security.* Springer, Cham, 205–246.

Somanathan, H., Krishna, S., Jos, E.M., Gowda, V., Kelber, A., and Borges, R.M. 2020. Nocturnal bees feed on diurnal leftovers and pay the price of day – night lifestyle transition. *Frontiers in Ecology and Evolution* 8: 566964.

Spitschan, M., Aguirre, G.K., Brainard, D.H., and Sweeney, A.M. 2016. Variation of outdoor illumination as a function of solar elevation and light pollution. *Scientific Reports* 6: 1–14.

Srinivasan, M.V. 2000. Honeybee navigation: Nature and calibration of the 'odometer'. *Science* 287: 851–853.

Steffan-Dewenter, I., and Kuhn, A. 2003. Honeybee foraging in differentially structured landscapes. *Proceedings of the Royal Society B. Biological Sciences/The Royal Society* 270: 569–575.

Tautz, J., Rohrseitz, K., and Sandemann, D.C. 1996. One-strided waggle dance in bees. *Nature* 382: 32.

Thom, C., Gilley, D.C., Hooper, J., and Esch, H.E. 2007. The scent of the waggle dance. *PLoS Biology* 5: 1862–1867.

Towne, W.F. 1985. Acoustic and visual cues in the dances of four honey bee species. *Behavioral Ecology and Sociobiology* 16: 185–187.

Towne, W.F., and Gould, J.L. 1988. The spatial precision of the honey bees' dance communication. *Journal of Insect Behavior* 1: 129–155.

Visscher, P.K., and Seeley, T.D. 1982. Foraging strategy of honeybee colonies in a temperate deciduous forest. *Ecology* 63: 1790–1801.

von Frisch, K. 1962. Über Die Durch Licht Bedingte 'Missweisung' Bei Den Tänzen Im Bienenstock. *Experientia* 18: 49–53.

von Frisch, K. 1967. *The dance language and orientation of bees.* Harvard University Press, Cambridge, MA.

Wenner, A.M. 1962. Sound production during the waggle dance of the honey bee. *Animal Behaviour* 10: 79–95.

Woyke, J., Wilde, J., and Wilde, M. 2012. Swarming and migration of *Apis dorsata* and *Apis laboriosa* honey bees in India, Nepal and Bhutan. *Journal of Apicultural Science* 56: 81–91.

Young, A.M., Kodabalagi, S., Brockmann, A., and Dyer, F.C. 2021. A hard day's night: Patterns in the diurnal and nocturnal foraging behavior of *Apis dorsata* across lunar cycles and seasons. *PLoS One* 16: e0258604.

Young, A.M., Kohl, P.L., Rutschmann, B., Steffan-Dewenter, I., Brockmann, A., and Dyer, F.C. 2021. Temporal and spatial foraging patterns of three asian honey bee species in Bangalore, India. *Apidologie* 52: 502–523.

9 Ecological Service Potential of *Apis dorsata* in Thailand

Atsalek Rattanawannee, Preecha Rod-im and Orawan Duangphakdee

9.1 INTRODUCTION

The lowland forests of Asia are dominated by the trees of the family Dipterocarpaceae. This forest type shows distinctive pollination ecology every 2 or 10 years by infrequent mass flowering events, in which all tree species flower simultaneously at a random times of year (Ashton 1988; Sakai et al. 1999). This forest type, particularly the canopy trees, appear to be adapted for pollination by migratory honeybees that can rapidly increase in size of their population by both reproductive and migratory swarming (Oldroyd and Nanork 2009; Corlett 2011). Additionally, because each tree species of dipterocarp forests tends to be largely spaced among individuals, pollen must be therefore transferred over long distances (Itioka et al. 2001). Thus, pollination of this forest type requires an animal vector that has species loyalty while foraging, a long-distance foraging range, and the tendency to visit multiple trees during foraging. The giant honeybee (*Apis dorsata* Fabricius, 1793), shows all these characteristics and is therefore likely to be a keystone species in dipterocarp forests.

The giant honeybee has a range that extends from the Indian subcontinent to Southeast Asia (Ruttner 1988). A massive single-comb nest is usually hanging beneath a tall tree branch, cliff overhang, or eaves of a building (Oldroyd and Wongsiri 2006). Throughout its range, the species is heavily hunted for honey, wax, and brood, which provides important sources of household income to local people of Asia (Strickland 1982; Lahjie and Seibert 1990; Soman and Kshirsagar 1991; Nath et al. 1994). In Thailand, colony hunting is usually done at night, which often results in the death of the harvested colonies because the queens and workers are killed or are otherwise unable to reform their colony (Oldroyd and Wongsiri 2006). Additionally, habitat loss and fragmentation is likely to have a significant impact on population viability (Oldroyd and Wongsiri 2006; Oldroyd and Nanork 2009; Rattanawannee et al. 2012b, 2013).

Beyond A. *dorsata*'s role as keystone insect pollinator and colony providing an importance source of household income, this chapter provides information on this honeybee species. The taxonomic status, a common morphology, ecological distribution, and population genetic diversity are defined and explained. Also, species behaviors, including the mating behavior, colony migration, and nesting biology are

DOI: 10.1201/9781003294078-9

briefly outlined. Additionally, because of its role as pollinator, *A. dorsata* has important economic, ecological, and social values throughout its range in Asia. Therefore, adaptation success in agricultural and urban areas, potential bee flora and pollination capacities, and impact by urbanization and anthropogenic activities of *A. dorsata* are summarized.

9.2 TAXONOMIC STATUS

In addition to their large size, the giant honeybee (*A. dorsata*) is distinguished from all other honeybees of the genus *Apis* by their wings, which are fuscous and more hairy (Oldroyd and Wongsiri 2006). The mean forewing and hind wing lengths of a typical *A. dorsata* worker are 12.84 ± 0.26 mm and 8.89 ± 0.21 mm, respectively. Furthermore, the development time from egg to adult is approximately 17–19 days for workers, 21–24 days for drones, and 13–16 days for queens (Qayyum and Ahmad 1967; Tan 2007). Caste dimorphism in the size of *A. dorsata* is less pronounced than that in other *Apis* species (Figure 9.1), especially those of dwarf honeybees (*A. andreniformis* and *A. florea*). The queen has a slightly larger thorax but smaller compound eyes than that of worker (Figure 9.1A), whereas drones have the largest compound eyes but slightly shorter abdomens compare to workers and queen (Figure 9.1B).

Previously, both Ruttner (1988) and Engels (1999) identified common giant honeybee (*A. dorsata*) and mountain giant honeybee (*A. laboriosa*) as the same species based on morphological characteristics. However, evidence supports the designation of these two giant honeybees as distinct species. First, the mating flight behavior of *A. laboriosa* drones occurs around noon (Underwood 1990), whereas *A. dorsata* drones perform mating flights just after dusk (Koeniger and Koeniger 1980; Rinderer et al. 1993). Second, the communication dances performed by *A. dorsata* and *A. laboriosa* are strikingly different (Oldroyd and Wongsiri 2006). Kirchner et al. (1996) reported that *A. dorsata* produces dance sounds, whereas *A. laboriosa*

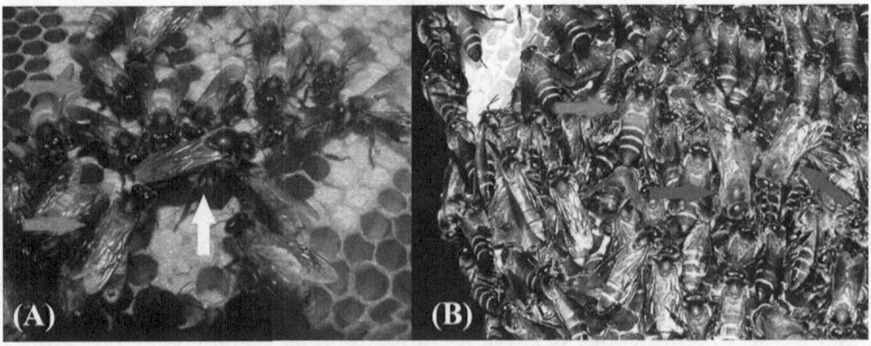

FIGURE 9.1 Morphology of *Apis dorsata*. (A) The queen (white arrow), slightly broader thorax, is surrounded by female workers (red arrows). (B) The drones (blue arrows) have larger compound eyes but slightly shorter abdomen than that of workers.

dances silently. Third, Arias and Sheppard (2005) demonstrated that DNA (deoxyribonucleic acid) sequence divergence between *A. dorsata* and *A. laboriosa* is 10.6%–11.5%, which is sufficient to support distinct species status. Finally, Raffiudin and Crozier (2000) revealed a 100% Bayesian consensus tree for grouping *A. dorsata* as a distinct group from *A. laboriosa*, supporting the recognition of *A. dorsata* and *A. laboriosa* as separate species.

A study by Lo et al. (2010), based on evidence from Bayesian and maximum parsimony phylogenetic analyses, supports the recognition of the giant Philippine honeybee (*A. breviligula* Maa, 1953), as a separate species from the more broadly distributed lowland *A. dorsata*. The species *A. breviligula* is found northwest of the Merrill line in Luzon, Philippines (Oldroyd and Wongsiri 2006). On the one hand, similar to *A. laboriosa* and *A. dorsata*, it never forms colony aggregations; on the other hand, *A. breviligula* is strikingly different from *A. dorsata* owing to the black rather than yellow coloration of the abdomen (Morse and Laigo 1969; Paar et al. 2004; Oldroyd and Wongsiri 2006; Rattanawannee et al. 2012a). Therefore, three species of giant honeybees in the subgenus *Megapis* of the genus *Apis* have been recognized.

9.3 DISTRIBUTION AND GENETIC DIVERSITY

Apis dorsata, a keystone pollinator within Asian lowland forests, is distributed over vast South and Southeast Asian geographies (Figure 9.2), including Thailand (Ruttner 1988; Oldroyd and Wongsiri 2006; Rattanawannee et al. 2012b). The farthest occurrence of *A. dorsata* toward the west is the Indus River. To the east, the *A. dorsata* areas are all in the Philippines on the other side of the Wallace line (Ruttner 1988; Oldroyd and Wongsiri 2006). It occurs at altitudes of 1,000–1700 m, or even up to 2,000 m during migration (Ruttner 1988).

To analyze the populations, Rattanawannee et al. (2012a) used geometric morphometric approaches, which are based on Cartesian coordinate shapes (Slice 2007), to examine the variation of the *A. dorsata* population in Thailand. The results of principal component analysis (PCA) and discriminant analysis demonstrated that *A. dorsata* populations in mainland Thailand showed the potential for a high degree of gene flow and were thus panmictic between *A. dorsata* populations across (mainland) Thailand. Consistent with this, *A. dorsata* colonies reportedly migrate up to a distance of 200 km (Koeniger and Koeniger 1980; Oldroyd and Wongsiri 2006). Thus, there is no major barrier to gene flow, and geographical obstacles seem to play no considerable role in directing the swarming migration of *A. dorsata* at the level of local scale and in mainland Thailand.

The geographical results are in close agreement with molecular DNA markers, and no substantial differences were found among *A. dorsata* populations throughout Thailand. The findings revealed the presence of one mitochondrial type at frequencies from 0.92–1.00 throughout the country, and nuclear microsatellites revealed no evidence of population structuring at broad scales (Insuan et al. 2007). At the local scale, no significant differences in the degree of genic differentiation between *A. dorsata* colony aggregations in disturbed and undisturbed habitats were observed,

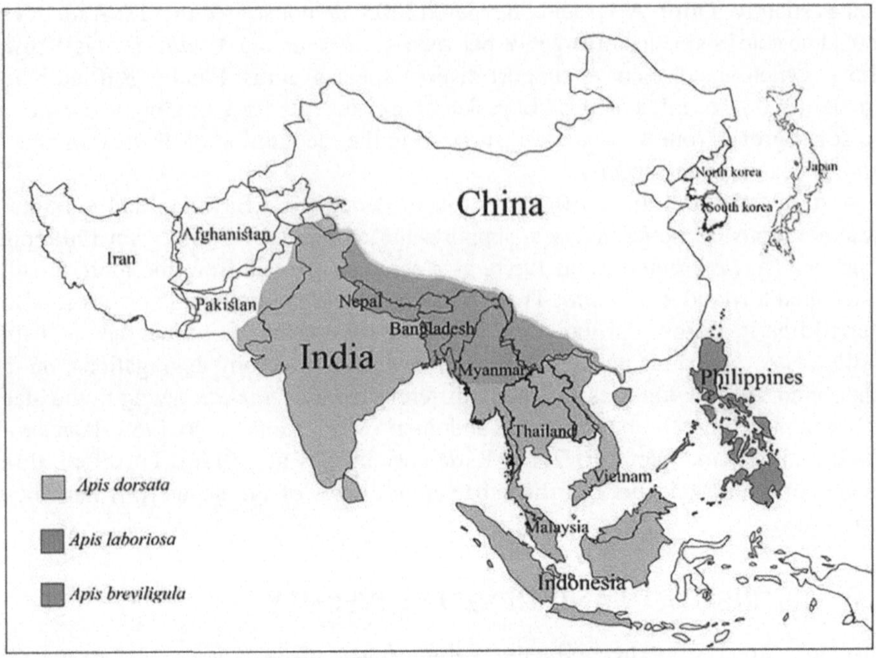

FIGURE 9.2 Approximate distribution of three giant honeybee species of the subgenus *Megapis*: *Apis dorsata*, *Apis laboriosa*, and *Apis breviligula* (amended in accordance with Ruttner [1988]; Oldroyd and Wongsiri [2006]; Lo et al. [2010]).

indicating a lack of barriers to gene flow and that geographical obstacles did not influence migrating swarms (Rattanawannee et al. 2012b). The authors also reported that habitat disturbance had no effect on the mating frequency or genetic diversity of this species. This finding suggests that the mating behavior of *A. dorsata* is robust to anthropogenic changes in the landscape. Additionally, Rattanawannee et al. (2013) found no genetic differentiation among colony aggregations of *A. dorsata* in northwestern Thailand based on microsatellite DNA markers. They found that the Thai *A. dorsata* population had high levels of heterozygosity, and the F_{st} values between aggregations were not significantly different from zero ($p > 0.05$). This suggests that despite formidable anthropogenic pressures endured by the *A. dorsata* population in northern Thailand, the species continues to enjoy a large effective population size and has high connectedness. The authors concluded that *A. dorsata* in Thailand can tolerate habitat fragmentation and annual harvesting. From these results, they speculated that the population was sustained by immigration from forested regions in Burma, northwest to our study sites (Rattanawannee et al. 2013).

The genetic structure of *A. dorsata* in Thailand is similar to that reported by Sahebzadeh et al. (2012), who showed high levels of genetic admixture in Malaysian *A. dorsata* populations using 15 single-locus DNA microsatellite markers (Table 9.1). The highest genetic diversity has been observed in Chinese *A. dorsata* populations

TABLE 9.1
Comparison of Genetic Diversity in the Populations of *Apis dorsata* in Thailand, Malaysia, and China from Microsatellite Data

	Thailand			Malaysia	China
	(Insuan et al. 2007)	(Rattanawannee et al. 2012b)	(Rattanawannee et al. 2013)	(Sahebzadeh et al. 2012)	(Cao et al. 2012)
N	154	18	54	20	16
AR (± SE)	7.800 ± 1.169	8.750 ± 3.304	10.000 ± 4.000	8.933 ± 0.345	10.167 ± 2.317
N_e (± SE)	4.904 ± 0.590	5.935 ± 1.885	4.855 ± 2.071	5.738 ± 0.223	9.371 ± 1.924
H_o (± SE)	0.715 ± 0.024	0.889 ± 0.078	0.749 ± 0.140	0.410 ± 0.025	Not reported
H_e (± SE)	0.732 ± 0.097	0.812 ± 0.082	0.757 ± 0.113	0.822 ± 0.007	9.208 ± 2.101

Notes: N = number of colonies sampled; AR = allelic richness; N_e = number of effective alleles; H_o = observed heterozygosity; H_e = expected heterozygosity; SE = standard error

(Cao et al. 2012). Qualitatively, the genetic diversity of *A. dorsata* in Thailand was similar in all three studies but lower than that observed in China (Cao et al. 2012) and Malaysia (Sahebzadeh et al. 2012) (Table 9.1)

9.4 MATING BEHAVIOR OF *APIS DORSATA*

Mating between honeybee queens and drones occurs in a specific area known as the drone congregation area (DCA), where many drones from nearby colonies gather (Baudry et al. 1998; Oldroyd and Wongsiri 2006). When a queen approaches a DCA, several drones rapidly copulate with the queen midair (Gries and Koeniger 1996) and then die immediately. The DCA persists year after year, regardless of a queen's presence (Baudry et al. 1998; Gries and Koeniger 1996). Although DCAs and mating behaviors of queens and drones have been extensively studied, it is still unclear why drones choose particular areas in which to congregate and how queens locate these areas. Baudry et al. (1998) examined the parentage of 142 *A. mellifera* drones collected from a DCA near Oberusel, Germany. They reported that the composition of the DCA contained an equal representation from the local colonies, approximately 240 in number. They suggested that most colonies within the recruitment parameter of a DCA delegated equal proportions of males to the DCA. Furthermore, they also found that the relatedness among the drones mated to a common queen is very low, indicating the genetic diversity among the different patrilines (paternal subfamilies) of a colony.

In two multiple-comb Asian honeybee species, *A. cerana* and *A. koschevnikovi*, the DCAs occur in the open air close to trees and under the cover of trees, respectively (Figure 9.3) (Koeniger and Koeniger 2000; Oldroyd and Wongsiri 2006). The DCAs of *A. dorsata* are found under the canopy of highly emergent trees, approximately 10–35 m above ground (Figure 9.3) (Koeniger et al. 1994). Mating time in *Apis* species seems to be a major behavioral barrier that increases reproductive

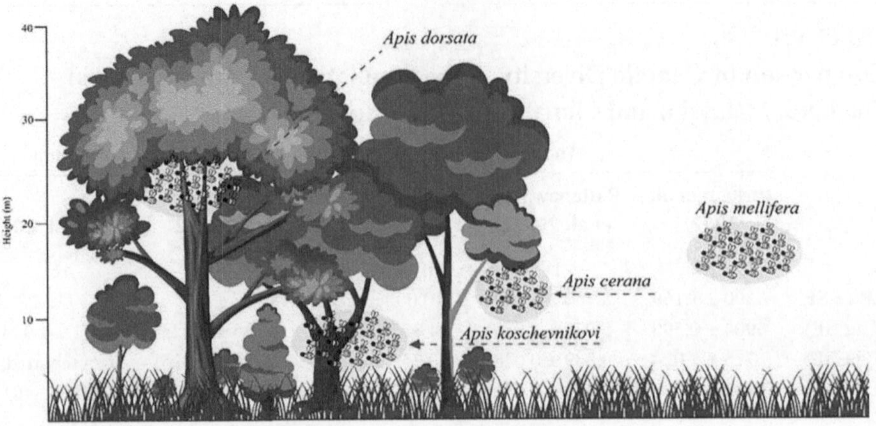

FIGURE 9.3 The drone congregation areas (DCA) of four honeybee species (Redrawn from Koeniger et al. [1994, 2005]; Oldroyd and Wongsiri [2006]).

isolation (Koeniger and Koeniger 2000). Observations of *A. dorsata* showed that the mating flight of this species occurs shortly after dusk (Koeniger et al. 1994; Koeniger and Wijayagunesekera 1976; Rinderer et al. 1993; Tan et al. 1999). Rinderer et al. (1993) observed the drone flight time of *A. dorsata* in Thailand and found that mating flights occurred after sunset between 18:15 hours and 18:45 hours. Similar patterns in drone flights have been reported for the species in Borneo (Koeniger et al. 1994). The mating flight of *A. dorsata* virgin queens has been reported to be shorter than that of drones. Queens return from mating flight after 15–30 minutes (Koeniger et al. 1994; Rinderer et al. 1993; Tan et al. 1999). This short mating period is similar to that observed in *A. andreniformis*, *A. florea*, and *A. koschevnikovi* (Koeniger and Koeniger 2000).

9.5 NESTING BIOLOGY IN *APIS DORSATA* AND TREE HOSTS

Apis dorsata displays substantial differences in nesting behavior and ecology compared to that of other *Apis* species. First, colonies are frequently found in dense aggregations (Kastberger and Sharma 2000; Paar et al. 2004; Oldroyd and Wongsiri 2006; Rattanawannee et al. 2012b, 2013), on a single tree, under the eaves of buildings, or otherwise on human-built structures (Figure 9.4). Second, colonies frequently undergo seasonal migration between alternate nesting sites. In this behavior, nest sites are occupied for three to four months. Toward the end of this period, brood rearing stops and stores of honey and pollen are depleted (Figure 9.5). Colonies are absent from alternative nest sites up to 200 km away (Koeniger and Koeniger 1980). The proximate cause of migration may be related to available food sources because *A. dorsata* swarms have been reported to travel between locations with different blooming seasons (Mahindre 2000). Additionally, colony absconding in *A. dorsata* may also help suppress the levels of parasitic mites (*Tropilaelaps clareae*) (Paar et al. 2004). Mites need a brood to reproduce (Kavinseksan et al. 2003). A colony may,

FIGURE 9.4 An aggregation of *Apis dorsata* colonies. (A) More than 20 colonies are on a single tree in Chiang Mai, Thailand. (B) *Apis dorsata* sometime aggregate on the human-made structures such as temple buildings or pagodas in Chiang Rai, Thailand.

FIGURE 9.5 Some broodless nests within a colony aggregation of *Apis dorsata* in Sakon Nakorn province, northeastern Thailand.

FIGURE 9.6 The returning colony of *Apis dorsata* is built next to the nest from previous season.

therefore, decrease infestation by *T. clareae* with a period of broodless migration (Rinderer et al. 1994) Third, nesting sites are reoccupied annually over periods of several decades (Oldroyd et al. 2000; Oldroyd and Wongsiri 2006). Interestingly, some returning colonies built their combs at exactly the same location that they occupied in the previous season (Figure 9.6) (Neumann et al. 2000; Paar et al. 2000).

Unlike the single comb of the dwarf honeybees (*A. florea* and *A. andreniformis*), the crown of the comb always encircles a small tree branch, while the massive nest of *A. dorsata* is always attached under the surface of a stout tree branch, an overhang of a rock face, or sometimes to the eves of buildings (Figure 9.7) or other urban structures (Paar et al. 2004; Rattanawannee and Chanchao 2011). The magnificent populous nest of a giant honeybee is awe-inspiring, with several thousand simultaneously visible individual workers. Trees where *A. dorsata* nests are found have supporting branches measuring 12–30 cm in diameter (Morse and Laigo 1969) or more (Oldroyd and Wongsiri 2006). A slightly sloping branch or support is preferred (Tan 2007).

The height and width of the *A. dorsata* combs range from 23–90 cm and 43–162 cm, respectively (Tan 2007). Approximately 3–4 weeks after nesting, during the blooming period, a colony of *A. dorsata* stores, on average, approximately 4 kg of honey in the comb, with the highest recorded quantity being 15.7 kg (Tan 2007). Honey is stored in the top corner of the comb in an area of approximately 10–20 cm

FIGURE 9.7 The massive nest of *Apis dorsata* is attached under (A) horizontal and (B) slightly sloping eves of buildings at Mae Fah Luang University, Chiang Rai, Thailand.

in a large nest. The number of individual workers can be over 50,000 in a large colony (Morse and Laigo 1969).

Many studies have demonstrated that *A. dorsata* prefers the same tree for nesting, which was used by them in the previous season (Parr et al. 2000; Misra et al. 2017). Oldroyd et al. (2008) reported that *A. dorsata* reoccupied the same host tree in Malaysia for more than 35 years. Misra et al. (2017) also reported that in India, the *A. dorsata* selected the same *Bombax ceiba* tree for more than 60 years to build their nest during blooming period. In Thailand, aggregations of *A. dorsata* colonies are mostly found in the northern part, including Chiang Mai, Chiang Rai, Mae Hong Sorn, Nan, and Tak provinces. Aggregation of *A. dorsata* colonies were also observed in Sakon Nakorn (northeastern region) and Ratchaburi (western region) (A.R. observation). Ton Yuan Puang (*Koompassia excelsa* Taub.) is a frequent host of *A. dorsata* colony aggregations. The reason for colony aggregation and host tree species preference remains unclear. However, Seeley et al. (1982) speculated that *A. dorsata* colony aggregation habits are a defensive strategy against vertebrate predators. The key point of this defensive strategy is the readiness of workers to create a massive attack (Seeley et al. 1982).

Large amounts of honey (up to 45 kg per colony per harvesting season) can be stored by a colony of *A. dorsata*. Thus, *A. dorsata* nests are harvested throughout its range by local people in Southeast Asia for honey, brood, and wax. These products, harvested from *A. dorsata* nests, are an important source of household income (Figure 9.8B). In Thailand, honey from *A. dorsata* is harvested between November

FIGURE 9.8 Honey and brood combs of *Apis dorsata* being sold in the local market.

and April depending on the area. In the northeastern region, honey, brood, and wax from *A. dorsata* nests were harvested from November–January. In northern Thailand, local bee hunters harvest *A. dorsata* nests from February–April. This period is the dry season of the northern and northeastern parts of Thailand; if it is harvested at this time, the honey will last longer as the moisture content is lower and fermentation is less likely (Waring and Jump 2004). In the local market, broods and honey from *A. dorsata* nests are offered for sale (Figure 9.8C). The brood is sold for Baht 200–250 (US$6–7) per kilogram, whereas the honey is sold for approximately Baht 400–500 (US$11–14) per liter. The average honey yield from each *A. dorsata* comb is estimated at 5 liters. Bee hunters also value the wax from the harvest. It is melted and sold for Baht 80–100 (US$2–3) per kilogram.

9.6 ADAPTATION SUCCESS OF *APIS DORSATA* IN AGRICULTURAL AND URBAN AREAS

Environmental threats have been reported to have relative effects on living organisms. *Apis dorsata* faces the greatest challenge in their survival success. As they are giant bees, forming large colonies with a single comb, nests are built under the overhangs of cliffs, branches of trees, or under the eaves of buildings. Upon reaching colony sizes of 100,000 individuals, they need a large amount of food resources, a strong nesting branching tree, or a cliff (Oldroyd and Wongsiri 2006). Existing nesting habitats originate from deep forests, where emergent high trees are abundant. It plays a major role as a pollinator in the rainforests of peninsular Malaysia and Southeast Asia. Honey hunting practices have traditionally been conducted by local villagers in forests since ancient times. This is a traditional activity that disturbs the growth of *A. dorsata* colonies. It is often regarded as "sustainable," whereas it clearly depends on the frequency of the harvesting. Wilson-Rich et al. (2014) hint that *A. dorsata* has survived in natural habitats for centuries. Recently, the threat of *A. dorsata* survival has shifted to larger scales relative to more unstable factors, i.e., landscape changes due to urbanization and expansion of infrastructure, climate change action, and natural disasters. Forest fires, flooding, and drought are now occurring unexpectedly worldwide. Increasing temperatures are the first challenge for their adaptation. Meteorological data for Thailand were obtained during 1951–1990, revealing a decline in the diurnal temperature range (Table 9.2). These declines are principally due to increases in nighttime minimum temperatures and may be related to urban influences at some sites; interestingly, most come from the largest cities in the country. The effects of urbanization in Thailand on the increasing temperature should be investigated. *A. dorsata* worker bees can regulate heat production during flight. The thoracic flight temperatures were higher at higher concentrations of sugar syrup than at lower concentrations, suggesting that they may vary their flight efforts according to the expected gain and associated costs (Underwood 1991). Hint: How have the bees adapted to this trend? Moreover, the differentiation of these adaptations between urban and rural sites must be further questioned and inspected.

Higher temperatures also raise concerns about the pathogen and disease management of the colony. Unlike cavity-nesting honeybees, *A. cerana* and *A. mellfera*,

TABLE 9.2
Change in Temperatures (°C), Using a Linear Trend, for the Four Countries Over Their Periods of Record

Thailand (1951–1990)			
DJF	0.12	1.75*	−1.63*
MAM	0.21	1.50*	−1.71*
JJA	0.39	1.14*	−1.71*
SON	0.36	0.93*	−0.58*
ANN	0.16	1.33*	−1.16*

Source: Jones (1995)
Note: Asterisks indicate significance at the 5% level

which remove dead pupae in sealed broods, Woyke et al. (2004b) reported that *A. dorsata* and *A. laboriosa* do not open undamaged cells with dead broods to prevent the spread of diseases and parasitic mites. Therefore, diseases that can be desolated by migration behavior will not be agitated because of the change in ambient temperature. However, another highly concerning disease is Nosema, caused by an intracellular microsporidian pathogen that lives in the midgut ventricular cells of bees. *Nosema ceranae*–infected *A. dorsata* workers showed significantly lower survival and higher energy stress and lower the capacities of other survival factors, such as trehalose levels, hypopharyngeal gland protein content, and midgut proteolytic enzyme activity (Ponkit et al. 2021). Energetic stress may consequently affect the foraging and defensive behavior of colonies. These pathogens may be gradually transmitted to the natural environment, awaiting a breakout crisis. The developmental cycle of *Nosema ceranae* is well maintained in the bee body at 33–34°C. The climate change crisis can be a risk factor for this case because it might change the rate of development of pathogens, including *Nosema*. It is noteworthy that *A. dorsata* nest products also play an important role in strengthening colony health and defense against pathogens and diseases. The antimicrobial activity of *Streptomyces* spp. agents has been isolated from *A. dorsata* combs against some phytopathogenic bacteria (Promnuan et al. 2020). *Apis dorsata* honey possesses antimicrobial properties that may contain an epinecidin homolog, an antimicrobial peptide, and can inhibit the growth of some bacteria (Chanchao 2009).

Antagogenic gut bacteria must also be taken into account in *A. dorsata* survival, as they play a relatively important role in several survival functions such as food processing, nutrient supplementation, immune boosting, and host protection. Midgut bacterial communities in the giant Asian honeybee *A. dorsata* have been identified from two bacterial phyla (Proteobacteria and Firmicutes) and four classes (Alphaproteobacteria, Betaproteobacteria, Gammaproteobacteria, and Bacilli). The frequency of occurrence of each group varies among different stages and locations (Saraithong et al. 2017). The bacterial community structure of honeybees culminates in host-bacterial and environmental interactions. The air pollution in Thailand, which dramatically increases the effect of anti-angiogenic gut bacterial profiles in honeybees, must be closely and continuously monitored.

The accomplishment of thermoregulation is also considered. Periodic mass flights (PMFs) in *A. dorsata* are an important adaptive strategy for open-nesting honeybees. The nests of *A. dorsata* vary in their periodic mass flights among different seasonal conditions. The duration and distribution of PMFs activities during the day were correlated with average daily temperatures. On the warmest days, more PMFs activities were performed after the hottest midday hours. These periodic mass flights minimize the thermal disruption of the protective bee curtain and consequently compromise colony thermoregulation (Woyke et al. 2004a). PMFs activities involve energy consumption and exploding colony material to the outside world. Implicitly, the bee curtain and thermoregulation are important enough for the colony to take this risk.

One captivating adaptation is foraging behavior. Most honeybee species are active during the day, whereas *A. dorsata* has additionally evolved foraging during the night. *Apis dorsata* start their activities early in the morning (06:30–09:25 hours) which accounts for 46.7–78.5% of all foraging activities of the day. The foraging activity continued even in the late afternoon (15:25), until the cessation of dust at approximately 19:00 (Figure 9.9). In contrast, for *A. florea* foraging on *Antigonon leptopus* and *Cosmos sulphureus*, foraging flights start early in the morning but terminate at 17:00–18:00 hours (Rod-im et al. 2015).

Young et al. (2021) further investigated the extent of nocturnal foraging behavior and the correlation between environmental factors. Strong evidence suggests that *A. dorsata* foraging is cathemeral and active throughout the day and night, when illumination is sufficient for nocturnal flight. The foraging activities during the dawn versus dusk twilight periods showed no differences. However, this shows a significant difference based on the seasonal period. *A. dorsata* colonies exhibited peak nocturnal activity in the hours before sunset (17:00–19:00 hours) and after sunrise (04:00–06:00 hours). Illumination at night had a significant positive effect on bee arrival. During the active lunar cycle, *A. dorsata* foraging activities can be active over the entire 24-hour day (Figure 9.10). The contribution of the nocturnal activity of *A. dorsata* permits this species to extend its ability to find extra nutrition and energy budgets; consequently, the roles of *A. dorsata* in pollination networks should be the focus of future studies.

Physiological characteristics of *A. dorsata* have evolved to support nocturnal foraging. Of these *Apis* groups, *A. dorsata* has the lowest spatial resolution and the most sensitive eyes, followed by the cavity-nesting bees which live in dim light, *A. mellifera* and *A. carana*, and finally by the open-nesting red dwarf bees, *A. florea* (Somanathan et al. 2009). In addition, *A. dorsata* showed a greater diversity of chemoreceptive antennal sensilla compared to that in dwarf honeybees, with eight types of antennal sensilla in *A. dorsata* compared to seven types of antennal sensilla found in *A. florea* and *A. andreniformis*. Accordingly, *A. dorsata* may have a better ability to sense a diversity of odors than species with fewer types of antennal chemoreceptive sensilla, which may support higher-quality foraging in the dark (Suwannapong et al. 2012)

Towne (1985) reported that in *A. dorsata*, sound signals were not emitted by the dancers, but later showed that dance sounds similar to those emitted by the

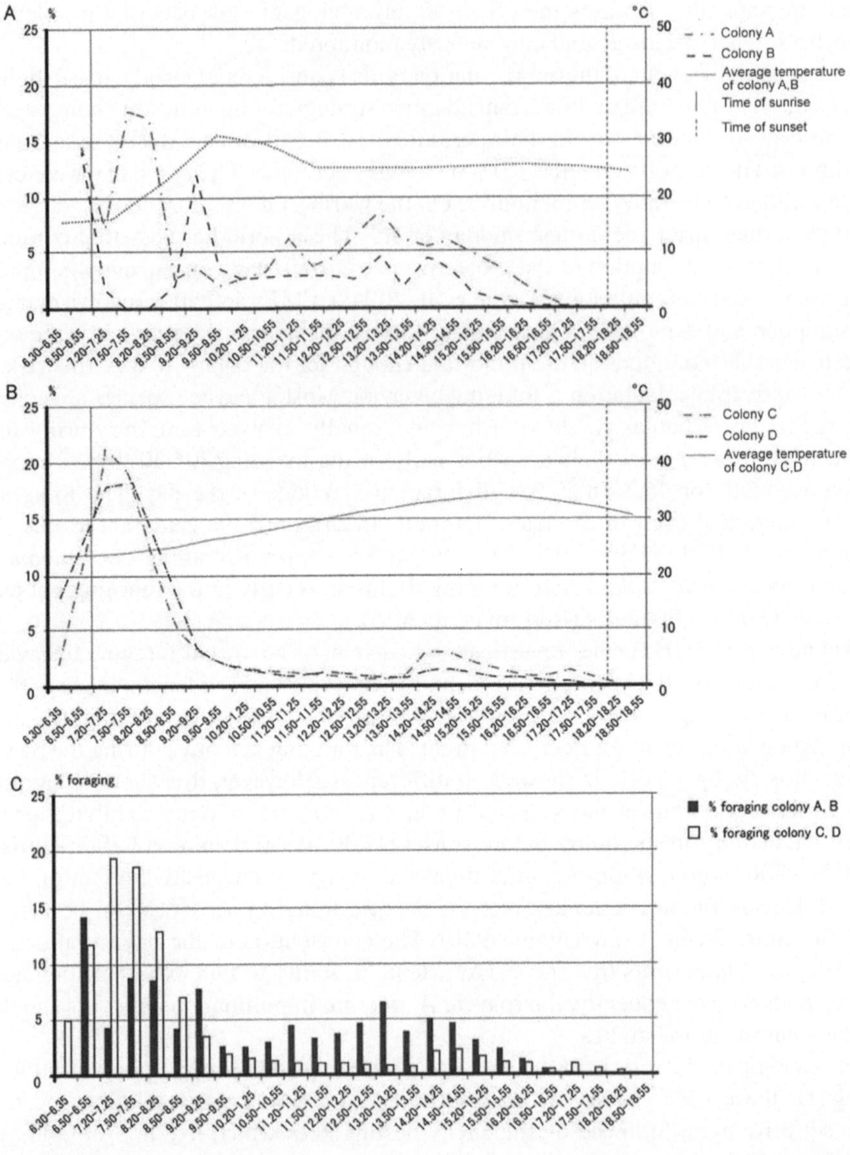

FIGURE 9.9 A, B: Averaged circadian daily flight profile for *A. dorsata* colonies. C: Summed averaged daily flight profile for colonies A and B and colonies C and D (Conforms from Saraithong and Burgett [2012]).

cavity-nesting bees, *A. mellifera*, were produced. This acoustic signal contains information regarding the distance, direction, and profitability of food sources (Kirchner and Dreller 1993). The acoustical transfer of information supports this competence in foraging during the low light intensity of *A. dorsata*. The significance of nocturnal

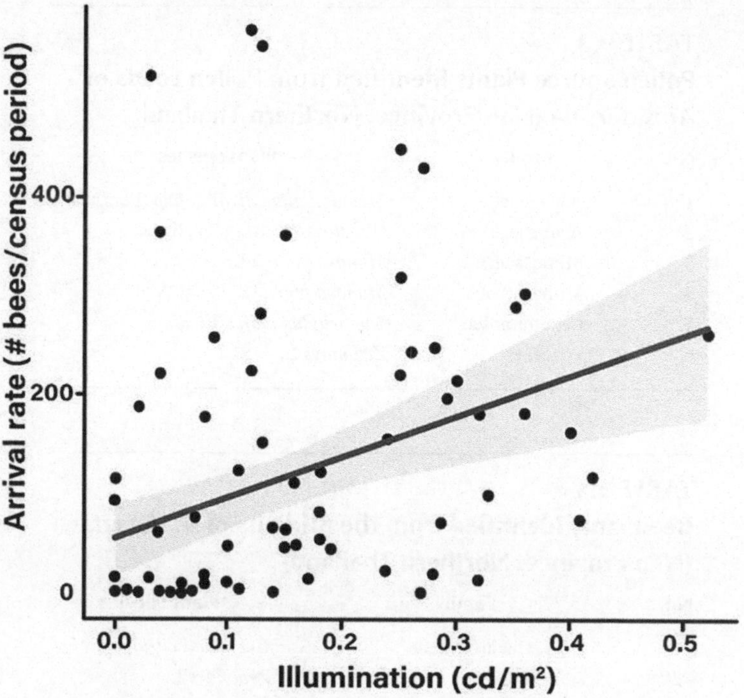

FIGURE 9.10 Relationship between nighttime illumination and *A. dorsata* arrival rates (Conform from Young et al. 2021).

activity in *A. dorsata* in the evolution of sound communication should be assessed further (Hepburn et al. 2014).

9.7 POTENTIAL BEE FLORA AND POLLINATION CAPACITIES

The giant honeybee is considered an important pollinator of crops and wild plants. It is one of the major pollinators in Southeast Asian lowland dipterocarp forests, which pollinate at least 15 species of emergent and canopy trees in Lambir Hills National Park, Sarawak, Malaysia (Momose et al. 1998). In Thailand, giant honeybees visit several plant species. Suwannapong et al. (2013) reported the bee flora utilized by *A. dorsata* in Nan province, northern Thailand, by identifying pollen grains from their pollen loads and midguts. The number of bee flora found from the pollen loads and in the midgut of *A. dorsata* were six (Table 9.3) and 11 species (Table 9.4), respectively (Suwannapong et al. 2013). Stewart et al. (2018) studied pollination networks in 52 green areas throughout Bangkok and found that *A. dorsata* visited 17 plant species. Wayo et al. (2018) revealed that *A. dorsata* appears to be the most legitimate and effective insect pollinator of durian ('Monthong' cultivar) in Southern Thailand since it commonly and consistently visited durian

TABLE 9.3
Pollen Source Plants Identified from Pollen Loads of *Apis dorsata* (Nan Province, Northern Thailand)

No.	Family	Plant species
1	Asteraceae	*Melampodium divaricatum* (Pers.) DC.
2	Asteraceae	*Wedelia trilobata* (L.) Hitchc.
3	Mimosaceae	*Mimosa pudica* L.
4	Mimosaceae	*Mimosa pigra* L.
5	Legumimosae	*Tamarindus indica* L.
6	Poaceae	*Zea mays* L.

TABLE 9.4
Bee Plants Identified from the Midguts of *A. dorsata* (Nan Province, Northern Thailand)

No.	Family	Plant Species
1	Cucurbitaceae	*Cucurbita citrullus* L.
2	Poaceae	*Zea mays* L.
3	Cucurbitaceae	*Momordica charantia* L.
4	Palmae	*Cocos nucifera* L.
5	Amaranthaceae	*Celosia argentea* L.
6	Asteraceae	*Wedelia trilobata* L.
7	Sapindaceae	*Dimocarpus longan* Lour.
8	Myrtaceae	*Syzygium malaccense* L.
9	Asteraceae	*Ageratum conyzoides* L.
10	Mimosaceae	*Mimosa pigra* L.
11	Mimosaceae	*Mimosa pudica* L.

flowers at dusk (Figure 9.11), which was also found in previous studies (Aziz et al. 2017; Bumrungsri et al. 2009; Sritongchuay et al. 2016). It should be noted that *A. dorsata* likely visited durian for nectar sources because the evidence indicates that bats play an important role in durian fruit setting. Moreover, *A. dorsata* is an important floral visitor of mangoes (Sritongchuay et al. 2022) and a vital pollinator of longan in Thailand (Sritongchuay 2016).

Apis dorsata is one of the species that well adaptive to new plants including the introduced species. The observation in the Queen Sirikit Botanical Park No. 9 (Rama Park 9) in Bangkok disclosed that *A. dorsata* pollinate the introduced plant *Ruellia simplex* C. Wright (Acanthaceae) (Hawkeswood and Sommung 2016). The introduced species, *Tecoma stans* (L.) Kunth (Bignoniaceae), which have now become a common ornamental plant in Thailand, is also beneficial for *A. dorsata* (Figure 9.12).

FIGURE 9.11 Durian flowers visited by *Apis dorsata* at dusk (Photo credit: K. Wayo).

9.8 *APIS DORSATA* AFFECTED BY URBANIZATION AND ANTHROPOGENIC ACTIVITIES

The giant honeybee is a wild honeybee species distributed from India to Southeast Asia, including Thailand. It builds a single open-nesting with a massive comb that can grow up to 2 × 2m comprising 80,000–100,000 individuals (Morse and Laigo 1969). Colonies in the wild are usually attached horizontally to nesting substrates, such as tree branches or below-rock overhangs on cliffs or roof beams or rafter construction. As it has big colonies, during resource vanquishment, *A. dorsata* frequently shows migratory behavior (Dyer and Seeley 1994; Oldroyd et al. 2000). In normal situations, when a colony faces food resource limitations or if the environment is unsatisfactory, honeybees tend to abandon the nesting site and migrate to areas of better quality. However, unlike other *Apis* species, *A. dorsata* exhibits a remarkable migratory behavior. Fundamentally, their migration is related to available forage, disturbance, or predation; however, they follow a seasonal pattern. During migration, colonies can stop over for temporary resting and preserve colony energy. After many stops, their travel distance can be up to 200 km (Crane et al. 1993) away from their original nesting sites. The ultimate target for seasonal migration is emergent

FIGURE 9.12 *Tecoma stans* flowers were visited by *Apis dorsata* at dusk (Photo credit: Preecha Rod-im).

trees, protruding from other surrounding trees in the virgin forest. Not long after the colony has set up on its host trees, the flower blooming season of the area gradually begins.

The "bee trees" *Ficus albipila* and "Yang pa/Yang na" *Kompassia* spp., are frequently aggregate sites for *A. dorsata*. The host trees confine unique characteristics that attain nesting sites for several colonies as it supports following activities, DCA proving genetic outbreeding, and increase their reproductive success and the collective mass defense behaviors that every colony joins to protect against the enemies. The lowland forests of Asia are dominated by trees belonging to the family Dipterocarpaceae. In Thailand, the trees in this type of forest started partial flowering during November–December. Later, the major flowering season begins in April–May, which is the most frequent time of the year when *A. dorsata* aggregations can be observed. Moreover, some groups of trees experience general flowering events every 2–10 years (Oldroyd and Wongsiri 2006; Sakai 2002). The spatial distribution of individual trees of each species tends to be widely spaced apart in the dipterocarp forests. During pollination, pollen must be transferred over long distances between trees. This requires migratory pollinators with large foraging ranges to visit multiple trees and a large population. *A. dorsata* has the required characteristics. Thus, *A. dorsata* is a key pollinator in this type of ecosystem.

Certain areas in Southeast Asia are known for high abundance of *A. dorsata*, such as the western part of Thailand, Tanassirim Mountain range, Suangphung, and

Banka district of Ratchaburi Province in Thailand. Relative to the traditional cultures, the name "Suanphung" (garden of bees) originates from the fact that hundreds of *A. dorsata* nests were commonly found in the district (Duangphakdee et al. 2013). In the past decade, huge forest areas, including thousands of bee trees, have been cut down to create landscapes for pineapple plantations and tourism activities, such as resorts and shops. Duangphakdee et al. (2013) estimated that the number of bee trees and other nesting sites remaining in the Suanphung area was approximately 21 regions. Consequently, the *A. dorsata* population in this area is seriously declining.

A total of 21 aggregation sites were investigated in the Suanphung district (Figure 9.13). Of these, 17 were *Ficus albipila* (Moraceae), one *Ficus microcarpa* (Moraceae), one *Ficus religeosa* (Moraceae), one *Dipterocarpus alatus* Roxb. ex, and one human-built structure (water tank). During 2013–2015, seven of these 21

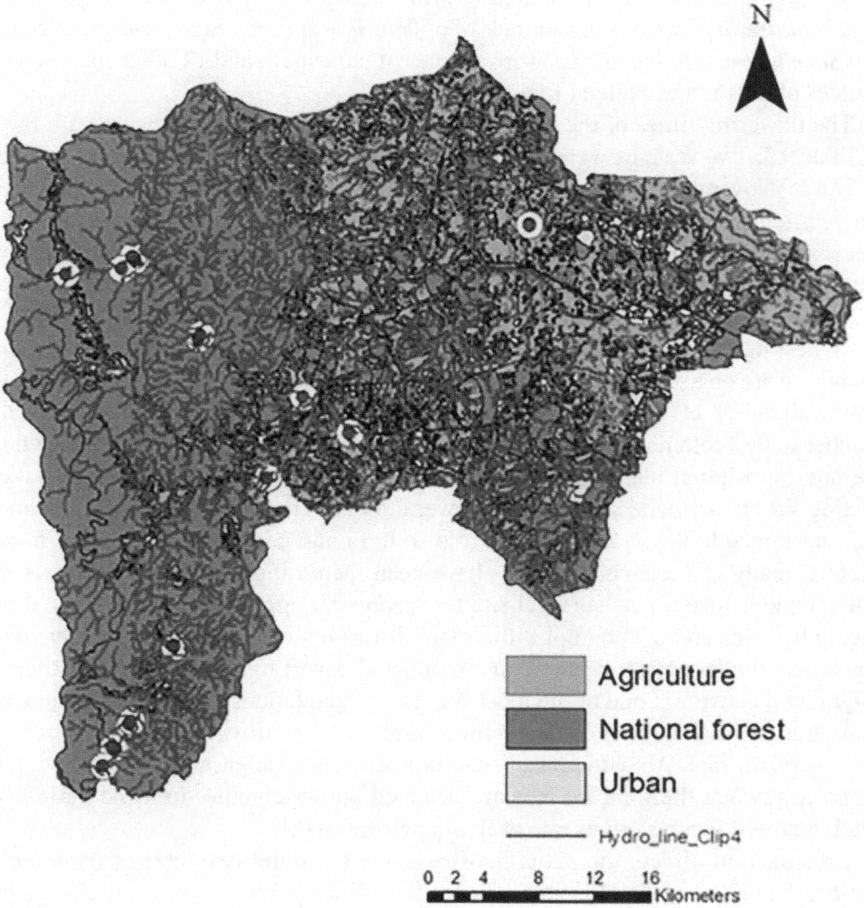

FIGURE 9.13 Distribution of 17 bee trees (red points) and four non-bee trees, which were not *Ficus albipila*, and one human-built structure (water tank) (blue points) at Suanphung district, Ratchaburi Province (some points obscured because they overlap) (Jansamut et al. 2017).

sites were observed to be active aggregation sites for giant honeybees. Four of the seven were *Ficus albipila*, two were other tree species, and one was a human-built structure (water tank). The average height of the seven aggregation nesting sites was 28.7 ± 4.6 meters (ranging from 21.7–33.2 meters). The width of the site canopy averaged 25.8 ± 1.9 meters, with a range of 22.9–28.5 meters. The average diameter was 5.6 ± 2.1 meters, ranging from 3.10–8.44 meters. The average tree circumference was 17.7 ± 6.7 meters, with a range of 6.91–26.50 meters (Jansamut et al. 2017).

In the present study, 29 common bee flora species in Suanphung were identified and recorded (Table 9.5). Thirteen species were significant sources of both nectar and pollen: *Cocos nucifera* L., *Eucalyptus camaldulensis* Dehn, *Psidium guajava* L., *Citrus aurantifolia* L., *Dimocarpus longan* Lour., *Lansium domesticum* Corr, *Bauhinia variegate* L., *Carica papaya* L., *Hevea brasiliensis* Muell, *Syzygium samarangense* (Blume) Merr. & L.M. Perry, *Albizia sama* (F.) Muell, *Benincasa hispida* (Thunb.) Cogn., and *Citrus maxima* (Burm.) Merr. Two species, *Zea mays* L. and *Zoysia matrella* Merrill, were sources of pollen. Two species, *Eleais guineensis* Jacq. and *Saccharum officinarum* L., were sources of honeydew, and 12 other species were sources of nectar and pollen (Table 9.5).

The flowering times of the common bee flora species in the study area are listed in Table 9.5. Two seasons were identified based on precipitation: wet (June–October) and dry (November–May). Of the 29 species, nine bloom throughout the year, whereas ten species bloom during the dry season, and two species bloom during the wet season. The other eight species bloom during both the dry and wet seasons, but not throughout the year (Table 9.5).

The study found that the aggregation sites of bee colonies declined from 21 to seven nesting sites in the Suanphung district. The bee trees that attracted *A. dorsata* colonies were on average tall (approximately 30 m) and had a large diameter (~1.8 m). The availability of 29 species of melliferous flora was highly correlated with the number of bee colonies occupying nesting sites. Approximately 50% of the bee flora are not the original plants in the area. Some migratory families have shifted their nesting sites to artificial structures. A tower tank is a common new adaptive human-built nesting side for *A. dorsata*. As it is so high and cannot be reached by honey hunters, many of the emergent trees have been increasingly abandoned every year. The temple is also a new survival site for *A. dorsata* aggregation. Buddhist culture deeply believes and acts by not killing and disturbing other animals; thus, the bees that live in the monastery area are an exceptional cavort for honey harvest activities. Migration behavior is one of the most successful adaptations for *A. dorsata* survival (Jansamut et al. 2017). The successful migration of *A. dorsata* is strongly related to its survival rate. Abandon nest behavior is a disease defense strategy. Moving to a new nest when the food sources are defeated allows colonies to avoid periods of death that will increase their rate of living performance.

Urbanization affects the patterns of resource flora and nest sites of *A. dorsata*. Artificial light at night (ALAN) also leads to light pollution on local and global scales. *Apis dorsata* has adapted to build nests in urban communities (Hepburn et al. 2014). Single colonies of *A. dorsata* are often found in cities and disturbed areas. This puts *A. dorsata* at risk of being affected by light pollution. The giant honeybee performs activities both during the day and night. The quantity of moonlight is the illumination that enhances nocturnal foraging activity. ALAN can directly affect

TABLE 9.5
List of Bee Flora Species Found and Sites Encountered (out of Seven Sampled) with Blooming Period and Potential Resource Provided to *A. dorsata* including Nectar (N), Pollen (P), and Honeydew (H) in Areas surrounding *A. dorsata* Nest Sites in the Suanphung District of Ratchaburi Province in Western Thailand.

No.	Local name	Scientific Name	Sites Encountered	Blooming Period	Resource		
					N	P	H
1	Bamboo	*Dendrocalamus asper* (Schult.) Backer	6	–	+	+	–
2	Banana	*Musa sapeantum* L.	7	Throughout the year	++	+	–
3	Cassava	*Manihot esculenta* Crantz	6	Throughout the year	+	++	–
4	Coconut	*Cocos nucifera* L.	6	Throughout the year	+++	+++	–
5	Maize (Corn)	*Zea mays* L.	4	Throughout the year	–	+++	–
6	Eucalyptus	*Eucalyptus camaldulensis* Dehn	4	Feb.–March	+++	+++	–
7	Giant sensitive plant	*Mimosa pigra* L.	2	Jun.–Nov.	+	+++	–
8	Manila grass	*Zoysia matrella* Merrill	5	Throughout the year	–	+	–
9	Guava	*Psidium guajava* L.	2	May–June	+++	+++	–
10	Lime	*Citrus aurantifolia* L.	3	Apr.–May	+++	+++	–
11	Longan	*Dimocarpus longan* Lour.	4	Jan.–Feb.	+++	+++	–
12	Longkong	*Lansium domesticum* Corr.	2	March–May	+++	+++	–
13	Mango	*Mangifera indica* L.	3	Nov.–Jan.	++	++	–
14	Orchid trees	*Bauhinia variegata* Linn.	1	June–July	+++	+++	–
15	Oil palm	*Eleais guineensis* Jacq.	4	Throughout the year	–	–	+++
16	Papaya	*Carica papaya* L.	2	Throughout the year	+++	+++	–
17	Para rubber	*Heavea brasiliensis* Muell	3	Throughout the year	+++	+++	–
18	Pineapple	*Ananas comosus* (L.) Merr.	2	Aug.–Sep.	++	+	–
19	Pine	*Pinus roxburghii* Sarg.	4	Feb.–April, Sept.–Oct.	–	+++	+
20	Rose apple	*Syzygium samarangense* (Blume) Merr. & L.M. Perry	1	Feb.–April	+++	+++	–
21	Sensitive plant	*Mimosa invisa* Mart ex Colla	3	Nov.–Dec.	+	++	–
22	Sugarcane	*Saccharum officinarum* L.	4	Dec.–Feb.	–	–	+++
23	Tamarind	*Tamarindus indica* L.	2	March–May	++	++	–
24	Teak	*Tectona grandis* L.	3	June–Oct.	+	+	–
25	Yard longbean	*Vigna unguiculata* (L.) Walp.	3	Throughout the year	+	++	–
26	Rain tree	*Albizia sama* F. Muell	1	Feb.–May	+++	+++	
27	Winter melon	*Benincasa hispida* (Thunb.) Cogn.	1	April–May	+++	+++	
28	Peacock crest	*Caealpinia pulcherrima* L.	1	March–June	+	++	–
29	Pomelo	*Citrus maxima* (Burm.) Merr.	2	Dec.–Jan., Aug.–Sept.	+++	+++	

Source: Jansamut et al. (2017)

Notes: + indicates low, ++ indicates medium, and +++ indicates high levels of potential resources provided worker foraging rhythms and behavior. However, it can also act as an inducing light in the flower blooming cycle of plants in the area. Giavi et al. (2021) reported that 19% of plant species were affected by light treatment and received either more or less diurnal pollinator visitation than the dark controls (Borges 2022). Comparison of the nocturnal activity in light-polluted urban environments and unpolluted natural environments requires further investigation. Different pollinator groups responded differently to disturbance gradients. The abundance of *A. dorsata*, one of the three local honeybee species, increased gradually from the village area towards the deep pristine forest (Akter et al. 2020). A similar assumption can be inferred for the case of big cities in Thailand, such as Bangkok and Chiang Mai. However, the direction of adaptation may vary depending on the bee species. *A. cerana* and *A. florea* have been reported in village sites and are completely absent in deep forests (Akter et al. 2020). The changes of future rainfalls under climate change emission scenarios play an important role and need to be investigated for *A. dorsata* survival. The study of rainfall characteristics under AR6 climate change scenarios was projected for the next 80 years based on a basin-scale in the Pasak River Basin using the quantile mapping method. The model revealed that extremely wet and dry years often occurred in 2051–2100 under SSP2–4.5 and SSP1–2.6, respectively. Rainfall may then affect the flowering pattern of trees and will eventually affect the migration success of *A. dorsata* in concern (Sittichok 2022).

Sustainable utilization and domestication can be alternative choices for maintaining and conserving *A. dorsata* populations under present circumstances. Even though this is an aggressive species that cannot be managed in hives or economic farming, it has long been exploited by honey hunters using the semi-domestication method called "rafter beekeeping" whereby colonies are kept on "rafters" made of wood fixed to poles. The rafters are provided by the beekeepers to attract the natural swarm of *A. dorsata* during the swarm season to build their nest on and develop the colonies.

A. dorsata has adapted to survive under various stress conditions created by human activities. We can encourage their success by promoting eco-friendly bee activities and landscapes. *A. dorsata* is closely connected to a wide range of ecological divisions, and its success can sustainably strengthen human natural resource security.

ACKNOWLEDGMENTS

This project is funded by National Research Council of Thailand (NRCT) and Kasetsart University (Project number: N42A650288); and Thailand Science Research and Innovation (TSRI) Basic Research Fund: Fiscal year 2021 under project number FRB640008.

REFERENCES

Akter, A., Biella, P., Batary, P., and Klecka, J. 2020. Changing pollinator communities along a disturbance gradient in the Sundarbans mangrove forest: A case study on *Acanthus ilicifolius, A. ilicifolius* and *Avicennia officinalis. Global Ecology and Conservation* 24: e01282.

Arias, M.C., and Sheppard, W.S. 2005. Phylogenetic relationships of honey bees (Hymenoptera: Apinae: Apini) inferred from nuclear and mitochondrial DNA sequence data. *Molecular Phylogenetics and Evolution* 37: 25–35.

Ashton, P.S. 1988. Dipterocarp biology as a window to the understanding of tropical forest structure. *Annual Review of Ecology and Systematics* 19: 347–370.

Aziz, S.A., Clements, G.R., McConkey, K.R., Sritongchuay, T., Pathil, S., Abu Yazid, M.N.H., Campos-Arceiz, A., Forget, P.M., and Bumrungsri, S. 2017. Pollination by the locally endangered island flying fox (*Pteropus hypomelanus*) enhances fruit production of the economically important durian (*Durio zibethinus*). *Ecology and Evolution* 7: 8670–8684.

Baudry, E., Solignac, M., Garnery, L., Gries, M., Cornuet, J., and Koeniger, N. 1998. Relatedness among honeybees (*Apis mellifera*) of a drone congregation. *Proceedings of the Royal Society B: Biological Sciences* 265: 2009–2014.

Borges, R.M. 2022. Impacts of artificial light at night on nocturnal and diurnal insect biology and diversity. *Indian Journal of Entomology* 84: 483–492.

Bumrungsri, S., Sripaoraya, E., Chongsiri, T., Sridith, K., and Racey, P. 2009. The pollination ecology of durian (*Durio zibethinus*, Bombaceae) in southern Thailand. *Journal of Tropical Ecology* 25: 85–92.

Cao, L.F., Zheng, H.Q., Fu-Liang Hu, F.L., and Hepburn, H.R. 2012. Genetic structure of Chinese *Apisdorsata* population based on microsatellites. *Apidologie* 43: 643–651.

Chanchao, C. 2009. Properties and antimicrobial activity of *Apisdorsata* honey from Thailand. *Pak J. Med. Sci.* 25: 313–318.

Corlett, R.T. 2011. Honey bees in natural ecosystems. In: Hepburn, H.R., and Radloff, S.E. (Eds.), *Honey bees of Asia*. Springer, Heidelberg, 215–225.

Crane, E., Luyen, V., Mulder, V., and Ta, T.C. 1993. Traditional management system for *Apisdorsata* in submerged forests in southern Vietnam and central Kalimantan. *Bee World* 74: 27–40.

Duangphakdee, O., Chidthaisong, A., Hepburn, R., Rodim, P., and Phunduang, P. 2013. Climatic cycle migrations of giant honeybees, *Apisdorsata*at their home Phung trees. *Ficus albipila* and *Kompassiamalaccensis*, 旭硝子財団助成研究成果報告 1–10.

Dyer, F.C., and Seeley, T.D. 1994. Colony migration in the tropical honey bee *Apisdorsata* Fabricius. (Hymenoptera: *Apidae*). *Biomedical and Life Sciences* 41: 129–140.

Engels, W. 1999. The taxonomy of recent and fossil honey bees (Hymenoptera: Apidae; *Apis*). *Journal of Hymenoptera Research* 8: 165–196.

Giavi, S., Fontaine, C., and Knop, E. 2021. Impact of artificial light at night on diurnal plant-pollinator interactions. *Nature Communication* 12: 1690.

Gries, M., and Koeniger, N. 1996. Straight forward to the queen: Pursuing honeybee drones (*Apis mellifera* L.) adjust their body axis to the direction of the queen. *Journal of Comparative Physiology* 179: 539–544.

Hawkeswood, T.J., and Sommung, B. 2016. Pollination of *Ruellia simplex* C. Wright (Acanthaceae) by the giant tropical bee, *Apis dorsata* (Fabr., 1793) (Hymenoptera: Apidae) in Bangkok, Thailand. *Calodema* 438: 1–5.

Hepburn, H.R., Pirk, C.W.W., and Duangphakdee, O. (Eds.). 2014. *The nest of the honeybees: Structure and functions*. Springer-Verlag, Berlin Heidelberg, 389.

Insuan, S., Deowanish, S., Klinbunga, S., Sittipraneed, S., Sylvester, H.A., and Wongsiri, S. 2007. Genetic differentiation of the giant honey bee (*Apis dorsata*) in Thailand analyzed by mitochondrial genes and microsatellites. *Biochemical Genetics* 45: 345–361.

Itioka, T., Inoue, T., Kaliang, H., Kato, M., Nagamitsu, T., Momose, K., Sakai, S., Yumoto, T., Mohamad, S.U., Hamid, A.A., and Yamane, S. 2001. Six-year population fluctuation of giant honey bee *Apis dorsata* (Hymenoptera: Apidae) in tropical lowland dipterocarp forest in Sarawak. *Annals of the Entomological Society of America* 94: 545–549.

Jansamut, P., Gale, G.A., and Duangphakdee, O. 2017. *The migration movement of the giant honeybees, (Apisdorsata) on particularly bee trees, (Ficus albipila) at Tanawsri Moutain Range, Thailand.* International Conference on Sustainable Agriculture and Bioeconomy, Bangkok International Trade & Exhibition Centre (BITEC), Bangkok, Thailand, February 27–March 2.

Jones, P.D. 1995. Maximum and minimum temperature trends in Irelabd, Italy, Thailand, Turkey and Bangladesh. *Atmospheric Research* 37: 67–78.

Kastberger, G., and Sharma, D.K. 2000. The predator-prey interaction between blue-bearded bee eaters (*Nyctyornis athertoni* Jardine and Selby, 1830) and giant honeybees (*Apis dorsata* Fabricius, 1798). *Apidologie* 31: 727–736.

Kavinseksan, B., Wongsiri, S., De Guzman, L.I., and Rinderer, T.E. 2003. Absence of *Tropilaelaps* infestation from recent swarms of *Apis dorsata* in Thailand. *Journal of Apicultural Research* 42: 49–50.

Kirchner, W.H., and Dreller, C. 1993. Acoustical signals in the dance language of the giant honeybee, *Apisdorsata. Behavioral Ecology and Sociobiology* 33: 67–72.

Kirchner, W.H., Dreller, C., Grasser, A., and Baidya, D. 1996. The silent dances of the Himalayan honeybee, *Apis laboriosa. Apidologie* 27: 331–339.

Koeniger, N., and Koeniger, G. 1980. Observations and experiments on migration and dance communication of *Apisdorsata* in Sri Lanka. *Journal of Apicultural Research* 19: 21–34.

Koeniger, N., and Koeniger, G. 2000. Reproductive isolation among species of the genus *Apis. Apidologie* 31: 313–339.

Koeniger, N., Koeniger, G., Gries, M., and Tingek, S. 2005. Drone competition at drone congregation areas in four *Apis* species. *Apidologie* 36: 211–221.

Koeniger, N., Koeniger, G., Tingek, S., Kalitu, A., and Mardan, M. 1994. Drones of *Apis dorsata* (Fabricius 1793) congregate under the canopy of tall emergent trees in Borneo. *Apidologie* 25: 249–264.

Koeniger, N., and Wijayagunesekera, H.N.P. 1976. Time of drone flight in the three Asian honeybee species (*Apis cerana, Apis florea, Apis dorsata*). *Journal of Apicultural Research* 15: 67–71.

Lahjie, A.M., and Seibert, B. 1990. Honey gathering by people in the interior of East Kalimantan. *Bee World* 71: 153–157.

Lo, N., Gloag, R.S., Anderson, D.L., and Oldroyd, B.P. 2010. A molecular phylogeny of the genus *Apis*suggests that the giant honey bee of the Philippines, *A. breviligula* Maa, and the plains honey bee of southern India, *A. indica* Fabricius, are valid species. *Systematic Entomology* 35: 226–233.

Mahindre, D.B. 2000. Developments in the management of *Apis dorsata* colonies. *Bee World* 81: 155–163.

Misra, T.K., Pahari, S., Murmu, S., and Raut, S.K. 2017. Nesting behaviour of the giant honeybees *Apisdorsata* occurring in Jhargram, West Bengal, India. *Proceedings of the Zoological Society* 70(2): 194–200.

Momose, K., Yumoto, T., Nagamitsu, T., Kato, M., Nagamasu, H., Sakai, S., Harrison, R.D., Itioka, T., Hamid, A.A., and Inoue, T. 1998. Pollination biology in a lowland dipterocarp forest in Sarawak, Malaysia. I. Characteristics of the plant-pollinator community in a lowland dipterocarp forest. *American Journal of Botany* 85: 1477–1501.

Morse, R.A., and Laigo, F.M. 1969. *Apis dorsata in the Philippines.* Philippine Association of Entomologists, Laguana.

Nath, S., Roy, P., Leo, R., and John, M. 1994. *Honeyhunters and beekeepers of Tamil Nadu.* A Survey Document Keystone, Pondicherry.

Neumann, P., Koeniger, N., Koeniger, G., Tingek, S., Kryger, P., and Moritz, R.F.A. 2000. Entomology—Home-site fidelity in migratory honeybees. *Nature* 406: 474–475.

Oldroyd, B.P., Gloag, R.S., Even, N., Wattanachaiyingcharoen, W., and Beekman, M. 2008. Nest-site selection in open-nesting honeybee *Apis florea. Behavioral Ecology and Sociobiology* 62: 1643–1653.

Oldroyd, B.P., and Nanork, P. 2009. Conservation of Asian honey bees. *Apidologie* 40: 296–312.

Oldroyd, B.P., Osborne, K.E., and Mardan, M. 2000. Colony relatedness in aggregations of *Apis dorsata* Fabricius (Hymenoptera, Apidae). *Insectes Sociaux* 47: 94–95.

Oldroyd, B.P., and Wongsiri, S. 2006. *Asian honey bees: Biology, conservation and human interaction.* Harvard University Press, Cambridge.

Paar, J., Oldroyd, B.P., Huettinger, E., and Kastberger, G. 2004. Genetic structure of an *Apisdorsata*populations: The significance of migration and colony aggregation. *Journal of Heredity* 95: 119–126.

Paar, J., Oldroyd, B.P., and Kastberger, G. 2000. Entomology—Giant honeybees return to their nest sites. *Nature* 406: 475–475.

Ponkit, R., Naree, S., Mayack, C.L., and Suwannapong, G. 2021. The pathological effects of a *Nosema ceranae* infection in the giant honey bee, *Apisdorsata*Fabricius, 1793. *Journal of Invertebrate Pathology* 185: 1–8.

Promnuan, Y., Promsai, S., and Meelai, S. 2020. Antimicrobial activity of *Streptomyces* spp. isolated from *Apis dorsata* combs against some phytopathogenic bacteria. *PeerJ* 8. https://doi.org/10.7717/peerj.10512.

Qayyum, A., and Ahmad, N. 1967. Biology of *Apisdorsata* F. *Pakistan Journal of Science* 19: 109–113.

Raffiudin, R., and Crozier, R.H. 2000. *Phylogenetic analysis of Apis based on inositol 1,4,5-triphosphate receptor gene sequences.* Seventh International Conference on Tropical Bees: Management and Diversity, International Bee Research Association, Cardiff, Wales, Chiang Mai, Thailand, 353–357.

Rattanawannee, A., and Chanchao, C. 2011. Bee diversity in Thailand and the applications of bee products. In: Grillo, O., and Venora, G.S. (Eds.), *Changing diversity in changing environment.* Intech, Rijeka, Croatia, 133–162.

Rattanawannee, A., Chanchao, C., Lim, J., Wongsiri, S., and Oldroyd, B.P. 2013. Genetic structure of a giant honey bee (*Apis dorsata*) population in northern Thailand: Implications for conservation. *Insect Conservation and Diversity* 6: 38–44.

Rattanawannee, A., Chanchao, C., and Wongsiri, S. 2012a. Geometric morphometric analysis of giant honeybee (*Apisdorsata*Fabricius, 1793) populations in Thailand. *Journal of Asia-Pacific Entomology* 15: 611–618.

Rattanawannee, A., Chanchao, C., Wongsiri, S., and Oldroyd, B.P. 2012b. No evidence that habitat disturbance affects mating frequency in the giant honey bee *Apisdorsata. Apidologie* 43: 761–770.

Rinderer, T.E., Oldroyd, B.P., Lekprayoon, C., Wongsiri, S., Boonthai, C., and Thapa, R. 1994. Extended survival of the parasitic honey bee mite *Tropilaelapsclareae* on adult workers of *Apis mellifera* and *Apisdorsata. Journal of Apicultural Research* 33: 171–173.

Rinderer, T.E., Oldroyd, B.P., Wongsiri, S., Potichot, S., Sheppard, W.S., and Buchmann, S. 1993. Time of drone flight in four bee species in south-eastern Thailand. *Journal of Apicultural Research* 32: 27–33.

Rod-im, P., Duangphakdee, O., Radloff, S., Hepburn, C., Pirk, C., and Hepburn, R. 2015. Azimuth-dependent waggle dances; flight and foraging activities of the red dwarf honeybee, *Apisflorea*Fabricius (1787). *Journal of Apicultural Research* 54(3): 246–254.

Ruttner, F. 1988. *Biogeography and taxonomy of honeybees.* Springer-Verlag, Berlin, 284.

Sahebzadeh, N., Mardan, M., Ali, A.M., Tan, S.G., Nur Azura, A., and Lau, W.H. 2012. Genetic relatedness of low solitary nests of *Apisdorsata* from Marang, Terengganu, Malaysia. *PLoS One* 7(7): e41020.

Sakai, S. 2002. General flowering in lowland mixed diterocarp forests of South-East Asia. *Biological Journal of the Linnean Society* 75: 233–247.

Sakai, S., Momose, K., Yumoto, T., Nagamitsu, T., Nagamasu, H., Hamid, A.A., and Nakashizuka, T. 1999. Plant reproductive phenology over four years including an episode of general flowering in a lowland dipterocarp forest, Sarawak, Malaysia. *American Journal of Botany* 86: 1414–1436.

Saraithong, P., and Burgett, M. 2012. *Apisdorsata* F.: Diurnal foraging patterns of worker bees in northern Thailand. *Journal of Apicultural Research* 51(4): 362–364.

Saraithong, P., Li, Y., Saenphet, K., Chen, Z., and Chantawannakul, P. 2017. Midgut bacterial communities in the giant Asian honeybee (*Apisdorsata*) across 4 developmental stages: A comparative study. *Insect Science* 24: 81–92.

Seeley, T.D., Seeley, R.H., and Akratanakul, P. 1982. Colony defense strategies of the honeybees in Thailand. *Ecological Monograph* 52: 43–63.

Sittichok, S. 2022. *Future rainfall characteristics under AR6 climate change scenarios, case study of Pasak River Basin, Thailand*. Engineering for Rural Development, Jelgava, May 25–27.

Slice, D.E. 2007. Geometric morphometrics. *Annual Review of Anthropology* 36: 261–281.

Soman, A.G., and Kshirsagar, K.K. 1991 Preliminary survey on the rockbee (*Apis dorsata* F.) and some observations on the traditional methods of honey hunting. *Indian Bee Journal* 53: 1–4.

Somanathan, H., Warrant, E.J., Borges, R.M., Wallén, R., and Kelber, A. 2009. Resolution and sensitivity of the eyes of the Asian honeybees *Apisflorea*, *Apiscerana* and *Apisdorsata*. *The Journal of Experimental Biology* 212: 2448–2453.

Sritongchuay, T. 2016. *Plant-pollinator networks in mixed fruit orchards in Southern Thailand*. PhD Thesis, Prince of Songkla University, Thailand, 132.

Sritongchuay, T., Dalsgaard, B., Wayo, K., Zou, Y., Simla, P., Tanalgo, K.C., Orr, M., and Hughes, A. 2022. Landscape-level effects on pollination networks and fruit-set of crops in tropical small-holder agroecosystems. *Agriculture Ecosystems and Environment* 339(2): 108–112.

Sritongchuay, T., Kremen, C., and Bumrungsri, S. 2016. Effects of forest and cave proximity on fruit set of tree crops in tropical orchards in Southern Thailand. *Journal of Tropical Ecology* 32(4): 269–279.

Stewart, A.B., Sritongchuay, T., Teartisup, P., Kaewsomboon, S., and Bumrungsri, S. 2018. Habitat and landscape factors influence pollinators in a tropical megacity, Bangkok, Thailand. *PeerJ* 6: e5335.

Strickland, S.S. 1982. Honey hunting by the Gurungs of Nepal. *Bee World* 63: 153–161.

Suwannapong, G., Maksong, S., Yemor, T., Junsuri, N. and Benbow, M.E. 2013. Three species of native Thai honey bees exploit overlapping pollen resources: Identification of bee flora from pollen loads and midguts from *Apis cerana*, *A. dorsata* and *A. florea*. *Journal of Apicultural Research* 52(5): 196–201.

Suwannapong, G., Noiphrom, J., and Benbow, M.E. 2012. Ultramorphology of antennal sensilla in Thai Single open nest honeybees (Hymenoptera: Apidae). *The Journal of Tropical Asian Entomology* 1: 1–12.

Tan, N.Q. 2007. Biology of *Apisdorsata*in Vietnam. *Apidologie* 38: 221–229.

Tan, N.Q., Maardan, M., Thai, P.H., and Chinh, P.H. 1999. Observations on multiple mating flights of *Apis dorsata* queens. *Apidologie* 30: 339–346.

Towne, W.F. 1985. Acoustical and visual cues in the dances of four honey bee species. *Behavioral Ecology and Sociobiology* 16: 185–187.

Underwood, B.A. 1990. Time of drone flight in *Apislaboriosa* Smith in Nepal. *Apidologie* 21: 501–504.

Underwood, B.A. 1991. Thermoregulation and Energetic decision-making by the honeybees *Apiscerana*, *Apisdorsata* and *Apislaboriosa*. *Journal of Experimental Biology* 157: 19–34.

Waring, C., and Jump, D.R. 2004. Rafter beekeeping in Cambodia with *Apisdorsata*. *Bee World* 85(1): 14–18.

Wayo, K., Phankaew, C., Stewart, A.B., and Bumrungsri, S. 2018. Bees are supplementary pollinators of self-compatible chiropterophilous durian. *Journal of Tropical Ecology* 34(1): 41–52.

Wilson-Rich, N., Allin, K., Carreck, N., and Quigley, A. 2014. *The bee: A natural history*. Princeton University Press, Princeton, 244.

Woyke, J., Kruk, C., Wilde, J., and Wilde, M. 2004a. Periodic mass flights of the giant honey bee, *Apisdorsata*. *Journal of Apicultural Research* 43(4): 181–186.

Woyke, J., Wilde, J., and Reddy, C.C. 2004b. Open-air-nesting honey bees *Apisdorsata* and *Apislaboriosa* differ from the cavity-nesting *Apis mellifera* and *Apiscerana* in brood hygiene behaviour. *Journal of Invertebrate Pathology* 86: 1–6.

Young, A.M., Kodabalagi, S., Brockmann, A., and Dyer, F.C. 2021. A hard day's night: Patterns in the diurnal and nocturnal foraging behavior of *Apisdorsata* across lunar cycles and seasons. *PLoS One* 16(10): e0258604.

10 Honey Harvesting for Sustainable Livelihoods and Agricultural Production

B.V. Shwetha, T. Neethu and N.S. Bhat

10.1 INTRODUCTION

Honeybees have the ability to elaborate honey. Honey is a sweet substance produced by honeybees from nectar of blossoms or other sources which they collect then transform by evaporating water and by action of enzymes they themselves secrete. There is a great diversity in honey based on flavor and aroma as different plants contribute their own characteristic constituents in the nectar that is converted into honey. Honey harvested from combs of honeybees is unique every time it is harvested owing to the variation in the sources. However, major constituents such as glucose and fructose remain more or less similar, constituting about 70% of the honey. Honey has more than 20 aroma constituents, 19 organic acids, five enzymes, 28 minerals, and 13 vitamins, apart from 22 types of sugars. Honey is therefore nature's concoction of more than 100 elements which is valuable to human health. Honey has diverse uses such as a flavoring agent in foods and drinks, in the confectionary industry, food products, alcoholic drinks, medicine and pharmacy, in preservation of biological materials etc.

India is endowed with five species of native honeybees, namely *Apis dorsata*, *Apis cerana*, *Apis florea*, *Apis laboriosa*, and *Apis andreniformis* (Radloff et al. 2010; Parr 2006; Hepburn and Radloff 2011). The Indian region is the center of the origin of honeybees and most commonly found are *A. dorsata*, *A. cerana*, and *A. florea*. Among these three, *A. dorsata* being most spectacular. They live in open in exposed positions, forming a cluster around a single comb. They are known for their aggressiveness and for higher honey yield. This species is distributed throughout the country but is most commonly found in forest areas with large trees.

It has been reported that India produces about 70,000 tons of honey annually (https://www.neist.res.in/stiner/bee_farming.php). The major part of this is said to come from wild honeybee, *Apis dorsata* but research on this species has seldom been done. However, the species is of great interest with regard to honey production, as well as pollination. It is reported to pollinate crops around its nest within a radius of about 10 km, which is an astounding coverage. Mostly the tribals dwelling in forest areas hunt for the honey from these colonies which provide economic

 DOI: 10.1201/9781003294078-10

security to these tribals. The honey hunters indulge in unhygienic and damaging modes of honey harvest, leading to loss of colonies in general and valuable pollinators in particular. The forest honey coming from wild bee colonies is totally free from pesticides. A large part of forest honey traded is not declared as organic and is sold as normal honey, accounting for huge financial loss to the honey hunters, traders, and nation. Organic declaration would enhance its global demand and help fetching increased amounts of foreign exchange. The forest honey will have better taste and aroma, which is therefore more suitable for value-added products of the dairy and bakery industries. Further, forest honey is multi-floral and has inherent medicinal properties which would go well with production of ayurvedic products and nutraceuticals.

Honey yield from *A. dorsata* is highly variable, and it depends on the floral situation around the nest and colony density in the area. Harvestable colonies will have on an average about 2.5–3 kg per harvest. The second harvest may yield on an average about 3–4 kg per colony. In certain areas of Karnataka, the colony yields up to 20 kg per colony, as recorded by some honey hunters.

10.2 TRADITIONAL METHOD OF HONEY HUNTING

Honey hunting is a traditional skilled activity practiced to harvest the honey from wild bee colonies—and it is one of the most ancient human activities, practiced for a thousand years. It is a source of income to the honey hunters. Local customs and traditions have become associated with honey hunting, and have been studied by anthropologists and social scientists. Most of the hives are located on high, steep, and inaccessible cliffs that face southwest in order to have increased exposure to direct sunlight and to avoid predators such as honey buzzards. This makes the honey hunting process challenging. It has been carried out very crudely by exposing the nest to smoldering fire so as to chase the bees away from the comb area and then comb is cut and the honey part is separated. Part of the comb containing larvae and egg stages of the honeybee is destroyed, resulting in the loss of huge numbers of bees. This crude method is unhygienic and also honey so obtained would not have acceptable appearance and gets spoiled soon due to mixing with immature stages and pollen grains. Also, the method damages and ultimately kills the colony.

The existing method of honey hunting is the same across Asia. People perform a religious ritual before harvesting the honey from wild bees to respect the hard work of honeybees and their contribution to the survival of human beings, and also to protect themselves from the attack of honeybees. The ritual involves worshipping the tree and offering flowers, grains, and sheep to God before hunting. A fire is lit under the bee tree or cliff to keep the bees away and get honey from the colonies of wild honeybees. In the traditional honey hunting system, smoking is considered crucial to disorientate the bees and save the honey hunters from being stung. After smoking off the bees from the comb, most honey hunters cut down the whole comb, destroying all the brood and food stores. A large number of young bees, some hundreds of adult bees and sometimes a queen and drones are also killed in this process.

The tribals in Yellapur, Karnataka, India harvest the honey from the *A. dorsata* colonies in the wild without using any protective clothing. The harvest is made during night hours. They use a torch made of plant materials (green and dry) to generate smoke and fire to chase the bees from the colonies, cut the entire comb by hand instead of a sharp spoon/knife. First, they cut off the brood area and let it fall down. Following this, they cut the honeycomb portion into a bowl without any filtering cloth. Further, after bringing down the collected combs, they hand-squeeze the comb piece containing the honey, resulting in turbid honey.

After careful study of the traditional method of honey harvest by the tribals in Yellapur, Karnataka, India, Bhat (2014) reported the items tribals used in wild bee honey harvest and gaps in their entire harvest exercise were worked out. During a training on sustainable honey harvest, all the shortcomings were met. Use of a bee suit gave confidence to handle bees during the daytime. Filtering cloths in the collection bowl facilitated clean honey collection. Removal of only the honey area from the comb helped to save the colony, along with brood, and helped repeated harvest of honey from the same colony 2–4 times (depending upon the floral availability). Cutting combs into pieces at the midrib and allowing honey to drain out helped to obtain clean honey. Use of a certain repellent plant material, genus *Mimosa*, helped to calm the temperament of the wild bees. Using a jumbo smoker helped easy chasing of bees from the comb area without any harm to the bees. The model is being followed by forest department and horticulture departments, as well.

Mulder et al. (2000) reported that honey is harvested at or around the new moon in West Kalimantan, Indonesia. In most cases, a group of local shamans, specialized bee hunters, gather for this activity. A few days before the harvest, they start making a ladder along the trunk of the bee tree up to the branches. Wooden pegs made of bamboo 30 cm long are hammered into the tree trunk at a distance of 1.5–2 m. A long pole is attached to the end of each peg by rattan. When the ladder is finished, the harvest can commence. Usually around 19:00 hours, one or two honey hunters ascend the ladder with a smoldering torch made of dried roots of jabai (*Ficusmicrocarpa*), a wooden knife and a basket attached to the hunter's waist by a long piece of rope. The hunters sing songs at various stages of the harvest. There appears to be a basic text formula, which is sung in five stages: (1) finishing the ladder; (2) clearing the bees from the nest; (3) cutting the comb; (4) hoisting the basket; and (5) descending the ladder. The songs are passed from fathers to sons, and are sung to the spirits of the trees to make them friendly.

Once the honey hunter reaches the branch above a comb, a wooden knife is used to cut the comb. The bees are brushed away from the comb with a smoldering torch, after which they disappear as sparks fall down below. It is believed that an iron knife should not be used to avoid damaging the tree bark, after which the bees would not return. In some cases, the brood comb is cut separately and thrown down. The honeycomb is then cut and put into the basket and lowered to the ground. Traditionally, these pieces are provided to appease any bad spirits.

10.3 SUSTAINABLE HARVESTING

Sustainable harvest involves removal of only honey in the comb area, still retaining comb area containing young bees and some honey. Paliwal et al. (2005) reported that in central India, the honey is harvested during evening hours as it is believed that darkness makes rock bees less active and reduces the nuisance of stinging. First, the nesting position of all the accessible colonies are pinpointed carefully using the torch. A thick rope is secured on the targeted tree where the colonies are nested. The honey hunters wear the bee suits and climb up the tree with the help of a rope. In the case of cliffs, water tanks, and buildings, a portable rope ladder is used. A bucket containing a sharp knife is pulled up on the rope and kept close. After climbing to the target place, the bees' curtain is gently cleared off from the elevated area of honeycomb. The evacuated bees fly around and settle temporarily on the nearby leaves. The evacuated portion is then smoothly cut, detached, collected in a bucket, and lowered. The whole process of bee evacuation and honeycomb separation requires only two minutes. The other accessible colonies are similarly "operated" in the next hour or two. Window-like holes are formed due to the removal of the elevated area of honeycomb since the ripened honey "in bulk" (70–80% of the honey stock of each nest) lies in the sealed, elevated cells (5–11 cm in depth), present at the corner area of the comb and in a few cases in the middle of the comb, extending for 20–28 cm. Unsealed flat cells, 2.5–5 cm in depth, containing unripe honey occur in the remaining portion of the honey area (length of 38–50 cm.). The harvested honeycombs are chopped into pieces along the mid-rib and are kept on a big sieve, allowing the honey to trickle into a lower pot. The de-honeyed combs on the sieve are melted, filtered, and condensed in water to separate crude beeswax. The harvest combs do not get detached from the supportive structure, as both the ends of their attachment remain intact with the substratum and all the evacuated bees settle back on the combs by the next morning. The detached portions are rebuilt and the honey is stored in the cells for the next 25–30 days, which facilitates subsequent harvests. In the second harvest, 6–7 kg of honey is obtained per nest, which is more than the first harvest of 2–3 kg. In each blooming season, two or three harvests may be obtained. By adopting this technique, the rock bee nests are not destroyed in the middle of their development period. The parent colonies multiply safely to add new daughter colonies, which ultimately helps to enrich the forest eco-system.

Harvesting honey crest alone, leaving the brood comb, was recommended to ensure sustainable honey harvest from wild bees nesting on trees in Satyamangalam, Tamil Nadu and Sundarbans, West Bengal (Bhat 2014). It was also attempted on colonies nesting on rock cliffs with limited success. Leaving a small portion of the honey crest induced the smoke-driven bees to resettle at the nesting site soon. They restarted comb building depending upon the availability of food sources. Provision of safety gadgets changed the mindset of the honey collectors and they applied sustainable honey harvest tactics for the colonies nesting on rock cliffs.

10.4 TIKUNG

Few honey hunters also collect honey using the *tikung* technique. Tikung is the name of a carved hardwood plank (approximately 0.8–2.5 m long by 25–40 cm wide); one side has a convex shape and the other side a concave shape. It is made of tembesu (*Fagraea fragrans*) or sometimes medang (*Litseas* p.). Such planks are attached to tree branches in the stunted swamp forests. The ends of the tikung planks are carved with notches (mainly rectangular, but sometimes V-shaped) to which a wooden peg is inserted, thus attaching it to a branch. Tikung planks are positioned with a slope of about 30°, with the upper part oriented toward the open sky. The concave side faces downward, so the upper convex side can facilitate rainwater runoff. Sometimes a pole is horizontally attached about 2 m below the tikung for the tikung owners to stand on while attaching or harvesting the tikung.

Although the *tikung* system is the most typical honey production method practised in the park, honey hunting from tall bee trees is also popular in this region. This technique is locally called *lalau* in Melayu language, or *tapang*, which is Iban language for *bee tree* (Bradbear 2009).

Honey collection from tikung resembles that of *lalau*. However, no songs are sung, as no spirits are believed to live in tikung trees. It is a collective practice. Harvest is done on moonless nights, usually from 21:00 hours until 04:00–05:00 hours, during which more than 20 tikung can be harvested. In discussions, beekeepers said that harvesting in daylight would be very dangerous as bees sting fiercely. However, in the village of Belibis, we were told that in recent years, a small group of tikung holders had started daytime collecting, using large quantities of smoke. They now seemed to favor daytime harvest as it is quicker due to better visibility. After harvest, the bees returned to the tikung for some days, after which they would leave.

The nests are approached in small boats. A man reaches up close to the tikung to smoke away the curtain of bees. All bees either fall into the water and drown or crawl up branches and leaves, as it is too dark to navigate and fly. In order to ensure floating bees do not crawl into the boat, other men in the boat use paddles (or their hands) to move the water away from the boat. Usually the brood comb (saranganak) is cut first and temporarily put on top of the tikung plank. This enables the honey collector to focus all his attention on the honeycomb, at the head of the tikung, which he then cuts and puts into the basket first. Bees are not likely to return to the tikung the following day, and are believed to return to the mountain area.

10.5 RAFTER BEEKEEPING

This method of "rafter beekeeping" was first reported by Fougères (1902). In *Apis dorsata*, there is a method of beekeeping called rafter beekeeping which was initially reported in the Mekong Delta of Vietnam (Crane et al. 1993; Chính et al. 1995). This beekeeping method is widespread in Asia. The beekeepers used to hang lengths of broad woods in the forest to attract the swarms so that they can have a control on individual nest and decrease the risk of climbing huge trees during night. Rafters are placed with an inclination of 25° facing into an opening (Tan et al. 1997). They can harvest the honey portion from the rafter comb without destroying the colony and there can be 3–4 harvests per year (Mulder 1992). Rafters help the honey hunters to have a control on more number of colonies (Oldroyd and Wongsiri 2006).

Tan and Ha (2002) reported that in Melaleuca forests of southern Vietnam, people use a traditional method of collecting honey and wax from *A. dorsata* colonies. The bees are chased away using a torch of dry leaves and *Melaleuca* leaves. This used to be the main cause of forest fires in the area, and therefore since 1993, smokers have been used. Honey is stored in the highest part of the comb and it is cut off without destroying the brood. Beekeepers usually cut a part of the brood from a big colony because they believe that if they do not, the next harvest will be smaller. This does not seem logical—the more bees, the more honey produced for harvesting. It is possible, however, that when the brood is cut, queen cells are removed which prevents swarming. Honey is squeezed, filtered and then sold in the local markets.

10.6 STEPS FOR NON-DESTRUCTIVE METHODS OF HONEY HUNTING

Joshi and Gurung (2015) suggested that honey hunting is best not done during daytime. They found that the best time to hunt honey is in the evening, before sunset. During dusk hours, the bees did not react aggressively or sting; instead, they became calm and started working on repairing the combs during the following day after the harvest. They reported that if the honey hunters assess the honey stores in the comb during the morning hours as the bees form a thicker layer over brood comb than over the honeycomb, an dif the comb is big enough and seems worth harvesting, then they proceed with honey hunting by spraying sugar solution over the comb and brush the bees when they are busy in sucking the solution and then cut the honey portion with a knife by wearing a bee veil.

10.7 BEE-REPELLENT PLANTS

Fifteen plant species have been listed to have repellent/calming ability against wild honeybees. However, eight have superior action against the bees. They are viz., *Amomum aculeatum*, *Amomum maximum*, *Zingiber zerumbet* (from Andamans), *Lobelia nicotianifolia*, *Zingiber nimonii*, *Anisomelos malabarica*, *Mimosa pudica*, and *Occimum* sp. (against A. florea) (Bhat 2014)

Of these, *Amomum maximum* had highest calming ability when fresh stem extracts were used against the wild bee, *A. dorsata*. First three species were studied for their chemical contents at National Chemical Laboratory, Pune. *Mimosa*, which is readily available in wild bee areas, is now used by tribals in smoke generation to induce calming behaviour. *Amomum* sp. extract has the proven property of silencing this honeybee species and therefore would be most ideal candidate to be used while handling this wild bee. The purification of active molecule would greatly help not only handling of the bees but also protecting the bees from pesticide hazards on crops.

10.8 CONCLUSION

Honey gathering from the wild is skilled job and the tribals undertaking this are unorganized. They are controlled by petty honey traders or local contractors, or sometimes by the forest department itself. They get meager payment for the honey they collect. The earnings from this profession can be easily doubled if they are organized and make a group approach in the profession. The proper training on sustainable honey harvest techniques of the tribals create awareness like use of bee suits can

TABLE 10.1
List of Plant Species which Have Action Calming/Repellent Action on Wild Honeybees

Sl. No.	Botanical Name	Common Name	Plant Part Used	Location
1	*Amomum aculeatum*	–	Stem	Andaman islands
2	*Amomum maximum*	–	Stem	Andaman islands
3	*Zingiber zerumbet*	–	Stem	Andaman islands
4	*Lobelia nicotianifolia*	–	Leaves	Mainland
5	*Zingiber nimonii,*	–	Stem	Mainland
6	*Anisomelos malabarica*	–	Leaves	Mainland
7	*Mimosa pudica*	Touch-me-not	Entire plant	Mainland
8	*Occimum* sp.	Tulsi	Leaves	Mainland
9	*Leucasaspera*		Entire plant	Mainland
10	*Strobilanthes reticulatus*	Persian Shield	Leaves	Mainland
11	*Albiziaamara*	Silk plant	Leaves	Mainland
12	*Lantana camera*	Spanish flag	Twigs/leaves	Mainland
13	*Cymbopogon martinii*	Lemon grass	Leaves	Mainland
14	*Cassia fistula*	Golden shower	Leaves/young branches	Mainland
15	*Synadenium grantii*		Leaves	Mainland

Note: Most of this information was learned through tribal interaction

gave confidence to handle bees during the daytime. Filtering cloths in the collection bowl facilitates clean honey collection. Removal of only honey area from the comb helps to save the colony, along with brood, and helped repeated harvest of honey from the same colony 2–4 times (depending upon the floral availability). Cutting combs into pieces at the mid-rib and allowing honey to drain out helped to obtain clean honey. Use of certain repellent plant material, genus *Mimosa*, helps to reduce the temperament of the wild bees. Jumbo smokers help easy chasing of bees from the comb area without any harm to the bees.

REFERENCES

Bhat, N.S. 2014. *A value chain on wild honey bees*. Final Report, National Agricultural Innovation Project. Indian Council of Agricultural Research, New Delhi, India, 130.

Bradbear, N. 2009. Chapter 10. Bees and their role in forest livelihoods: A guide to the services provided by bees and the sustainable harvesting, processing and marketing of their products. In: Bradbear, N. (Ed.), *Production and trade of beeswax*. Food and Agriculture Organization of the United Nations (FAO). Rome, Italy.

Chính, P.H., Thái, P.H., Minh, N.H., and Tan, N. Q. 1995. Rafter beekeeping in melaleuca forests in Vietnam. *Bees for Development Journal* 36: 8–9.

Crane, E., van Luyen, V., Mulder, V., and Ta, T.C. 1993. Traditional management system for *Apisdorsata*in submerged forests in southern Vietnam and central Kalimantan. *Bee World* 74: 27–40.

Fougères, M. 1902. Rapport surl'apiculturecoloniale. *III Congrèsinternationalel'apiculture* 53–58.

Hepburn, H., and Radloff, S. 2011. Biogeography of the dwarf honeybees, *Apisandreniformis* and *Apisflorea*. *Apidologie* 42(3): 293–300.

Joshi, S., and Gurung, M. 2015. Non-destructive method of honey hunting. *Bee World* 86(3): 63–64. https://doi.org/10.1080/0005772X.2005.11417313.

Mulder, V. 1992. Honey and wax production from *Apis dorsata* and *Apis florea* in submerged Melaleuca forests in Vietnam. In: Beetsma, J. (Ed.), *Bees and forests in the tropics. Proceedings of the 2nd NECTAR seminar.* International Agricultural Centre, Wageningen, 27–41.

Mulder, V., Heri, V., and Wickham, T. 2000. Traditional honey and wax collection from *Apisdorsata*in West Kalimantan, Indonesia. *Borneo Research Bulletin* 31: 246–261.

Oldroyd, B., and Wongsiri, S. 2006. *Asian honey bees: Biology, conservation, and human interactions.* Harvard University Press, Cambridge, MA.

Paar, J. 2006. Genetic structure of an *Apisdorsata* population: The significance of migration and colony aggregation. *Journal of Heredity. A* 12(2): 119–126.

Paliwal, G.N., Paliwal, S., and Tembhare, D.B. 2005. Practical beekeeping Eco-friendly harvesting of rock bees. *Journal of Bees for Development* 77: 1–4.

Radloff, S.E., Hepburn, C., Randall Hepburn, H., Fuchs, S., Hadisoesilo, S., Tan, K., Engel, M.S., and Kuznetsov, V. 2010. Population structure and classification of *Apiscerana. A pidologie* 41(6): 589–601. https://doi.org/10.1051/apido/2010008.

Tan, N.Q., Chinh, P.H., Thai, P.H., and Mulder, V. 1997. Rafter beekeeping with *Apisdorsata:* Some factors affecting occupation of rafters by bees. *Journal of Apicultural Research* 36: 49–54.

Tan, N.Q., and Ha, D.H. 2002. Socio-economic factors in traditional rafter beekeeping with *Apisdorsata*in Vietnam. *Bee World* 83: 165–170.

11 Distribution and Nest Site Preference of *Apis dorsata* Fabricius

B.V. Shwetha, T. Neethu, A.K. Bharath Kumar and N.S. Bhat

11.1 INTRODUCTION

Apis dorsata Fab. (Hymenoptera: Apidae), also known as rock bees, is the second largest honeybee species. In India, *A. dorsata* plays an important role as the potential pollinator of various agricultural, horticultural crops and forest trees producing multifloral honey in various agro-ecosystems. This contributes a huge share to the overall honey production in India (Bradbear and Reddy 1998). These giant honeybees are native bees of South and Southeast Asia, found mainly in forested areas such as the Terai of Nepal, parts of Pakistan and all the way to the eastern part of the Indonesian chain of islands (Ruttner 2013; Smith 2020; Guerin 2019). Nearly 60%–70% of the wax and honey harvested in India is from *A. dorsata*, and it has its own distinct pollination importance (Sihag 1986, 1988, 2000a, 2000b, 2014).

The colonies are defensive and demonstrate seasonal migration. Therefore, the domestication of this species, in the sense of the western honeybee *Apis mellifera* L. or the eastern honeybee *Apis cerana* F., is often thought to be impossible (Tan et al. 1997, 2004; Chinh et al. 1995). Nevertheless, it is always best to admire them from afar, as rock bees are known to be aggressive (Wu 2019).

Foraging behaviour of *A. dorsata* is peculiar as they forage during both the day and the night (Young et al. 2021). Foragers visit more flowers per foraging due to the high energetic demand during flight, thus making them more efficient pollinators (Sihag and Kapil 1983). *A. dorsata* F. is an important pollinator of many cultivated and wild flora. Conservation of this species is crucial. Nesting sites have profound implications on the conservation aspects of rock bees. The migrating colonies return and settle back on the trees where impressions of the previous season's colonies are present. Hence, it is really important to know the nest site characteristics which will, in turn, help us conserve them–thereby conserving the bees.

11.2 NESTING SITE PREFERENCES

Nest site selection is critical for social insects, as poor choices can be vulnerable to predation risks and result in reproductive failures (Franks et al. 2002). Social bees vary in their nest site requirements—among the *Apis* bees in South Asia, *Apis*

DOI: 10.1201/9781003294078-11

dorsata and *A. florea* nest in open combs, whereas *Apis cerana* nests in cavities (Crane 1999). *Apis dorsata* often nests in aggregations, and the large open nests can be about 1.5 m wide and are located at such sites as cliff faces or on the underside of branches of tall trees, making them inaccessible to most predators except skilled fliers and climbers (Crane 1999; Seeley et al. 1982).

Selection of a nest site is very important because the wrong choice can increase the risk of predation, threat of bad weather and reproductive failure, causing harm to colony of social insects such as bee colonies (Franks et al. 2002; Neupane et al. 2013). Kahono et al. (1999) reported that nests of *Apis dorsata* were found at heights of less than 10 m from the ground surface, contradictory to the conclusion of Weihmann et al. (2014) that *Apis dorsata* prefer high trees to make their nests. *Apis dorsata* tend to prefer places with a height of more than 10 m to build nests (Starr et al. 1987; Hadisoesilo 2001). Kahono et al. (1999) reported that the colony of *A. d. dorsata* found in West Java, Indonesia tend to nest on a branch with a 50° slope.

Nest design of *Apis dorsata* species is different compared to cavity-nesting bees, namely *A. mellifera* L. and *A. cerana* L., which build large, single, open comb at inaccessible places unlike the cavity nesting species which hold multiple combs. Seeley et al. (1982) observed that all rock bee combs were in open vegetation. They do not tend to hide themselves, as large numbers of bees can be seen forming a protective curtain around the nests and are commonly aggregated (Starr et al. 1987) and show very strong nest site fidelity (Neumann et al. 2000; Paar et al. 2000).

Apis dorsata construct nests on varieties of nesting structures that provide suitable habitat for leading successful life. The colonies have a strong tendency to aggregate on a single structure. Colony aggregation is a unique characteristic of this taxon. Paar et al. (2004) reported that preferential nesting sites may be shared by more than 150 colonies.

Neupane et al. (2013) found that greater preference of *A. dorsata* colonies towards strong-cemented water tanks and residential buildings for nesting. These colonies are often found in aggregations on suitable nesting structures and their numbers vary from a few colonies to many hundreds colonies. *Apis dorsata* build nests by orienting towards suitable compass directions. They are found to orient their nests in large numbers towards the north–south axis to minimize exposure to strong winds and bright sunlight.

Deodikar et al. (1977) observed 1861 colonies and reported that 45% of them preferred terrestrial support for nesting while 55% were arboreal. The terrestrial supports preferred in general are caves, ledges, temples, high-rise buildings, bridges, huge water tanks and weather sheds (Venkata Rau 1946; Singh 1962; Morse and Laigo 1969; Chakrabarti and Chaudari 1972). In the Philippines (Morse and Laigo 1969) and the Sunderbans (India) (Chakrabarti and Chaudari 1972), colonies were found to be nesting on slanting branches of 12–30 cm in diameter. They noticed that the same supports were repeatedly used by the bees, as the remnants of deserted combs were obvious. The reason for the repeated selection of a particular nesting site is yet to be deciphered.

Rock bees face a wide range of predators from birds, bears and humans; this makes nest location crucial for better survival (Crane 1999; Seeley et al. 1982). Trees with more branches, well spread, will provide space for more colonies to nest in aggregations. The nest site located in high trees was aimed to avoid the vertebrate predators (Starr et al. 1987) and nests located in low altitude from the ground surface indicated the lack or absence of predators. Weihmann et al. (2014) reported that *A. d. dorsata* tends to build nests in open circumstances so that the predators, such as wasp *Polistes olivaceus* can be more easily visible. The defence by weaker colonies put their larvae and colonies at higher risk of diseases. The wild honeybees usually build grouped nests in the same tree by hanging on a tree branch that is clean and free of epiphytic plants (Hadisoesilo and Kuntadi 2007; Roy et al. 2011). Colonies of *A. d. dorsata* never built the nests on old buildings, weak branches or dead trees because they are not strong enough to support the load of the nest (Neupane et al. 2013). In the Nilgiri Biosphere Reserve of India, *A. dorsata* was found to prefer cliffs to human-built structures and tall trees for nesting (Roy et al. 2011). There was significant variation in nest densities between sites; the Sathyamangalam area, which had more cliffs than human-built structures and tall trees, witnessed the highest nest densities. Results of studies carried out by Basavarajappa and Raghunandan (2013) from the same Karnataka state of South India were, however, entirely different. Here, *Apis dorsata* colonies were found to select 20 tree species to nest with 1,646 colonies (68.38%) at an elevation ranging from 5–80 feet; 580 colonies (24.10%) nested on human-built structures at heights between 8 feet and 75 feet, and 181 colonies (7.5%) nested on rock cliffs at heights between 10 m and 30 m. In Chitwan (Nepal), the annual mean number of *Apis dorsata* nests was highest on water towers, followed by residential buildings; the lowest number were in trees, and they were rarely found on rocks (Neupane et al. 2013). The colonies stayed for longer at residential buildings (6.2 months per year) followed by water towers (5.8 months per year) and trees (3.3 months per year). Previous year's sites were given the highest preference. However, the bees never built nests on the remnants of a previous colony if it had been burnt, treated with chemicals or painted with enamels.

The colonies chose the same site year after year for decades (Oldroyd et al. 2000). Often, some colonies return exactly to the same nesting structure they occupied the previous season (Neumann et al. 2000; Paar et al. 2000). Though workers probably live for less than two months (Otis et al. 1990), it is the queen that could have direct experience of the previous nest structure. There has been a colony documented to have returned after an absence of two years (Paar et al. 2000). Such precise nesting behaviour has not been reported for any other insect but is only known for migratory vertebrates, stock animals, seals and toads (reviewed in Papi 1992).

11.3 COMB ARCHITECTURE

In general, *Apis dorsata* tends to nest high in the air, usually from 3–25 m above the ground. Nests of *Apis dorsata* may occur singly or in groups; it is not uncommon to find 10–20 nests in a single tall tree, known locally as a "bee tree". In India and Thailand, trees harbouring more than 100 nests are occasionally seen in or near the tropical forest (Gupta et al. 2014). The nests of *Apis dorsata* were distributed over different heights. Seeley et al. (1982) described 15 trees in Thailand with *A. dorsata* nests; 14 of these were straight, smooth-barked and limbless for at least 13 m. He said that most nests located between 13 m and 27 m. They tend to nest at heights as low as 5 m to as high as 15 m above the ground level, and at times they even nest at 40 m above the ground level. This may be related to the incidence of sun rays on combs, the convenience of foragers and/or protection from pests, and predators (Deodikar et al. 1977).

The combs are generally directed towards north–south rather than east–west. This holds true even for arboreal colonies. Least resistance towards seasonal winds and exposure to sun rays from east or west can be probable reasons for such peculiar orientation. Further, earth's magnetic field may also be operative in deciding the angle of the comb since the polar axis and magnetic axis coincide with each other with a marginal difference (Deodikar et al. 1977).

Once a particular spot has been selected by the scout bees after preliminary survey, the swarm arrives and settles there, commencing the formation of wax pellets on the supporting surface. At times, new combs are built over the remnant of the colonies of previous seasons. The comb construction starts with the midrib. Cells are attached secondarily to the midrib with their openings slightly facing upwards. They generally build combs that are semicircular or cuneiform in shape with the width of the comb slightly larger than the length (Ruttner 1988). The surface of the comb is found to vary from 0.1–1 m^2. Like all other species, the comb is differentiated into a thick honeycomb in the highest section with a diameter of up to 15–20 cm, to a brood comb of 3.5 cm thickness (Roepke 1930; Muttoo 1956; Morse and Laigo 1969; Seeley et al. 1982). By simply extending or reducing them, the walls each cell can be used for storing or brood rearing (Koeniger and Koeniger 1980). The uniform diameter of the cells is 5.35–5.64 mm and the depth of brood cells is 16 mm (Muttoo 1956; Thakar and Tonapi 1961). The weight of the bees (6,500 per kilogram; that is, 5–10 kg per colony; Muttoo 1956) is mainly held by the support. A colony has 80%–90% of the bees incorporated into a thick "pelt" covering the comb (Morse and Laigo 1969). The "protective curtain" of bees is detached from the comb by 1–2 cm (Morse and Laigo 1969). This bee space serves as a moving area for the "nest service", for ventilation. The bees forming the curtain hang upwards with their wings spread. The body has an inclination of 5° towards the comb, with a covered head and exposed abdomen. Unless disturbed, these bees are completely motionless. In the low part of the curtain, there is a portion distinct with increased activity and bees with their heads directed downwards. there is a continuous movement of alighting field bees, and this active area is termed the "mouth" of the colony (Morse and Laigo 1969).

Nest size of *Apis dorsata* varies in length as well as height. According to Sattigi et al. (1995–1996), average comb length varied from 38.75–82.5 cm, height from 34.5–92.5 cm and thickness from 3.2–7.3 cm. Honey and brood cells were 4.0 cm and 1.7 cm in depth, respectively. However, Basavarajappa and Raghunandan (2013) reported different statistics of these parameters from the same region of India. They found that nest length was 54 cm, nest height 35 cm and nest thickness for brood and honeycomb cells was 1.74 cm and 2.03 cm, respectively. Neupane et al. (2013) reported nest height from 2–30 cm and nest length from 5–150 cm, with a shape variation from round to elliptical. The thickness of the comb at the attachment point and bottom end was 7 cm and 3 cm, respectively.

Of the total immigrated swarm colonies (N = 796), the majority (50.76%) in this region attained nest length (horizontal or base dimension) of 100–120 cm, followed by those having nest length 80–100 cm (26%), 120–140 cm (21.36%) and 60–80 cm (1.88%). Minimum nest length recorded was 63 cm and maximum was 139 cm. Likewise, there was a difference in the height (vertical dimension) of the nest of two types of colonies: the majority of the immigrated swarm colonies had nest height of 40–50 cm (51.76%), followed by those having nest height of 50–60 cm (25.25%), 30–40 cm (15.58%) and least were those with 20–30 cm (7.41%).

The basal nest thickness (just below the attachment) of *Apis dorsata* varied greatly; in the non-honey region, there was narrower nest thickness (mean ± SD = 2.04 ± 0.6 cm, range = 1.7–3.1 cm, N = 150) whereas in the honey region, the nests had broader basal thickness (mean ± SD = 5.7 ± 1.2 cm; n = 150); the difference between the two was significant ($p < 0.05$, df = 148, independent sample t-test). The nest shape of *Apis dorsata* varied greatly depending on the angle of the nesting support. There were usually three types of shapes of the nests attached to the horizontal supports: (1) vertical semi-ellipse; (2) semi-circle; and (3) horizontal semi-ellipse. Of the 308 colonies that nested on the horizontal surfaces of buildings (offices and residences), 262 (85.1%) had horizontal semi-ellipse shape with 2.37 ± 0.19 (mean ± SD) nest shape index. Colonies with other shapes were low in number: 29 (9.4%) had semi-circle shape with 2.04 ± 0.05 (mean ± SD) nest shape index and 17 (5.5%) had vertical semi-ellipse shape with 1.73 ± 0.03 (mean ± SD) nest shape index; among these, one colony had a curved (concave) nest. This was due to the non-availability of enough nesting space on the nesting site (Woyke et al. 2016).

The diameter (thickness) of the tree branch seemed to influence the nesting behaviour of *Apis dorsata*—degrees of incident nesting angles and directions of the nests. *Apis dorsata* made nests in different directions and on various inclined supports giving an initial impression of their role in the nesting site preferences. More than 95% of nests were found to be attached on supports making angles from 0–45°; there were small differences among the groups. Nevertheless, an inclination of nesting support of 15–30° seemed to be maximally preferred (25.25%) over others. This was followed by 0–15° (20.38%), 0° (19.20%), 30–45° (17.78%) and 45–60° (14.96%). Nest support with >60° inclination was least preferred and had only 39 nests (2.43%). Thus, the angle of nesting support too seemed to guide the nesting preference of *Apis dorsata*. Of the 1604 colonies studied, 780 (48.6%) constructed their nests in an east–west direction, followed by 472 (29.4%) in a northeast–southwest direction and 215 (13.4%) in a northwest–southeast direction. Only a small proportion of the nests were constructed in other directions; for example, 30 (1.8%) nests were in north–south direction, 58 (3.7%) in east–southwest direction and 49 (3.1%) in west–southeast direction, respectively. Therefore, in the semi-arid environments of northwest India, *Apis dorsata* seemed to prefer those nesting sites where it could construct its nests in east–west and northeast–southwest directions.

11.4 MIGRATION BEHAVIOUR

Apis dorsata colonies can perform seasonal long-distance migrations in order to exploit the nectar and pollen resources available at different times throughout many parts of Southern Asia (Oldroyd et al. 2000). The colony leaves behind an empty comb when it departs from a nest site. Interestingly, the same colony has been to return to the exact same observed branch six months later, even though those bees that knew of the old nesting location would have died long before. In South India, during the months of February–April, they construct nests in forest and move to plains land in the months of September–December, according to the availability of flora and also to escape from heavy rain.

In Punjab, colonies of *A. dorsata* desert the comb during dearth period, i.e., May–June. They abscond and migrate to the safer places with thick vegetation and favourable conditions (Abrol 1997; Atwal 2000). Absconded bees migrated towards safer places. Bee colonies which were in favourable conditions (good bee flora, fresh water, shadow and not attacked by wax moth) did not abscond. Queens of *A. dorsata* are not much larger than workers, which is an adaptation for migration of queens along with the colony over larger distances. *Apis dorsata* is a migratory species, travelling up to 200 km through a series of shorter steps to follow currently available resources (Koeniger and Koeniger 1980; Dyer and Seeley 1994; Robinson 2012). In Thailand, colonies of *Apis dorsata* reside at the nesting sites between November and April. After April, during the rainy season, *Apis dorsata* migrates to forest areas. Newly arrived swarms in the subsequent dry season form colonies, forage and reproduce until the beginning of the rainy season (May–June) (Wongsiri et al. 1996). Nagaraja (2017) reported that *A. dorsata* colonies tend to be found within Bangalore during the dry season (which lasts from October–May), migrating into Bangalore in November and December then migrating out of Bangalore in May and June before the start of monsoon season (which lasts from June–September).

Colonies migrate seasonally. Though details vary with region, colonies typically live in lofty communal nest sites of <20–100 colonies at relatively high elevations (1,000 m) during the dry season. There they build <1.5 m–diameter single wax combs. The colonies grow quickly and reproduce by fission (swarming). As forage decreases towards the end of the season, colonies abandon their combs and migrate to lower elevations, establishing new nests there for the mass flowering of the monsoon season. Entire colonies of *Apis dorsata* migrate regularly, with changes in the monsoon season, back and forth between highland and lowland nesting sites. While nesting, a colony typically builds a single, exposed, 1–1.5 m–wide wax comb on which its population may grow to more than 50,000 individuals (Morse and Laigo 1969) before reproducing by swarming (colony fission). *Apis dorsata* F. colonies are frequently found in dense aggregations (Kastberger and Sharma 2000; Koeniger and Koeniger 1980; Lindauer 1956; Ruttner 1988). In general, the colonies are found nesting in aggregations or groups including 50 colonies. There have been 200 colonies that occur on a single tree (Oldroyd et al. 2000) and those are often only a few centimetres apart. The newly arrived colonies build their combs over the remnants of the previous season's combs.

11.5 COLONY AGGREGATION

Colony aggregation is a unique characteristic of this taxon. Preferential nesting sites may be shared by more than 150 colonies (Paar et al. 2004). Seasonal migration of 100–200 km is a common behaviour of *Apis dorsata*, and queens appear to return to their previously occupied sites (Koeniger and Koeniger 1980; Dyer and Seeley

1994). Home-site fidelity was recently reported in *Apis dorsata* based on microsatellite analysis (A14, A76 and A88; Neumann et al. 2000).

Apis dorsata often forms large aggregations of up to 150 unrelated colonies (Oldroyd et al. 2000). They construct large (up to 2 m–wide) single comb colonies on the underside of rocky outcrops or branches of smooth-barked trees (Oldroyd and Wongsiri 2006). Colonies are known to return to the same roosting locations every season, and seem to use the presence of old comb fragments as a cue to indicate a good nesting location (Liu et al. 2007; Neumann et al. 2000; Paar et al. 2000). Like the nest sites of the open-nesting *A. florea*, trees and rock surfaces that are suitable for *Apis dorsata* nest sites are relatively common, but, owing to *Apis dorsata*'s preference to nest in aggregations, the choices deemed by the scout bees to be of higher quality are no doubt restricted.

The giant honeybee differs substantially in behaviour and ecology from other *Apis* species (Morse and Laigo 1969). First, colonies are frequently found in dense aggregations (Kastberger and Sharma 2000; Koeniger and Koeniger 1980; Lindauer 1956; Roepke 1930; Ruttner 1988). Up to 200 colonies may occur on a single tree (Oldroyd et al. 2000) and these are often separated by only a few centimetres. Second, colonies often undergo seasonal migration between alternate nesting sites. In these populations, nest sites tend to be occupied for 3–4 months. Toward the end of this period, brood rearing ceases and the honey and pollen stores are depleted (Ruttner 1988). Eventually, the colonies abscond to alternative nest sites which may be 200 km distant (Crane et al. 1993; Koeniger and Koeniger 1980). The proximate cause of migration may be related to available forage. Nesting sites are reoccupied year after year over periods of several decades or more (Oldroyd et al. 2000). Amazingly, some returning colonies find their way back to exactly the same nesting structure they occupied the previous season (Neumann et al. 2000; Paar et al. 2000). This is despite the fact that workers probably live for less than two months (Otis et al. 1990), so only the queen could have direct experience of the former nest site. Dozens—even hundreds—of nesting colonies may aggregate at close quarters in a single tall tree or other lofty structure, such as a water tower (Oldroyd et al. 2000; Oldroyd and Wongsiri 2006; Hawkeswood et al. 2019). In one tree, more than ten nests of *Apis dorsata* can be found, and usually clustered in the nesting tree (Kahono et al. 1999; Hadisoesilo 2001; Hepburn and Radloff 2011; Mead 2013).

Although the occurrence of isolated colonies is not uncommon, colonies can be found nesting in aggregations or groups in both inhabited and uninhabited areas. In abandoned forts and forests, colonies are found nesting in aggregations at single spots. However, in the Philippines, they are found only as isolated colonies. Sister (twin) combs with a bee space in between have also been reported occasionally (Morse and Laigo 1969). There are several reasons for having such colonies in aggregation which include protection from predators, shelter, availability of floral resources and the presence of water in the vicinity.

11.6 HOST PLANTS

It is clear that the majority of *Apis dorsata* nesting sites were located on forest trees. This implies that *A. dorsata* construct their nest on forest tree branches for nesting support. They prefer to build their nests in close proximity to the pollen source. The preference of *A. dorsata* toward forest trees is possibly related to the tree architecture of the forest species (Izzat et al. 2012). The existence of bee colony in habitat is supported by the availability of food resources and nesting conditions. *A. dorsata* also utilize wild plants and crops around the forest as their food source (Bawa 1990; Kahono et al. 1999).

Pattazhy (2020) reported common tree species used as nesting sites were *Ficus* sp. (Moraceae), *Adenanthera* sp. (Fabaceae), *Spondias pinnata* (Anacardiaceae), *Artocarpus sericoarpus* (Moraceae), *Alstonia scholaris* (Apocynaceae), *Knema cinerea* (Myristicaceae), *Palaquium obovatum* (Sapotaceae) and *Litsea mappacea* (Lauraceae). Nagir et al. (2016) found 34 species belonging to 27 genera and 17 families of trees used as nesting site of *A. d. binghami*.

Hadisoesilo and Kuntadi (2007) also stated that in Riau, nesting trees of *Apis dorsata* were in banyan (Ficus sp.), kempas (*Koompassia excelsa*), lumbuai (*Metroxylon* sp.), siluang (*Polythia shipolenca*), jangkang (*Dellenia exinia*), and mahang (*Macaranga* sp.), whereas in Sulawesi, nesting trees of *A. d. binghami* were in durian (*Durio zibethinus*), mango (*Mangifera* sp.), guava (*Syzygium* sp.), and *Ceiba pentandra*. Some studies have also found *A. d. dorsata* tend to like kempas (*K. excelsa*) for nesting (Starr et al. 1987; Hadisoesilo 2001). In Bogor Botanical Garden, West Java, *Apis dorsata* tend to choose *Ficus albipila* as nesting trees (up to 72%) (Kahono et al. 1999). The colonies of *A. d. binghami* build nests in trees depending on the species of trees that dominate in these areas (Hadisoesilo and Kuntadi 2007).

Ram (2017) found that preferences for nesting sources and sites were big trees such as Indian rosewood (*D. sissoo*), eucalyptus (*Eucalyptus hybrida*), Indian banyan (*Ficus benghalensis*) and religious peepal (*Ficus religeosa*). Besides these, horticultural trees such as guava (*Psidium guajava*) or plum (*Prunus domestica*) were also present. Eucalyptus was the most abundant (~1,000), followed by ber (~800) and Indian rosewood (~500); buildings were fewer in number than the trees.

Out of 796 immigrated swarm colonies, 440 constructed nests on trees, with the remaining 356 on the buildings. Likewise, out of the 808 reproductive swarm colonies, 807 made nests on the trees and only one on a building. If the number of nests present on the nesting sources is taken as a measure of nesting preference, at Hisar, *Apis dorsata* seemed to show distinct preference for trees over human-built structures—77.74% of colonies constructed nests on trees. Indian rosewood (*D. sissoo*), attracted maximal number of bee colonies (57.29% of the total colonies), followed by eucalyptus (*Eucalyptus hybrid*) (8.29%), religious peepal, *Ficus benghalensis/Ficus religiosa* (3.86%), *Acacia nilotica* (2.81%), *Z. mauritiana* (2.37%), *Kigelia africana* (2.12%) and *Mangifera indica* (1.00%) (Ram 2017). In the Sunderbans area (West Bengal), a large number of colonies (39%) were found on *Excoecaria agallocha* and 16% nested on *Avicennia* species (Chakrabarti and Chaudari 1972).

Apis dorsata seemed to show distinct preference for trees over human-built structures. The majority of the colonies nesting on buildings made nests on sites having relics of old nests. However, trees did not provide such a clue for nesting site reoccupation by this honeybee. Smoothness of the surface may be the reason for nesting site reoccupation in the majority of the latter cases, as the surface of the occupied sites were smooth; the unevenly grooved or highly rough surfaces were seldom occupied.

That was the reason why only those tree branches having smooth surfaces were selected for fresh nesting.

REFERENCES

Abrol, D.P. 1997. *Bees and beekeeping in India.* Kalyani Publishers, Ludhiana, India, 450.

Atwal, A.S. 2000. *Essentials of beekeeping and pollination.* Kalyani Publishers, Ludhiana, India.

Bawa, K.S. 1990. Plant-pollinator interactions in tropical rain forests. *Annual Review of Ecology and Systematics* 399–422.

Bradbear, N., and Reddy, M.S. 1998. *Existing apicultural practices within Karnataka.* Mission Report, FAO and UN, Pune, 1–20.

Chakrabarti, W.B., and Chaudari, A.B. 1972. Honey production and behaviour pattern of the honey bee. *Sci. & Cult.* 38: 257–290.

Chính, P.H., Thái, P.H., Minh, N.H., and Tan, N.Q. 1995. Raftering, a traditional technique for honey and wax production from Apisdorsata in Vietnam. *Beekeeping and Development* 36: 8–9.

Crane, E. 1999. *The world history of beekeeping and honey hunting.* Routledge, London.

Crane, E., van Luyen, V., Mulder, V., and Ta, T.C. 1993. Traditional management system for *Apis dorsata* in submerged forests in southern Vietnam and central Kalimantan. *Bee World* 74: 27–40.

Deodikar, G.B., Ghatge, A.I., Phadke, R.P., Mahindre, D.B., Kshirsagar, K.K., Muvel, K.S., and Thakar, C.V. 1977. Nesting behaviour of Indian honeybees III. Nesting behaviour of *Apis dorsata* Fab. *Indian Bee Journal* 39(1–4): 1–12.

Dyer, F.C., and Seeley, T.D. 1994. Colony migration in the tropical honey bee *Apis dorsata* F. (Hymenoptera: Apidae). *Insects Soc.* 41(2): 129–140.

Franks, N.R., Pratt, S.C., Mallon, E.B., Britton, N.F., and Sumpter, D.J. 2002. Information flow, opinion polling and collective intelligence in house–hunting social insects. *Philosophical Transactions of the Royal Society of London. Series B: Biological Sciences* 357(1427): 1567–1583.

Guerin, E. 2019. *Rafter beekeeping—sustainable management with Apis dorsata.* Training manual. https://www.wwf.or.th/en/scp/upcoming_event_/reports_/rafter_beekeeping/.

Gupta, R.K. et al. (Eds.). 2014. *Beekeeping for poverty alleviation and livelihood security.* Springer, Dordrecht. https://doi.org/10.1007/978-94-017-9199-1.

Hadisoesilo, S. 2001. *Diversity in traditional techniques for enticing Apisdorsata colonies in Indonesia.* Proceedings of the 37th International Congress, Apimondia, Durban.

Hadisoesilo, S., and Kuntadi, S. 2007. *Local wisdom in "beekeeping" of a forest honeybee (Apisdorsata).* Forest Department. Research and Development of Forestry, Center Research and Development for Forestry and Conservation (translated).

Hawkeswood, T.J., Sommung, B., and Sommung, A. 2019. A record of the giant honey bee, *Apis dorsata* Fabricius, 1793 (Insecta: Hymenoptera: Apidae) from Sisaket Province, Thailand. *Calodema* 720: 1–4.

Hepburn, H.R., and Radloff, S.E. 2011. *Honeybees of Asia.* Springer Science & Business Media, Berlin.

Izzat, F., Balasundram, S.K., Abdullah, N.A.P., Sood, A.M., Mardan, M., and Saberioon, M.M. 2012. The spatial distribution of *Apis dorsata* host plants using an integrated geographical information system-remote sensing approach. *American Journal of Agricultural and Biological Sciences* 7(4): 396–406.

Kahono, S., Nakamura, K., and Amir, M. 1999. Seasonal migration and colony behavior of the tropical honeybee Apis dorsa Ta F. (Hymenoptera: Apidae). *Treubia* 31(3): 283–297.

Kastberger, G., and Sharma, D.K. 2000. The predator-prey interaction between blue-bearded bee-eaters (Nyctyornis Atherton Jardine and Selby 1830) and giant honeybees (*Apisdorsata* Fabricius 1798). *Apidologie* 31(6): 727–736.

Koeniger, N., and Koeniger, G. 1980. Observations and experiments on migration and dance communication of *Apis dorsata* in Sri Lanka. *J Apic. Res.* 19(1): 21–34.

Lindauer, M. 1956. Über die Verständigung bei indischen Bienen. *Zeitschrift für Vergleichende Physiologie* 38: 521–557.

Liu, F., Roubik, D.W., He, D., and Li, J. 2007. Old comb for nesting site recognition by *Apis dorsata*? Field experiments in China. *Insectes Sociaux* 40: 963–964. https://doi.org/10.1007/s00040-007-0963-4.

Mead, D. 2013. *A guide to some bees and wasps of Indonesia.* Sulang Language Data and Working Papers: Topics in lexicography, no. 11. Sulawesi Language Alliance. http://sulang.org/.

Morse, R.A., and Laigo, F.M. 1969. *Apis dorsata in Philippines.* Monograph of the Philippines Association of Entomologists Inc., Laguna.

Muttoo, R.N. 1956. Facts about beekeeping in India-1. *Bee World* 37(7): 125–133.

Nagaraja, N. 2017. Population fluctuation of giant honeybee, *Apisdorsata* (Hymenoptera: Apidae) colonies in Bengaluru, Karnataka, India. *J Entomol Res.* 41(3): 307–310.

Nagir, M.T., Atmowidi, T., and Kahono, S. 2016. The distribution and nest-site preference of *Apisdorsatabinghami* at Maros Forest, South Sulawesi, Indonesia. *Journal of Insect Biodiversity* 4(23): 1–14.

Neumann, P., Koeniger, N., Koeniger, G., Tingek, S., Krygerll, P., and Moritz, R.F.A. 2000. Homesite fidelity in migratory honeybees. *Nature* 406: 474–475.

Neupane, K.R., Woyke, J., and Poudel, S.M. 2013. *Behavior of giant honeybee Apisdorsata.* Proceedings of International Beekeeping Congress of APIMONDIA, Kyiv, Ukraine, September 29–October 4.

Oldroyd, B.P., Osborne, K.E., and Mardan, M. 2000. Colony relatedness in aggregations of *Apisdorsata* Fabricius (Hymenoptera, Apidae). *Insectes Sociaux* 47(1): 94–95.

Oldroyd, B.P., and Wongsiri, S. 2006. *Asian honey bees: Biology, conservation, and human interactions.* Harvard University Press, Cambridge. https://doi.org/10.4159/9780674041622.

Otis, G.W., Mardan, M., and McGee, K. 1990. Age polytheism in *Apisdorsata.* In: Veeresh, G.K., Mallick, B., and Viraktamath, C.A. (Eds.), *Social insects and the environment. Proceedings of the 11th international congress of IUSSI.* Oxford & IBH Publishing, New Delhi, 378.

Paar, J., Oldroyd, B.P., Huettinger, E., and Kastberger, G. 2004. Genetic structure of an *Apis dorsata* population: The significance of migration and colony aggregation. *Journal of Heredity* 95(2): 119–126.

Paar, J., Oldroyd, B.P., and Kastberger, G. 2000. Giant honeybees return to their nest sites. *Nature* 406(6795): 475.

Papi, F. 1992. General aspects. In: *Animal homing.* Springer, Dordrecht, 1–18.

Pattazhy, S. 2020. First record of the nest orientation of *Apis dorsata* in relation to cell tower. *Int J Zoo Animal Biol* 3(2): 000218.

Ram, C.S. 2017. Nesting behaviour and nest site preferences of the giant honey bee (*Apis dorsata* F.) in the semi-arid environment of north west India. *Journal of Apicultural Research.* https://doi.org/10.1080/00218839.2017.1338443.

Robinson, W.S. 2012. Migrating giant honey bees (*Apis dorsata*) congregate annually at stopover site in Thailand. *PLoS One* 7(9): e44976. https://doi.org/10.1371/journal.pone.0044976 PMID: 23028715.

Roepke, W. 1930. Beobachtungen an Indischen Honigbienen, insbesondere an *Apis dorsata* F. *Meded Landbouwhogesch Wageningen* 34: 1–28.

Roy, P., Leo, R., Thomas, S.G., Varghese, A., Sharma, K., Prasad, S., Bradbear, N., Roberts, S., Potts, S.G., and Davidar, P. 2011. Nesting requirements of the rock bee Apisdorsata in the Nilgiri Biosphere Reserve, India. *Tropical Ecology* 52(3): 285–291.

Ruttner, F. 1988. *Apisdorsata* Fabricius 1793: 328. In: *Biogeography and taxonomy of honeybees*. Springer, Berlin, Heidelberg, 103–119.

Ruttner, F. 2013. *Biogeography and taxonomy of honeybees*. Springer Science & Business Media, Berlin.

Seeley, T.D., Seeley, R.H., and Akratanakul, P. 1982. Colony defense strategies of the honeybees in Thailand. *Ecol. Monogr.* 52(1): 43–63.

Sihag, R.C. 1986. Insect pollination increase seed production in cruciferous and umbelliferous crops. *Journal of Apicultural Research* 25: 121–126.

Sihag, R.C. 1988. Characterization of the pollinators of cultivated cruciferous and umbelliferous crops of sub-tropical Hisar (India). *Bee World* 69: 153–158. https://doi.org/10.1080/0005772X.1988.11098972.

Sihag, R.C. 2000a. Eco-biology of the little honey bee (*Apis florea* F.) in semi-arid subtropical climates of India. In: Matsuka, M., Verma, L.R., Wongsiri, S., Shrestha, K.K., and Partap, U. (Eds.), *Asian bees and beekeeping: Progress of research and development*. Oxford & IBH Publishing Company Private Limited, New Delhi, 46–49.

Sihag, R.C. 2000b. Eco-biology of the giant honey bee (*Apis dorsat*a F.) in semi-arid subtropical climates of India. In: Matsuka, M., Verma, L.R., Wongsiri, S., Shrestha, K.K., and Partap, U. (Eds.), *Asian bees and beekeeping: Progress of research and development*. Oxford & IBH Publishing Company Private Limited, New Delhi, 50–52.

Sihag, R.C. 2014. Phenology of migration and decline in colony numbers and crop hosts of giant honey bee (*Apis dorsata* F.) in semiarid environment of Northwest India. *Journal of Insects* 2014: 1–9. https://doi.org/10.1155/2014/639467.

Sihag, R.C., and Kapil, R.P. 1983. Foraging strategies of honey bees as determined by quality and quantity of nectar. In: Tasei, J.N. (Ed.), *Proceedings of 5th international symposium on pollination*. Les Colleques de INRA, France, Versailles, 51–59.

Singh, S. 1962. *Beekeeping in India*. ICAR, New Delhi.

Smith, D.R. 2020. Biogeography of honey bees. In: Starr, C. (Eds.), *Encyclopedia of social insects*. Springer, Cham.

Starr, C.K., Schmidt, P.J., and Schmidt, J.O. 1987. Nest-site preferences of the giant honey bee, Apisdorsata (Hymenoptera: Apidae), in Borneo. *Pan-Pacific Entomologist* 63(1): 37–42.

Tan, K., Ge, F.C., Zhao, R., and He, S.Y. 2004. Morphometric characters of Changbaishan A. cerana. *Journal of Bee* 869(6): 8–9. https://doi.org/10.3969/j.issn.1003-9139.2004.06.002.

Tan, N.Q., Chinh, P.H., Thai, P.H., and Mulder, V. 1997. Rafter beekeeping with *Apisdorsata*: Some factors affecting the occupation of rafters by bees. *Journal of Apicultural Research* 36(1): 49–54.

Thakar, C.V., and Tonapi, K.V. 1961. Nesting behaviour of Indian honeybees I. Differentiation of worker, queen and drone cells on the combs of Apisdorsata Fab. *Bee World* 42(3): 61–62.

Venkata Rau, R.N. 1946. My final report of the observations made on the *Apisdorsata* hive built in the Ramakrishna Mission students home in Mylapore, Madras. *Indian Bee Journal* 8: 160–163. Keywords: Beekeeping/dorsata/India.

Weihmann, F., Waddoup, D., Hötzl, T., and Kastberger, G. 2014. Intraspecific aggression in giant honey bees (Apisdorsata). *Insects* 5(3): 689–704.

Wongsiri, S., Thapa, T., Oldroyd, B., and Burgett, D.M. 1996. A magic bee tree: Home to Apis dorsata. *American Bee J.* 136: 796–799.

Woyke, J., Wilde, J., and Wilde, M. 2016. The reasons for the different nest shapes of Megapis bees. *Journal of Apicultural Science* 60(1): 121.

Wu, Y. 2019. *Apis (Megapis) dorsata* Fabricius, 1793. Wikipedia: Common Giant Honey Bee (Hymenoptera: Apidae: Apini). https://idtools.org/tools/1078/index.cfm?packageID=1183&entityID=9007.

Young, A.M., Kodabalagi, S., Brockmann, A., and Dyer, F.C. 2021. A hard day's night: Patterns in thediurnal and nocturnal foraging behaviour of *Apisdorsata* across lunar cycles and seasons. *PLoS One* 16(10): e0258604. https://doi.org/10.1371/journal.pone.0258604.

12 Giant Honeybees Exploit Multiple Floral Resources in Natural and Agricultural Landscapes

B.M. Rathna Kumari, N. Nagaraja and Dharam P. Abrol

12.1 INTRODUCTION

Bees and flowering plants exhibit a distinct relationship in which both benefit from each other by co-evolving during the middle Cretaceous period (Michener 2007). Honeybees play an important role in pollination of tropical agricultural and natural ecosystems (Hung et al. 2018; Requier et al. 2019). The bee flora is pre-requisite for development and sustainability of honeybee colonies (Dimou et al. 2006). The qualities of bee flora are assessed based on the long flowering period, density of flowers in a unit area, sugar concentration of the nectar and accessibility of nectar to the honeybees. Honeybees visit flowers for collection of nectar and pollen by recognizing their colour, shape and odour. After locating flowers, they respond to floral odours, including olfactory nectar guides (Hopkins et al. 1969). Bees are attracted to flowers that are available in large numbers with readily accessible pollen or nectar (Kevan and Baker 1983). Honeybees forage on flowers and sometimes by restricting their visits to a single plant (Sihag 2000). The quantity of nectar and the concentration of sugars in floral nectars determine attractiveness to bees. The proportion of pollen to nectar that collected is largely determined by the nutritional needs of the colonies. Honeybees communicate the direction, distance, odour and source of food to other bees by various kinds of dances (von Frisch 1967). This communication system allows them to become efficiently informed about nectar and pollen plants in their flight range. The foraging biology of honeybees is important in tropical ecosystems, where most forest trees and agricultural crops depend upon bee pollinators (Dainese et al. 2019; Thimmegowda et al. 2020).

12.2 FLORAL RESOURCES

Similar to other insects, honeybees require carbohydrates, proteins, fats and other nutrients including water for growth and development of colonies (Loper and Berdel 1980). Nectar is a major carbohydrate source and is secreted by flowers to attract

 DOI: 10.1201/9781003294078-12

bee pollinators. It is the major energy source of the colony, providing food for the activities of all colony members. The major sugars of nectar are sucrose, fructose and glucose, which differ in sugar concentration from 15%–75%. The nectar secretion is greater during summer than other seasons, as the nectar sugars are products of photosynthesis, which are in turn influenced by sunlight. Similarly, the pollen is an important source of proteins and fats to both adult honeybees and developing larvae. It is obvious that adequate nectar and pollen resources are critical in maintaining honeybee health (Michener 2007). Deficiency of pollen and nectar can lead to decline in colony populations, followed by absconding of bee colonies. Production of pollen and nectar in the plants varies with plant species, and also during different seasons (Free 1970).

12.3 FORAGING BEHAVIOUR OF *APIS DORSATA*

Honeybees live in complex societies and are characterized by rearing several generations in a colony with cooperative brood rearing. Honeybee workers exhibit age-related biases in task performance as temporal polyethism whereby young individuals performing tasks inside the colony and older worker bees specializing in foraging (Dukas 2008). Individual worker bees forage across varied locations and can carry substantial pollen and nectar loads during flight (Harrison and Fewell 2002). Each colony consists of a large single nest with about 30,000–60,000 worker bees. *A. dorsata* is one of the major honeybee species on the Indian sub-continent. It produces large quantities of honey and is an efficient pollinator of both agricultural crops and wild flora (Chantawannakul et al. 2018; Sihag 2014). *A. dorsata* builds single nests on tall trees, buildings, rock cliffs, water tanks, flyovers, etc. It is well distributed in plains and hilly regions, but the number of colonies are greater in dense forest regions (Nagaraja and Rajagopal 2009). Congregation of *A. dorsata* colonies is quite common and their number may vary from 50–200 on a tree or rock where the bee forage sources are quite abundant (Rajagopal and Nagaraja 1999; Oldroyd et al. 2000) (Figure 12.1). The pollen and nectar from varieties of ornamental plants is a major resource for these bees during summer.

Pollination in the tropics is often dominated by *A. dorsata* because of its wide foraging activity on most of the crops, and also for its seasonal migration from the regions of floral dearth to floral rich by rapidly increasing their population in new nest sites (Stewart et al. 2018; Mukherjee et al. 2019). The workers of these species are larger in size than those of other *Apis* species (Underwood 1991), and such variation in body size may coincide with differences in their tongue length (Cariveau et al. 2016), foraging range (Greenleaf et al. 2007), sensory capabilities (Streinzer et al. 2013) and metabolic rate (Dyer and Seeley 1987). The nests of these bees have comparatively less comb surface area than cavity-nesting species such as *A. mellifera* and *A. cerana*, as *A. dorsata* build single open comb.

Honeybees not only forage to meet their own nutritional demands, but also to fulfil the needs of the entire colony (Page and Erber 2002). They are generalist feeders and visit several plant species for gathering nectar and pollen (Michener 1990), and they prefer the flowers with exposed nectar, abundant pollen, zygomorphic nature

FIGURE 12.1 Congregation of *Apis dorsata* colonies on a banyan tree near Hosakote, Karnataka, India.

and compact umbels (Koul et al. 1989). Even though they are polytrophic, sometimes the foragers may restrict to single flower in a foraging trip. *A. dorsata* foragers prefer relatively large sucrose-dominated flowers (Abrol 2011, 2016). The foraging activity of honeybees is influenced by the weather conditions and availability of nectar and pollen resources (Neupane and Thapa 2005). These bee species collect floral resources from the distance of 2–3 km radius. *A. dorsata* continues foraging even after sunset in bright moonlight (Wongsiri et al. 1991), due to large concentrations of visual pigment in the retinular cells of their ommatidia (Suwannapong and Wongsiri 1999).

12.4 NATURAL AND AGRICULTURAL LANDSCAPES

Natural landscapes consist of landforms such as mountains, hills, plains and plateaus; lakes, streams and natural vegetation are other characteristics of natural landscapes. Agricultural landscapes are the territories designed for agricultural production. These habitats may be field margins, hedge rows, woodlands, ponds, ditches and fallow farm fields (Corbet 1995). However, most landscapes are characterized by varying proportions of natural, semi-natural, agricultural and urban areas. The transformation of landscapes and increase in human land use made negative impacts on landscape biodiversity due to intensive agriculture (Shackelford et al.2013) and threaten crop pollinators (Martin et al. 2013). The size of semi-natural habitats has been reported to affect bee abundance in adjacent agricultural landscapes (Liow et al. 2001).

The biodiversity and populations of insect pollinators including honeybees are under decline in agricultural landscapes (Freitas et al. 2009), causing serious threat to the stability and yield of food crops (Aizen and Harder 2009). The decline in the bee population may be due to degradation and fragmentation of bee habitats and use of agricultural pesticides (Johnson et al. 2010). However, information on nesting biology and floral resources of honeybees in general and *A. dorsata* in particular in natural and agricultural landscapes are meagre despite these bee species being seasonally well distributed in both landscapes by continuous exploitation of available floral resources. Further, an understanding of foraging biology *A. dorsata* is important in tropical ecosystems (Bawa 1990) where the majority of the forest trees and agricultural crops depend on these bee species for pollination (Dainese et al. 2019).

12.5 FLORAL RESOURCES IN NATURAL LANDSCAPES

Apis dorsata is well distributed in southern Asia, covering both natural and agricultural landscapes. They are found abundantly in rural, urban and suburban plains across the Indian sub-continent (Nagaraja 2019). *A. dorsata* immigrates to plains during onset of monsoon season during July–August and rapidly establish their colonies. The widespread distribution of *A. dorsata* colonies in rural regions which are agricultural landscapes is due to availability of pollen and nectar from agricultural and horticultural bee flora during late rainy and winter seasons. The major floral resources of *A. dorsata* in both natural and agricultural landscapes are presented in Table 12.1. During winter, *A. dorsata* foragers collect huge amounts of pollen and nectar from the agricultural and horticultural crop plants. The guava (*Psidium guajava*)

TABLE 12.1
Pollen and Nectar Flora of Giant Honeybee *Apis dorsata* from Natural and Agricultural Landscapes in India

Common Name	Scientific Name	Family	Flowering Period
Onion	*Allium cepa* L	Liliaceae	Jan.–March
Cashew nut	*Anacardium occidentale* L.	Anacardiaceae	Feb.–April
Coral creeper	*Antigonon leptopus* Hook and Arn.	Polygonaceae	April–May
Toria	*Brassica campestris* L. var. toria)	Cruciferae	Oct.–Nov.
Indian mustard	*Brassica juncea* Czern. & Coss)	Cruciferae	Dec.–Feb.
Rape	*Brassica napus* L.	Cruciferae	Dec.–Feb.
Cauliflower	*Brassica oleracea* L. var. *botrytis*	Cruciferae	Oct.–Dec.
Turnip	*Brassica rapa* L.	Cruciferae	Oct.–Dec.
Black mustard	*Brassica nigra* L.	Cruciferae	Oct.–Dec.
Pigeon pea	*Cajanus cajan* (L.) Millsp	Fabaceae	Sept.–Oct.
Papaya	*Carica papaya* L.	Caricaceae	Feb.–March
Safflower	*Carthamus tinctorius* L.	Compositae	June–July

(*Continued*)

TABLE 12.1 ***(Continued)***
Pollen and Nectar Flora of Giant Honeybee *Apis dorsata* from Natural and Agricultural Landscapes in India

Common Name	Scientific Name	Family	Flowering Period
Chicory	*Cichorium intybus* L.	Compositae	June–Sept.
Watermelon	*Citrullus vulgaris* Schrader	Cucurbitaceae	July–Aug.
Lemon	*Citrus limon* (L) Burm. f.)	Rutaceae	March–April
Coconut	*Cocos nucifera* L.	Arecaceae	Jan.–Dec.
Coffee	*Coffea arabica* L.	Rubiaceae	April–May
Coriander	*Coriandrum sativum* L.	Umbelliferae	Jan.–Dec.
Gherkin	*Cucumis anguria* L	Cucurbitaceae	Jan.–Dec.
Musk melon	*Cucumis melo* L	Cucurbitaceae	Jan.–Dec.
Cucumber	*Cucumis sativus* L.	Cucurbitaceae	Jan.–Dec.
Pumpkin	*Cucurbita pepo* L	Cucurbitaceae	Jan.–Dec.
Shisham	*Dalbergia sissoo* Dc	Fabaceae	March–June
Carrot	*Daucus carota* L.	Apiaceae	June–Aug.
Cardamom	*Elettaria cardamomum* (L).	Zingiberaceae	July–Sept.
Eucalyptus	*Eucalyptus* spp	Myrtaceae	Oct.–Dec.
Niger	*Guizotia abyssinica* Cass	Compositae	Sept.–Nov.
Sunflower	*Helianthus annuus* L.	Compositae	Jan.–Dec.
Litchi	*Litchi chinensis* Sonner	Sapindaceae	Aug.–Sept.
Mahua	*Madhuca longifolia* (Koenig.)	Sapotaceae	Feb.–April
Apple	*Malus domestica* Borkh	Rosaceae	April–May
Mango	*Mangifera indica* L.	Anacardiaceae	Feb.–March
Bitter gourd	*Momordica charantia* L.	Cucurbitaceae	Aug.–Oct.
Drumstick	*Moringa oleifera* Lamk.	Moringaceae	Feb.–April
Alfalfa	*Medicago sativa* L.	Fabaceae	March–Oct.
Copper pod	*Peltophorum pterocarpum* (Dc)	Fabaceae	April–June
Pongamia	*Pongamia pinnata* (L) Pierre	Fabaceae	Feb.–May
Almond	*Prunus amygdalus* Batsch	Rosaceae	Feb.–March
Apricot	*Prunus armeniaca* L.	Rosaceae	Mar–April
Plum	*Prunus domestica* L.	Rosaceae	Feb–April
Guava	*Psidium guajava* L.	Myrtaceae	March–May
Pomegranate	*Punica granatum* L.	Punicaceae	April–June
Pear	*Pyrus communis* L.	Rosaceae	March–April
Peach	*Prunus persica* (L.) Stokes	Rosaceae	Feb.–March
Radish	*Raphanus sativus* L.	Cruciferae	Jan.–Aug.
Sesame	*Sesamum indicum* L	Pedaliaceae	Aug.–Oct.
Brinjal	*Solanum melongena* Aubergine	Solanaceae	July–Aug.
Tamarind	*Tamarindus indica* L.	Fabaceae	May–July
Arjun tree	*Terminalia arjuna* (Roxb.) Wight & Arn	Combretaceae	Feb.–April
Clover	*Trifolium alexandrinum* L.	Fabaceae	March–May

FIGURE 12.2 Foraging of *Apis dorsata* worker bees on guava flowers.

is pollinated by different species of honeybees including *A. dorsata* (Figure 12.2). The foraging activity of honeybees was found to be maximum during morning hours of the day, followed by evening hours. The increased foraging activity during morning hours was due to the availability of moist pollen and large quantities of nectar.

Apis dorsata can persist and even thrive in urban and suburban areas with support of bee flora from the gardens, parks, lawns and other green spaces. Availability of abundant suitable nesting structures such as high-rise buildings, water tanks and large trees attract huge number of *A. dorsata* colonies to urban regions during October–December. In metropolitan cities like Bengaluru, India, plenty of floral resources are available throughout the year from the ornamental plants grown in gardens. The trees such as Pelptophorum and Eucalyptus on the roadsides provide pollen and nectar during March–June. Even a single tree of Peltophorum can produce thousands of flowers and copious quantities of pollen and nectar with high sugar content (Somme et al. 2016). The important forest flora found in urban regions was *Eucalyptus* spp., *Tamarindus indica*, *Peltophorum pterocarpum*, etc. (Table 12.1). The mango is a major horticultural crop in rural plains of Karnataka, and is an important source of nectar for honeybees (Figure 12.3). Greater protection from wind, slashing rains and sunlight also results in greater abundance of nests on buildings in urban regions (Nagaraja 2012). Establishing plants of suitable woody landscapes to provide better food for bees should be part of any strategy to conserve and restore urban pollinators.

Roy et al. (2011) found that bee forage and availability of nesting sites are major factors influencing the distribution of *A. dorsata* colonies. The giant honeybee nests were most abundant in the areas of agro-forestry in contrast to nests found in forest fragments (Pavageau et al. 2018). Abrol (1985) found that *A. dorsata* visited sucrose

FIGURE 12.3 Blooming of mango flowers during summer in rural plains.

dominated flowers such as *Cajanus cajan* and *Pongamia pinnata* which provide large amounts of nectar with caloric rewards. *A. dorsata* spends more energy during foraging, and its foraging cost stands higher than *A. florea*. The reward system developed by the flowers enable pollinators to discriminate between closely related plant species. Various species of weeds are also excellent sources of pollen and nectar for honeybees during dearth periods when agro-horticultural crops were not flowering.

12.6 FLORAL RESOURCES IN AGRICULTURAL LANDSCAPES

According to an estimate, about 35% of world food production is derived from agricultural products that are dependent on pollinators (Klein et al. 2007). Agricultural landscapes are characterized by high spatial and temporal dynamics of cropping patterns whereby floral resources are often available only for short periods. However, the tropical agro-forestry landscapes are often characterized by a high spatial heterogeneity of land covers with limited agricultural intensification. These heterogenous landscapes are known to provide a wide variety of habitat conditions and resources for honeybees (Williams and Kremen 2007). Crop pollination by honeybees is one of the important ecosystem services in agricultural landscapes in food production (Bentrup et al. 2019). To increase pollination in agricultural landscapes, it is critical to understand how wild bees like *A. dorsata* are affected by habitat quality and landscape structure (Olsson et al. 2015).

Migration allows honeybee colonies to track seasonally varying resources in different regions. *A. dorsata* colonies typically migrate to take advantage at different altitudes of the flowering plants that provide their diet of nectar and pollen. *A. dorsata* seems to provide more effective and stable pollination services in agricultural landscapes (Garibaldi et al. 2013). They migrate to agricultural fields during flowering season of agricultural crops. However, the abundance and diversity of *A. dorsata* in agricultural regions depends on the distance between their nesting sites and patches of foraging regions (Kline and Joshi 2020). In the plains of Karnataka, these bees gather large quantities of pollen and nectar from varieties of agricultural crops and build their colonies in a rapid way. These crops provide floral resources to bees from September–December. Later, they move to forest habitats which provide a high abundance and diversity of resources for *A. dorsata*.

In south Indian coffee landscapes, short forest patches were found around agro-forest coffee plantations. These forest fragments provide substantial amounts of nesting habitat for *A. dorsata* colonies (Thomas et al. 2009). It is obvious that more nests of giant honeybees are found in agro-forest regions than in natural forest patches (Krishnan et al. 2012). Forest fragments experience lower anthropogenic disturbance and have higher density and diversity of suitable large nesting trees than do coffee plantations (Bhagwat et al. 2005). The diversity and abundance of trees in coffee plantations and forest fragments might ensure spatio-temporal stability of pollen and nectar availability. *A. dorsata* nests were found close to natural forest patches, water bodies and built-up areas around coffee plantations. They respond positively to spatially separated resources and are not affected by fragmentation of land cover.

The nests of *A. dorsata* tend to be few when the tree-covered patches at broader scale are insufficient, whereas the nests are numerous when the amount of agro-forest patches are present within the bees' foraging range. In addition, unlike forest fragments, coffee plants are characterized by temporary pulses of floral resources over few weeks in February and March. This readily available and abundant resource has been shown to coincide with the migration and settlement of *A. dorsata* nests between January and June (Krishnan 2011). Similar studies have shown the importance of the availability of rich food sources for large aggregations of *A. dorsata* during the migration and nest establishment period (Singh et al. 2007). Even though application of agricultural pesticides on coffee plants have only limited impact on coffee pollination by honeybee pollinators, Decourtye et al. (2010) reported on serious losses of honeybee populations in Europe and North America were attributed to lack of diverse nectar and pollen resources in intensively farmed agricultural landscapes. The same tendency may be observed in pollination of varieties of crops grown in agricultural landscapes.

When the nesting structures in a landscape are limited, fragmentation forces pollinators to devote more energy for foraging (Maurer et al. 2020). Landscapes with high amounts of habitat and low degrees of fragmentation have shown higher pollinator visitation. Joshi et al. (2016) showed that higher number of patches increased pollination efficiency because these patches provide more nesting habitat and floral resources for the bees. Saturni et al. (2016) showed that forest cover had a positive effect on the diversity and abundance of bees and that the diversity and abundance of bees increased with increasing area of forest patches. If the coffee estates are away from more than 800 m from forest patches, the abundance of pollinators and visiting rates were reduced (Ricketts et al. 2006). Bee diversity, visiting rate and pollen accumulation rates were significantly higher in coffee fields near forest patches. Similarly, Saturni et al. (2016) showed that the effects of landscape configuration on bee visiting rate on small scales were more detectable for apple orchards. Optimal foraging theory states that, in a landscape, with uniform distribution of resource patches, bees forage near their nesting areas to reduce energy and increase foraging efficiency (Heinrich 2004; Rahimi et al. 2021).

12.7 CONCLUSIONS

Apis dorsata exploits pollen and nectar resources more or less equally from both natural and agricultural landscapes through seasonal migrations. To maintain large bee populations by repeated migrations, it is important to ensure the conservation and management of bee flora in agricultural and surrounding natural landscapes (Zhang et al. 2007). Habitat loss due to deforestation and traditional honey hunting practices leads to declines in diversity and abundance of *A. dorsata* colonies. Therefore, preserving natural areas with diverse bee flora may enhance the population of *A. dorsata* colonies in natural and agricultural landscapes (Aslan 2016). Furthermore, creating new forest patches around agricultural fields also increases the pollination rate and crop production by these giant honeybee species.

ACKNOWLEDGEMENTS

The authors are grateful to Bangalore University, Bengaluru, for providing research facilities.

REFERENCES

Abrol, D.P. 1985. *Analysis of biophysical interactions in causing foraging behaviour of some bees-A study in bioenergetics*. Ph.D thesis, Haryana Agricultural University, Hisar, 287.

Abrol, D.P. 2011. Foraging. In: Hepburn, H.R., and Radloff, S.E. (Eds.), *Honeybees of Asia*. Springer-Verlag, Berlin, 257–292.

Abrol, D.P. 2016. Foraging strategies in honeybees, *Apisdorsata* F. and *Apisflorea* F. in relation to availability of energy rewards. *Journal of Apiculture* 31(1): 9–18.

Aizen, M.A., and Harder, L.D. 2009. The global stock of domesticated honeybees is growing slower than agricultural demand for pollination. *Current Biology* 19: 915–918.

Aslan, C.E. 2016. The role of honeybees as pollinators in natural areas. *Natural Areas Journal* 36(4): 478–488.

Bawa, K.S. 1990. Plant-pollinator interactions in tropical rain forests. *Annual Review of Ecology and Systematics* 21: 399–422.

Bentrup, G., Hopwood, J., Adamson, N.L., and Vaughan, M. 2019. Temperate agro-forestry systems and insect pollinators: A review. *Forests* 10(11): 981.

Bhagwat, S.A., Kushalappa, C.G., Williams, P.H., and Brown, N.D. 2005. The role of informal protected areas in maintaining biodiversity in the Western Ghats of India. *Ecological Society* 10(1): 8.

Cariveau, D.P. et al. 2016. The allometry of bee proboscis length and its uses in ecology. *PLoS One* 11: e0151482.

Chantawannakul, P., Williams, G., and Neumann, P. 2018. Asian *beekeeping in the 21st century*. Springer Nature, Singapore, 325.
Corbet, S.A. 1995. Insects, plants and succession: Advantages of long-term set-aside. *Agricultural Ecosystem and Environment* 55: 61–67.
Dainese, M., Martin, E.A., Aizen, M.A., Albrecht, M., Bartomeus, I., Bommarco, R., and Steffan-Dewenter, I.A. 2019. Global synthesis reveals biodiversity-mediated benefits for crop production. *Science Advances* 5: eaax0121.
Decourtye, A., Mader, E., and Desneux, N. 2010. Landscape enhancement of floral resources for honeybees in agro-ecosystems. *Apidologie* 41: 264–277.
Dimou, M., Thrasyvoulou, A., and Tsirakoglou, V. 2006. Efficient use of pollen traps to determine the pollen flora used by honey bees. *Journal of Apicultural Research and Bee World* 45: 42–46.
Dukas, R. 2008. Evolutionary biology of insect learning. *Annual Review of Entomology* 53: 145–160.
Dyer, F.C., and Seeley, T.D. 1987. Interspecific comparisons of endothermy in honeybees (*Apis*); deviations from the expected size related patterns. *Journal of Experimental Biology* 127: 1–26.
Free, J.B. 1970. *Insect pollinators of crops*. Academic Press, London and New York, 544.
Freitas, B.M., Imperatriz-Fonseca, V.L., Medina, L.M., Kleinert, A.D.P., Galetto, L., Nates-Parra, G., and Quezada-Euan, J.J.G. 2009. Diversity, threats and conservation of native bees in the neotropics. *Apidologie* 40: 332–346.
Garibaldi, L.A., Steffan-Dewenter, I., Winfree, R., Aizen, M.A., Bommarco, R., Cunningham, S.A., and Klein, A.M. 2013. Wild pollinators enhance fruit set of crops regardless of honey bee abundance. *Science* 339: 1608–1611.
Greenleaf, S.S., Williams, N.M., Winfree, R., and Kremen, C. 2007. Bee foraging ranges and their relationship to body size. *Oecologia* 153(3): 589–596.
Harrison, J.F., and Fewell, J.H. 2002. Environmental and genetic influences on flight metabolic rate in the honeybee, *Apis mellifera. Journal of Comparative Biochemistry and Physiology A* 133: 323–333.
Heinrich, B. 2004. *Bumblebee economics*. Harvard University Press, Cambridge.
Hopkins, C.Y., Jevans, A.W., Evans and Boch, R. 1969. Occurrence of octadeca-trans-2, cis-9, cis-12-trienoic acid in pollen attractive to the honey bee. *Canadian Journal of Biochemistry* 47: 433–436.
Hung, K.L., Kingston, J.M., Albrecht, M., Holway, D.A., and Kohn, J.R. 2018. The worldwide importance of honey bees as pollinators in natural habitats. *Proceedings of Royal Society B: Biological Sciences* 285(1870): 20172140.
Johnson, R.M., Ellis, M.D., Mullin, C.A., and Frazier, M. 2010. Pesticides and honey bee toxicity, USA. *Apidologie* 41: 312–331.
Joshi, N.K., Otieno, M., Rajotte, E, G., Fleischer, S.J., and Biddinger, D.J. 2016. Proximity to woodland and landscape structure drives pollinator visitation in apple orchard ecosystem. *Frontiers in Ecology and Evolution* 4: 38.
Kevan, P.G., and Baker, H.G. 1983. Insects as flower visitors and pollinators. *Annual Review of Entomology* 28: 407–453.
Klein, A.M., Vaissiere, B.E., Cane, J.H., Steffan-Dewenter, I., Cunningham, S.A., Kremen, C., and Tscharntke, T.D. 2007. Importance of pollinators in changing landscapes for world crops. *Proceedings of Royal Society B: Biological Sciences* 274: 303–313.
Kline, O., and Joshi, N.K. 2020. Mitigating the effects of habitat loss on solitary bees in agricultural ecosystems. *Agriculture* 10(4): 115.
Koul, A.K., Hamal, I.A., and Gupta, S.K. 1989. Pollination mechanism in *Coriandrum sativum* Linn. *Proceedings of Indian Academy of Science (Plant Science)* 99: 509–515.
Krishnan, S. 2011. *Pollinator services and coffee production in a forested landscape mosaic*. Ph.D thesis, ETH, Zurich.

Krishnan, S., Kushalappa, C.G., Shaanker, R.U., and Ghazoul, J. 2012. Status of pollinators and their efficiency in coffee fruit set in a fragmented landscape mosaic in South India. *Basic and Applied Ecology* 13: 277–285.

Liow, L.H., Sodhi, N.S., and Elmquist, T. 2001. Bee diversity along a disturbance gradient in tropical lowland forests of south-east Asia. *Journal of Applied Ecology* 38: 180–192.

Loper, G.M., and Berdel, R.L. 1980. A nutritional bioassay of honeybee brood-rearing potential. *Apidologie* 11: 181–189.

Martin, E.A., Reineking, B., Seo, B., and Steffan-Dewenter, I. 2013. Natural enemy interactions constrain pest control in complex agricultural landscapes. *Proceedings of National Academy of Sciences USA* 110: 5534–5539.

Maurer, C., Bosco, L., Klaus, E. et al. 2020. Habitat amount mediates the effect of fragmentation on a pollinator's reproductive performance, but not on its foraging behaviour. *Oecologia* 193: 523–534.

Michener, C.D. 1990. Classification of the Apidae (Hymenoptera). Appendix: *Trigona genalis* Friese, a hitherto unplaced New Guinea species. *University of Kansas Science Bulletin* 54: 75–163.

Michener, C.D. 2007. *The bees of the world*. 2nd edn. The Johns Hopkins University Press, Baltimore and London.

Mukherjee, R., Deb, R., and Devy, S.M. 2019. Diversity matters: Effects of density compensation in pollination service during rainfall shift. *Ecology and Evolution* 9(17): 9701–11.

Nagaraja, N. 2019. Nesting patterns of giant honeybee, *Apisdorsata* in plains of Karnataka, India. *Journal of Entomological Research* 43(4): 503–508.

Nagaraja, N., and Rajagopal, D. 2009. *Diseases, parasites, pests, predators and their management*. MJP Publishers, Chennai, India, 210.

Nagaraja, N. 2012. Asian honeybees: Biology, threats and their conservation. In: Florio, R.M. (Ed.), *Biology, threats and colonies*. Nova Publishers Inc., New York, 99–123.

Neupane, K.R., and Thapa, R.B. 2005. Pollen collection and brood production by honeybees (*Apis mellifera* L.) under Chitwan condition of Nepal. *Journal of Institute of Agriculture and Animal Science* 26: 143–148.

Oldroyd, B.P., Osborne, K.E., and Mardan, M. 2000. Colony relatedness in aggregations of *Apis dorsata* Fabricius (Hymenoptera: Apidae). *Insectes Sociaux* 47: 94–95.

Olsson, O., Bolin, A., Smith, H.G., and Lonsdorf, E.V. 2015. Modelling pollinating bee visitation rates in heterogeneous landscapes from foraging theory. *Ecological Modelling* 316: 133–143.

Page, R.E., and Erber, J. 2002. Levels of behavioral organization and the evolution of division of labor. *Naturwissenschaften* 89: 91–106.

Pavageau, C., Gaucherel, C., Garcia, C., and Guazoul, J. 2018. Nesting sites of giant honeybees modulated by landscape patterns. *Journal of Applied Ecology* 55: 1230–1240.

Rahimi, E., Barghjelveh, S., and Dong, P. 2021. Using the Lonsdorf model for estimating habitat loss and fragmentation effects on pollination service. *Ecological Process* 10(2).

Rajagopal, D., and Nagaraja, N. 1999. Beekeeping status in Karnataka, India. *Asian Bee Journal* 1: 50–59.

Requier, F., Jowanowitsch, K., Kallnik, K., and Steffan-Dewenter, I. 2019. Data from: Limitation of complementary resources affects colony growth, foraging behavior and reproduction in bumble bees. *Ecology* 101(3): e02946.

Ricketts, T.H., Williams, N.M., and Mayfeld, M.M. 2006. Connectivity and ecosystem services: Crop pollination in agricultural landscapes. In: Crooks, K., and Sanjayan, M. (Eds.), *Connectivity conservation*. Cambridge University Press, New York, 255–290.

Roy, P., Leo, R., Thomas, S.G., Varghese, A., Sharma, K., Prasad, S., and Davidar, P. 2011. Nesting requirements of the rock bee *Apisdorsata* in the Nilgiribiosphere reserve, India. *Tropical Ecology* 52: 285–291.

Saturni, F.T., Jafe, R., and Metzger, J.P. 2016. Landscape structure influences bee community and coffee pollination at different spatial scales. *Agriculture, Ecosystems and Environment* 235: 1–12.

Shackelford, G., Steward, P.R., Benton, T.G., Kunin, W.E., Potts, S.G., Biesmeijer, J.C., and Sait, S.M. 2013. Comparison of pollinators and natural enemies: A meta-analysis of landscape and local effects on abundance and richness in crops. *Biological Reviews* 88: 1002–1021.

Sihag, R.C. 2000. Ecobiology of the giant honeybee (*Apisdorsata* F.) in semi-arid subtropical climates of India. In: Matsuka, M. (Ed.), *Asian bees and beekeeping: Progress of research and development*. Oxford University Press, IBH, New Delhi, 50–52.

Sihag, R.C. 2014. Phenology of migration and decline in colony numbers and crop hosts of giant honeybee (*Apisdorsata* F.) in semiarid environment of Northwest India. *Journal of Insects* 9.

Singh, R.P., Singh, A.K., and Singh, R.P. 2007. The effect of the availability of bee forage and environmental conditions on the nesting of *Apisdorsata* Fabr. *Journal of Apicultural Research* 46: 276–281.

Somme, L., Moquet, L., Quinet, M. et al. 2016. Food all in row: Urban trees offer valuable floral resources to pollinating insects. *Urban Ecosystems* 19: 1149–1161.

Stewart, A.B., Sritongchuay, T., Teartisup, P., Kaewsomboon, S., and Bumrungsri, S. 2018. Habitat and landscape factors influence pollinators in a tropical megacity, Bangkok, Thailand. *Peer J.* 6: e5335.

Streinzer, M., Brockmann, A., Nagaraja, N., and Spaethe, J. 2013. Sex and caste-specific variation in compound eye morphology of five honeybee species. *PLoS One* 8(2): e57702.

Suwannapong, G., and Wongsiri, S. 1999. Ultrastructure of the compound eyes of the giant honey bee queens, *Apisdorsata* Fabricius, 1793. *Journal STREC* 7(1–2): 60–68.

Thimmegowda, G.G., Mullen, S., Sottilare, K., Sharma, A., Mohanta, S.S., Brockmann, A., Dhandapany, P.S., and Olsson, S.B. 2020. A field-based quantitative analysis of sublethal effects of air pollution on pollinators. *Proceedings of National Academy of Sciences USA* 117(34): 20653–20661.

Thomas, S.G., Varghese, A., Roy, P., Bradbear, N., Potts, S.G., and Davidar, P. 2009. Characteristics of trees used as nest sites by *Apisdorsata* (Hymenoptera, Apidae) in the Nilgiribiosphere reserve, India. *Journal of Tropical Ecology* 25: 559–562.

Underwood, B.A. 1991. Thermo regulation and energetic decision-making by the honeybees *Apiscerana*, *Apisdorsata* and *Apislaboriosa*. *Journal of Experimental Biology* 157(1): 19–34.

von Frisch, K. 1967. *The dance language and orientation of bees*. Harvard University Press, Cambridge, MA.

Williams, N.M., and Kremen, C. 2007. Resource distributions among habitats determine solitary bee offspring production in a mosaic landscape. *Ecological Applications* 17: 910–921.

Wongsiri, S., Rinderer, T.E., and Sylvester, H.A. 1991. *Biodiversity of honeybees in Thailand*. Prachachon, Bangkok, 50–63.

Zhang, W., Ricketts, T.H., Kremen, C., Carney, K., and Swinton, S.M. 2007. Ecosystem services and dis-services to agriculture. *Ecological Economics* 64: 253–260.

13 Safety of Giant Honeybee *Apis dorsata* in Relation with Agricultural Pest Management

Saeed Mohamadzade Namin, Tekalign Bergna and Chuleui Jung

13.1 INTRODUCTION

Pollination is a key ecological service that is crucial for sustainable food production. Among insects, wild and domesticated honeybees play an important role in pollination of wild flowering plants and agricultural crops (Gill et al. 2016; Ballantyne et al. 2017). Crop yield decline of more than 90% was reported mainly in some fruit, seed, and nut crops in the absence of pollinators (Klein et al. 2007). The giant honeybee *Apis dorsata* F. is native to South and Southeast Asia and distributed from Pakistan in the west, the Indian subcontinent and Sri Lanka to Indonesia, southern parts of China, and the Philippines in the east (Oldroyd and Wongsiri 2006). It is similar to the Himalayan giant honeybee *A. laboriosa* in their body size, but differs in their thoracic hair color which is tawny yellow in *A. laboriosa* and mostly dark in A. dorsata. In addition, the first two abdominal tergites in *A. laboriosa* are black (gray in callow adults) whereas they are orange-brown (pale yellow in callows) in *A. dorsata* (Sakagami et al. 1980; Kitnya et al. 2020). Even though *A. dorsata* is not domesticated and cannot be maintained for honey harvest or pollination purposes, it plays an enormous role in the pollination of tropical rainforest plants and local crops (Suwannapong et al. 2012; Wongsiri et al. 2000; Corlett 2011; Partap 2011; Ali et al. 2021).

The availability of natural insect pollinators—specifically, honeybees—is in decline, rapidly raising concerns about the sustainability of the food security and biodiversity of flowering plants. Although some factors—such as habitat degradation and fragmentation, land-use changes, climate changes, air pollution, and appearance of invasive pests and diseases—have significant effect in reduction of honeybee populations, bee poisoning due to the extensive use of pesticides remains one of the most important factor in the decline of pollinators (Johansen 1977; Atkins 1992). Pesticides are agrochemicals widely used to control a wide range of damaging pests, pathogens, and weeds (Lehmann and Camp 2021). While collecting nectar and pollens to maintain the colony, foragers are exposed to toxic materials directly in

DOI: 10.1201/9781003294078-13

agricultural or natural ecosystems. Collection of pollen and nectar from pesticide contaminated flowers is another route of exposure which usually have sub-lethal effect on adults and immature stages of honeybees.

Managed honeybee *Apis mellifera* is one of the model organism to assess the negative impact of pesticides on the ecological environment. However, while the shortage of availability of data on the impact of pesticides like neonicotinoids is suspected to be the cause of honeybee colony decline for *A. dorsata*, dependence on the *A. mellifera* as a replacement for risk assessment has some drawbacks. Bearing in mind that with the differences in size, lifestyle, pesticide sensitivity, and exposure between *A. mellifera* and *A. dorsata*, it is very important to consider the safety of this giant honeybee for greater honey production and effective cross-pollination (Nagaraja 2016). Therefore, we evaluate in this chapter the risk assessment of agrochemical application in agroecosystem and its adverse effect on giant honeybees. Additionally, the relatedness of *A. dorsata* biological features which are important in their toxicity, the detoxification mechanisms, and the integrated pest and pollinator management (IPPM) strategies to protect and conserve *A. dorsata* against agrochemicals are discussed.

13.2 INSECTICIDES AND THEIR MODES OF ACTION

About two million tons of pesticides are being used annually worldwide (Sharma et al. 2019). The misuse of pesticides against a pest species developed resistance that results from the Darwinian selection of resistant forms of the pest and the consequent evolution of populations that are resistant to those pesticides (Sparks and Nauen 2015). To control the damaging pests effectively and protect the environment from contamination, pesticide management according to their target site and modes of action is very crucial. Insecticides are chemicals formulated to control one or more pest species that packed in sprays, dusts, gels, or bait forms. Insecticides possess different modes of action that poses various risk to non-target pollinator.

Nicotinic acetylcholine receptor (nAChR) modulators: Neonicotinoids, sulfoximines, butenolides, and mesoionics target nAChR of insect pests. Neonicotinoids (NNIs) are synthetic analogs of nicotine that activate the nicotinic acetylcholine receptor (nAChR), classified into nitro- ($-NO_2$) or cyano- ($-CN$) depending on their toxicity. The nitro-NNIs—imidacloprid, clothianidin, thiamethoxam, dinotefuran and nitenpryram—are highly toxic, with acute contact LD_{50} ranges from 0.004–0.14 μg/bee, while the cyano-NNIs—thiacloprid and acetamiprid—are much less toxic with acute contact LD_{50} values of 7.1–14.6 μg/bee. The difference is likely related to rapid cytochrome P450 detoxification (Iwasa et al. 2004). Sulfoximines and butenolides are relatively new class of insecticides targeting nAChR. Sulfoxaflor is the initial compound commercialized from sulfoximine insecticide. Flupyradifurone is another new insecticide class first represented novel butenolide. Both Sulfoxaflor and flupyradifurone are relatively less toxic than some of nitro-NNI with LD_{50} of 0.05 μg/bee and 1.2 μg/bee (Azpiazu et al. 2021; Campbell et al. 2016).

Acetylcholinesterase inhibitors: Relatively old classes of insecticides target the nervous system by inhibiting acetylcholinesterase (AChE) activity. These include carbamates and organophosphates, which have a broad spectrum of effects that poses high risk to honeybees. They are also absorbed and accumulated in the beeswax and pose a greater risk for entire honeybee populations. Benfuracarb, carbaryl, carbosulfan, ethiofenca, and formetanate from carbamate and dimethoate, acephate, chlorpyrifos, and coumaphos from organophosphate are some examples.

GABA-gated Cl– channel agonists: Organochlorines and fiproles insecticide bind to the GABA-gated Cl– channel in insect neurons and lead to hyperpolarization of the neuron and an inability to transmit action potentials (Casida and Durkin 2013). Examples are endosulfan and fipronil. Fipronil is a broad-spectrum insecticide highly toxic to honeybees (oral LD_{50} = 4 μg/bee, Li et al. 2010).

Sodium channel modulator insecticides: Pyrethroids, pyrethrins, and dichlorodiphenyltrichloroethane (DDT) target voltage-gated sodium channels (VGSCs) to intoxicate insects. They differ in action on VGSCs that have no effect at hyperpolarized membrane potentials, but cause a voltage-dependent, nearly irreversible block as the membrane potential is depolarized. This class of insecticides also highly toxic to honeybees, *A. mellifera*. For example, permethrin and etofenprox have LD_{50} = 0.024 μg/bee and 0.015 μg/bee (Piccolomini et al. 2018).

Insect growth regulators (IGRs): IGRs are insecticides related to juvenile hormone (JH) or the ecdysteroids or chitin synthesis. They are more toxic to immature bees than adult bees. For instance, diflubenzuron—a chitin synthesis inhibitor—affects larval survival at concentrations as low as 1 ppm. Whole colonies treated with a field rate of fenoxycarb immediately experienced high mortality of immature bees and reduced brood abundance a year following treatment (Thompson et al. 2005).

Other modes of action of insecticides targeting the nerve action of insects are glutamate-gated chloride channel (GluCl) allosteric modulators and octopamine receptor agonists. Those insecticides are commonly used to control pests in hives or in agricultural crops. Emamectin benzoate, glutamate-gated chloride channel (GluCl) allosteric modulators insecticide, is highly toxic to *A. mellifera*, possessing very low contact LD_{50} value of 0.00006 μg a.i/bee (Abdu-Allah and Pittendrigh 2018). Amitraz is commoly used acaricides moderately toxic to *A. mellifera* with LD_{50} value of 1.99 μg/bee (Gashout et al. 2018).

13.3 ROUTES OF PESTICIDE EXPOSURE TO HONEYBEES

Many different classes of insecticides are used mainly to control pests that damage agricultural crops. Depending on target organisms, pesticides can be applied in a different ways like foliar application, seed covering, and root drench. Spraying is one of the main application methods, which can be applied in agroecosystems as well as forest ecosystems to control agricultural and forest pests. Honeybees can be exposed to pesticides orally or by contacting directly during application to plants, or drinking contaminated water, or collecting pollen or nectar from sprayed plants. Neonicotinoids are broad-spectrum, systemic compounds which exhibit activity against phloem-feeding pests (e.g. aphids, whiteflies, leafhoppers). These pesticides are mostly applied by seed dressing or root drench. Due to their systemic nature, they are readily absorbed and move through the vascular system to growing part of plants to provide long-term protection against piercing-sucking pests (Iwasa et al. 2004). Soil and seed application of pesticides were considered safe methods for non-target organisms such as pollinators and natural enemies; however, due to their systemic nature, they can be found in sub-lethal concentrations in plant nectar and pollen, which can indirectly affect both immature and adults stages.

13.4 PESTICIDE EFFECTS ON HONEYBEES

13.4.1 Lethal Effects

Pesticides sprayed on agricultural crops against different pests and pathogens could affect natural pollinators like *A. dorsata*. Evaluation of toxicity of those pesticides in laboratorys has been very crucial in assessing toxicological safety of pollinators and their relation with plant flowers. In this regard, the acute toxicity of some classes of pesticides with their toxicological endpoint (LD_{50}) is presented in Table 13.1. Except for organochlorine (endosulfan), which is moderately toxic insecticides with LD_{50} value of contact 4.825 μg/bee and oral 3.903 μg/bee, all the remaining insecticides are lethal to *A. dorsata* having LD_{50} value of <1μg/bee (Table 13.1).

13.5 SUSCEPTIBILITY OF HONEYBEES TO PESTICIDES

According to Table 13.1, for both contact and oral exposure, *A. dorsata* was more susceptible than was *A. mellifera* to oranophosphate (dimethoate, malathion, monocrotophos, oxydemiton-methyl, phosphamidon), organochlorine (endosulfan), carbamate (carbaryl), and pyrethriod (cypermethrin, deltamethrin, fenvalerate). Additionally, *A. dorsata* was about 14.7, 320.5, and 31.4 times more tolerant to neonicotinoid (thiametoxam, imidacloprid, and clothianidin, respectively), than *A. mellifera* and 29.3 and 962 times more tolerant to thiametoxam and imidacloprid, respectively, than *A. cerena* (Lee et al. 2016). The contact and oral toxicity of carbaryl is equal in *A. dorsata*, while the oral toxicity for *A. mellifera* is several times more than its contact toxicity (Lee et al. 2016). On the other hand, dimethoate is equally toxic in contact and oral methods to *A. mellifera* but is several times more toxic orally for *A. dorsata*. Variation in the vulnerability among honeybee species can be attributed to different body biomass, fat deposit levels, hemolymph pH, the penetration ability and lipophilic property of bee cuticle, and different detoxification mechanisms due variation in the number of detoxification enzymes (da Silva et al. 2020).

TABLE 13.1
Acute Contact and Oral Toxicity of Different Classes of Pesticides to *A. dorsata* in Laboratory Tests and Relative Tolerance Index Compared with *A. mellifera* (Relative Tolerance Index; $RTI = \frac{LD_{50} \text{ or } LD_{50} \text{ of Apis dorsata}}{LD_{50} \text{ or } LD_{50} \text{ of Apis mellifera}}$)

		LD_{50} (µg/Bee)		LC_{50} (ppm)	RTI		References
Insecticides	Active Ingredients	Contact	Oral	Oral	Contact	Oral	Sihag et al. 1999, 2004
Organophosphate	Dimethoate	0.0549	0.0104		0.42	0.086	Sihag et al. 1999, 2004
	Malathion	0.1165	0.0553		0.47	0.35	Sihag et al. 1999, 2004
	Monocrotophos	0.2580	0.1309		0.78	0.62	Sihag et al. 1999, 2004
	Oxydemiton-methyl	0.2804	0.2103		0.51	0.64	Sihag et al. 1999, 2004
	Phosphamidon	0.0847	0.0299		0.60	0.28	Sihag et al. 1999, 2004
Organochlorine	Endosulfan	4.825	3.9030		0.54	0.64	Sihag et al. 1998, 2004
Carbamate	Carbaryl	0.2364	0.2317		0.13	0.63	Sihag et al. 1998, 2004
Pyrethriod	Cypermethrin		0.0059			0.16	Sihag et al. 2004
	Deltamethrin		0.0027			0.10	Sihag et al. 2004
	Fenvalerate		0.0913			0.49	Sihag et al. 2004
	Thiamethoxam			0.088		14.7	Lee et al. 2016
Neonicotinoids	Imidacloprid			1.923		320.5	Lee et al. 2016
	Clothianidin			0.220		31.4	Lee et al. 2016

13.6 SUB-LETHAL EFFECTS OF PESTICIDES

Pesticide effect on honeybees has been evaluated by short-term acute contact and oral toxicity study, as well as long-term exposure at low doses. The acute toxicity of neonicotinoids and other pesticides has been extensively identified on *A. mellifera*, and the probability of bees exposed to high dose to harm them is very low in the field. However, exposure to mixtures of pesticides can lead to severe sub-lethal effects to honeybees exposed chronically at low doses. The effects from field realistic concentration or doses of pesticide that may have behavioral or physiological impact without causing statistically significant mortality in the experimental population is sub-lethal effect. Effects like motor activity, orientation in the field, feeding, development, learning ability, and memory, weakening the immune and reproductive system, and triggering the antioxidant and detoxification mechanisms are some examples of sub-lethal effects observed in honeybee due to low-dose pesticides exposure. For instance, the total haemocyte counts increases immediate in response to ethofenprox, diafenthiuron, and imidacloprid and decreased for all ethofenprox, bifenthrin, diafenthiuron, endosulfan, and imidacloprid after 60 minutes at field-realistic concentrations on *A. dorsata* (Perveen and Ahmad 2017).

13.7 RESIDUE ANALYSIS

A. dorsata colony death from insecticide poisoning is continuous process throughout the year. However, greater number of colonies mortality was recorded between March–April and August–September. Nagaraja (2016) reported brutal killings of thousands of *A. dorsata* colonies due to pesticides application to remove the colonies from frequent attack of bees in urban regions. The misuse of pesticides in agriculture affects *A. dorsata* and may result in the contamination of the environment. However, the residue analysis from multifloral honey of *A. dorsata* showed pesticide detection limit was 0.01 ppm and most of the honey samples did not show higher concentrations of different pesticides like organochlorine, organophosphorus, pyrethroids, and herbicides. Thus, the combs of *A. dorsata* in the wild are not contaminated by chemical pesticides, and their residues level in multifloral honey was below detection limit and safe for consumption (Raghunandan and Basavarajappa 2013). It is very important to point out the urgent need to establish inclusive monitoring programs in *A. dorsata* colony products due to increasing demands for pesticide use in agriculture.

13.8 INSECTICIDE RESISTANCE AND DETOXIFICATION MECHANISMS

Insecticide resistance is a survival strategy to escape the adverse effect of agrochemicals which is based on either an increase in the levels of detoxification enzymes or is related to reduced target-site sensitivity (Feyereisen 1995; Ffrench-Constant 1999). Insecticide resistance by behavioral escaping has been documented in *A. mellifera* against commonly used pesticides including fungicides, acaricides, and insecticides (Kang and Jung 2017). In this type of resistance, honeybees leave the insecticide-treated area only after making physical contact with the chemical.

Cytochrome P450 monooxygenases (P450), glutathione transferases (GST), and carboxylesterases (COEs) are the major enzyme families that participate in the detoxification process of pesticides in insects (Bergé et al. 1999; Hemingway 2002; Hemingway et al. 2004; Mao et al. 2011). Multiple genes from these families are responsible for detoxification of different pesticides; however, high-throughput microarray analysis is required to understand which gene(s) is associated with the resistance to a particular class of insecticide (Naidoo et al. 2005). Recent studies demonstrated that honeybees and bumblebees harbor limited number of genes encoding for xenobiotic detoxification enzyme families in comparison to other hymenopterans or insect pests of agricultural and medical importance (Claudianos et al. 2006; Berenbaum and Johnson 2015; Sadd et al. 2015). The genome of giant honeybees has not been fully studied yet. Genome analysis of the European honeybee *A. mellifera* indicated high similarity with the genome of bumblebees in number of detoxification-related genes, but hormone- and semiochemical processing–associated CCE genes are reduced, while detoxification-associated P450 genes are increased (Sadd et al. 2015; Jung et al. 2020). Recent toxicology studies revealed the low toxicity of neonicotinoid pesticides on giant honeybees compared with domesticated honeybees, *A. mellifera* and *A. cerana*, which can be possibly attributed to the larger size, higher number of detoxification enzymes, or the diversity of their gut microbiome community.

13.9 BIOLOGICAL FACTORS AND BEE POISONING

Unlike cavity-nesting honeybees, *A. mellifera* and *A. cerana*, giant honeybee colonies are generally found as a single comb about 1–2 m long and approximately 0.5 m high under thick tree branches in rainforests or under cliffs, but they also can be occasionally found in building ledges of urban areas (Lindauer 1961). By the fact that *A. dorsata* make their nests in open areas, they are more vulnerable to environmental disturbances such as climate change, invasive pests, diseases, etc. In addition, increasing usage of agrochemicals more adversely affect honeybees with open nests then cavity-nesting honeybees. Colonies that are close to the agricultural landscape can be exposed to pesticides directly due to the pesticide drift through foliar application. In cavity-nesting honeybees, only foragers can be exposed to the high concentration of pesticides directly; however, due to the open nest structure in *A. dorsata*, all stages of giant honeybee (immature, newly emerged adults and queen, and foragers) are at the risk of being exposed by higher concentrations making them more vulnerable to pesticide application.

Due to the negative consequences of pesticides such as their dramatic effect on non-target organisms, it has been recommended to apply pesticide in late afternoon or at night. However, unlike other species of honeybees, *A. dorsata* has the unusual habit of continuing to forage after sunset on bright moonlit nights (Young et al. 2021), a nocturnal ability that is not reported in other honeybee species (Suwannapong and Wongsiri 1999; Suwannapong et al. 2010). *A. dorsata* foragers can fly at night in part because of large concentrations of visual pigment in the retinular cells of their ommatidia (Suwannapong and Wongsiri 1999). This ability makes them more sensitive to the pesticides in comparison to other honeybee species.

The much larger body size provides the opportunity of having up to 3.8 km foraging distance, approximately double that of other honeybee species such as *A. florea* (~ 1.3 km) or *A. cerana* (1–2.5 km) (Dyer and Seeley 1991; Dhaliwai and Sharma 1974; Hisashi 2010). Having this biological feature makes them capable of collecting pollen and nectar from distant flowering plant resources; however, this makes them susceptible to exposure to pesticides in a larger landscape, as well.

13.10 IPPM STRATEGIES FOR PROTECTING *A. DORSATA*

Due to the hazardous effects of insecticides on non-target organisms—specifically, natural enemies and pollinators—implementation of eco-friendly strategies are crucial to mitigate the adverse effect of pesticides. These scientifically based strategies—include cultural, physical, biological, and chemical control methods—are called integrated pest management (IPM) (Barzman et al. 2015). Although many IPM strategies are pollinator-friendly, some specific curation of pollinators are required which is beyond the target of IPM programs. To meet these requirements, integrated pest and pollinator management (IPPM) was introduced and a systematic framework for this concept was recently proposed (Biddinger and Rajotte 2015; Egan et al. 2020;

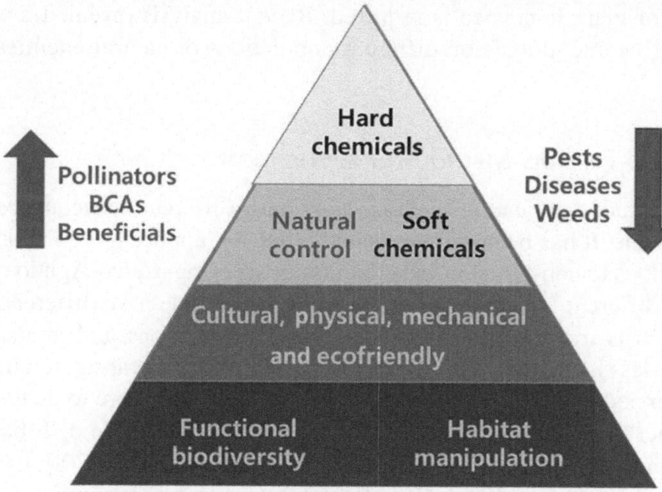

FIGURE 13.1 Conceptual presentation of integrated pollinator-pest management system adopted from Jung 2021 and Egan et al. 2020.

Jung 2021; Lundin et al. 2021). For this, ecological engineering of creating habitats for pollinators, landscape management and agroecosystem diversification—as well as selective soft pesticide uses on time guided by pest monitoring and phenology, and targeting delivery—are further proposed (Figure 13.1). The recent shift of agroecosystems from climate change and new pest outbreaks requires new paradigm of pest management for sustainable agricultural production. Any management strategy leading to decrease of either the amount or number of applications of pesticides will have positive effect on domesticated and wild pollinators.

13.10.1 Reduction of Pesticide Application

Using alternative eco-friendly and selective pest management strategies is recommended to mitigate the adverse effects of pesticides. Regular monitoring of the pest population and application of pesticide when the population level is above economic threshold level is helpful in reduction of pesticide application (Furlan and Kreutzweiser 2015). On the other hand, more studies on the cost-effectiveness of using pesticides are crucial, since recent analysis demonstrated that seed-coating by neonicotinoids is often not necessary and may not always contribute effectively to yield gain (Goulson 2013).

13.10.2 Low-Toxic Pesticides

Pesticides are extremely variable in their toxicity on different organisms, their durability, and the range of plants and animals that are affected in the application. More study of the lethal and sub-lethal toxicity of different pesticides on different populations of giant honeybee is required. Recent analysis revealed a wide range of toxicity of neonicotinoids on different populations of natural enemies (Balanza et al. 2019).

13.10.3 Eco-Friendly Methods of Application

Different pesticide application methods also have different consequences on non-target organisms. It has been shown recently that soil application on neonicotinoids is less hazardous to non-target insects than foliar spraying (Calvo-Agudo et al. 2019). In addition, different formulations of the same insecticide have different effects on honeybees. Dusts are more hazardous than liquid formulations, and emulsifiable concentrates are less hazardous than wettable powders. Microencapsulated insecticides, which are carried by honeybees to their colonies, are much more toxic to honeybees than any other formulation (Sanford 2003; Abrol 2011). Pesticide drift has dramatic effect on honeybee foragers (Barnett et al. 2007; Mineau et al. 2008). Since giant honeybees can be more or less active after sunset, evening spraying may not always reduce the negative effect of insecticides on *A. dorsata*; however, other pollinator-friendly strategies such as avoiding spraying during windy weather, low-pressure spraying, higher water volumes during application, and spraying close to the crop surface would reduce the risk of exposure (Craig et al. 1998; Beekman and Ratnieks 2000; Sanchez-Bayo and Goka 2016).

ACKNOWLEDGEMENT

This study was partly supported by the BSRP through the National Research Foundation of Korea (NRF), Ministry of Education (NRF-2018R1A6A1A03024862), and Rural Development Administration agenda (PJ01577802) on honeybee and pesticide interaction.

REFERENCES

Abdu-Allah, G.A., and Pittendrigh, B.R. 2018. Lethal and sub-lethal effects of select macrocyclic lactones insecticides on forager worker honey bees under laboratory experimental conditions. *Ecotoxicology* 27(1): 81–88.

Abrol, D.P. 2011. *Pollination biology: Biodiversity conservation and agricultural production*. Springer, Dordrecht, The Netherlands.

Ali, H., Iqbal, J., Raweh, H.S., and Alqarni, A.S. 2021. Proboscis behavioral response of four honey bee Apis species towards different concentrations of sucrose, glucose, and fructose. *Saudi Journal of Biological Sciences* 28(6): 3275–3283.

Atkins, P.W. 1992. *General chemistry*. Scientific American Books, New York.

Azpiazu, C., Bosch, J., Bortolotti, L., Medrzycki, P., Teper, D., Molowny-Horas, R., and Sgolastra, F. 2021. Toxicity of the insecticide sulfoxaflor alone and in combination with the fungicide fluxapyroxad in three bee species. *Scientific Reports* 11(1): 1–9.

Balanza, V., Mendoza, J.E., and Bielza, P. 2019. Variation in susceptibility and selection for resistance to imidacloprid and thiamethoxam in Mediterranean populations of Orius laevigatus. *Entomologia experimentalis et Applicata* 167(7): 625–635.

Ballantyne, G., Baldock, K.C.R., Rendell, L., and Willmer, P.G. 2017. Pollinator importance networks illustrate the crucial value of bees in a highly speciose plant community. *Scientific Reports* 7: 8389.

Barnett, E.A., Charlton, A.J., and Fletcher, M.R. 2007. Incidents of bee poisoning with pesticides in the United Kingdom, 1994–2003. *Pest Management Science* 63(11): 1051–1057.

Barzman, M., Bàrberi, P., Birch, A.N.E., Boonekamp, P., Dachbrodt-Saaydeh, S., Graf, B., Hommel, B., Jensen, J.E., Kiss, J., Kudsk, P. et al. 2015. Eight principles of integrated pest management. *Agron. Sustain. Dev.* 35: 1199–1215.

Beekman, M., and Ratnieks, F.L.W. 2000. Long-range foraging by the honey-bee, Apis mellifera L. *Functional Ecology* 14(4): 490–496.

Berenbaum, M.R., and Johnson, R.M. 2015. Xenobiotic detoxification pathways in honey bees. *Current Opinion in Insect Science* 10: 51–58.

Bergé, J.-P., Feyereisen, R., and Amichot, M. 1999. Cytochrome P450 monooxygenases and insecticide resistance. In: Denholm, I., Pickett, J.A., and Devonshire, A.L. (Eds.), *Insecticide resistance: From mechanisms to management.* CABI Publishing, Wallingford.

Biddinger, D.J., and Rajotte, E.G. 2015. Integrated pest and pollinator management-adding a new dimension to an accepted paradigm. *Curr. Opin. Insect Sci.* 10: 204–209.

Calvo-Agudo, M., González-Cabrera, J., Picó, Y., Calatayud-Vernich, P., Urbaneja, A., Dicke, M., and Tena, A. 2019. Neonicotinoids in excretion product of phloem-feeding insects kill beneficial insects. *PNAS* 116(34): 16817–16822.

Campbell, J.W., Cabrera, A.R., Stanley-Stahr, C., and Ellis, J.D. 2016. An evaluation of the honey bee (Hymenoptera: Apidae) safety profile of a new systemic insecticide, flupyradifurone, under field conditions in Florida. *Journal of Economic Entomology* 109(5): 1967–1972.

Casida, J.E., and Durkin, K.A. 2013. Neuroactive insecticides: Targets, selectivity, resistance, and secondary effects. *Annual Review of Entomology* 58: 99–117.

Claudianos, C., Ranson, H., Johnson, R.M., Biswas, S., Schuler, M.A., Berenbaum, M.R. et al. 2006. A deficit of detoxification enzymes: Pesticide sensitivity and environmental response in the honeybee. *Insect Molecular Biology* 15: 615–636.

Corlett, R.T. 2011. Honeybees in natural ecosystems. In: Hepburn, H.R., and Radloff, S.E. (Eds.), *Honeybees of Asia*. Springer-Verlag, Berlin, Germany, 215–226.

Craig, I., Woods, N., and Dorr, G.A. 1998. Simple guide to predicting aircraft spray drift. *Crop Protection* 17(6): 475–482.

da Silva, I.M., Zanuncio, J.C., Brügger, B.P., Soares, M.A., Zanuncio, A.J.V., Wilcken, C.F., Tavares, W.D.S., Serrão, J.E., and Sediyama, C.S. 2020. Selectivity of the botanical compounds to the pollinators *Apis mellifera* and *Trigona hyalinata* (Hymenoptera: Apidae). *Scientific Reports* 10(1): 1–8.

Dhaliwai, H.S., and Sharma, P.L. 1974. Foraging range of the Indian honeybee. *J. Apic Res.* 13: 137–141.

Dyer, F.C., and Seeley, T.D. 1991. Dance dialects and foraging range in three Asian honey bee species. *Behav. Ecol. Sociobiol.* 28: 227–233.

Egan, P.A., Dicks, L.V., Hokkanen, H.M.T., and Stenberg, J.A. 2020. Delivering integrated pest and pollinator management (IPPM). *Trends Plant Sci* 25: 577–589.

Feyereisen, R. 1995. Molecular biology of insecticide resistance. *Toxicol Lett* 82–83: 83–90.

Ffrench-Constant, R.H. 1999. Target site mediated insecticide resistance: What questions remain? *Insect Biochem Mol Biol* 29: 397–403.

Furlan, L., and Kreutzweiser, D. 2015. Alternatives to neonicotinoid insecticedes for pest control: Case studies in agriculture and forestry. *Environ Sci Pollut Res* 22: 135–147.

Gashout, H.A., Goodwin, P.H., and Guzman-Novoa, E. 2018. Lethality of synthetic and natural acaricides to worker honey bees (Apis mellifera) and their impact on the expression of health and detoxification-related genes. *Environmental Science and Pollution Research* 25(34): 34730–34739; Germany 2011, 227–255.

Gill, R.J., Baldock, K.C., Brown, M.J., Cresswell, J.E., Dicks, L.V., Fountain, M.T., Garratt, M.P., Gough, L.A., Heard, M.S., Holland, J.M., and Ollerton, J. 2016. Protecting an ecosystem service: Approaches to understanding and mitigating threats to wild insect pollinators. *Advances in Ecological Research* 54: 135–206.

Goulson, D. 2013. An overview of the environmental risks posed by neonicotinoid insecticides. *J Appl Ecol* 50: 977–987.

Hemingway, J. 2002. The molecular basis of two contrasting metabolic mechanisms of insecticide resistance. *Insect Biochem Mol Biol* 30: 1009–1015.

Hemingway, J., Hawkes, N.J., McCarroll, L., and Ranson, H. 2004. The molecular basis of insecticide resistance in mosquitoes. *Insect Biochem Mol Biol* 34: 653–665.

Hisashi, F. 2010. Profitable beekeeping with *Apis cerana. Bees Dev. J.* 94: 8–10.

Iwasa, T., Motoyama, N., Ambrose, J.T., and Roe, R.M. 2004. Mechanism for the differential toxicity of neonicotinoid insecticides in the honey bee, *Apis mellifera. Crop Protection* 23: 371–378.

Johansen, C.A. 1977. Pesticides and pollinators. *Annual Review of Entomology* 22: 177–192.

Jung, C. 2021. Integrated pollinator-pest management (IPPM) strategy as future Apple IPM. *Korean J. Appl. Entomol.* 60(1): 145–154.

Jung, C., Noor-ul-Ane, M., and Mohamadzade Namin, S. 2020. Safety of dwarf honeybee, Apis florea in relation with agricultural pest management. In: Abrol, D.P. (Ed.), *The future role of dwarf honeybees in natural and agricultural systems.* CRC Press, Boca Raton.

Kang, M., and Jung, C. 2017. Avoidance behavior of honey bee, Apis mellifera from commonly used fungicides, acaricides and insecticides in apple orchards. *Journal of Apiculture* 32(4): 295–302.

Kitnya, N., Prabhudev, M.V., Bhatta, C.P., Pham, T.H., Nidup, T., Megu, K., Chakravorty, J., Brockmann, A., and Otis, G.W. 2020. Geographical distribution of the giant honey bee Apis laboriosa Smith, 1871 (Hymenoptera, Apidae). *ZooKeys* 951: 67–81.

Klein, A.M., Vaissiere, B.E., Cane, J.H., Steffan-Dewenter, I., Cunningham, S.A., Kremen, C., and Tscharntke, T. 2007. Importance of pollinators in changing landscapes for world crops. *Proceedings of the Royal Society B: Biological Sciences* 274(1608): 303–313.

Lee, C., Jeong, S., Jung, C., and Burgett, M. 2016. Acute oral toxicity of neonicotinoid insecticides to four species of honey bee, *Apis florea*, *Apis. cerana*, *Apis. mellifera*, and *Apis dorsata. Journal of Apiculture* 31(1): 51–58.

Lehmann, D.M., and Camp, A.A. 2021. A systematic scoping review of the methodological approaches and effects of pesticide exposure on solitary bees. *PLoS One* 16(5): e0251197.

Li, X., Bao, C., Yang, D., Zheng, M., Li, X., and Tao, S. 2010. Toxicities of fipronil enantiomers to the honeybee *Apismellifera* L. and enantiomeric compositions of fipronil in honey plant flowers. *Environmental Toxicology and Chemistry* 29(1): 127–132.

Lindauer, M. 1961. *Communication among social bees.* Harvard University Press, Cambridge, 346.

Lundin, O., Rundlöf, M., Jonsson, M., Bommarco, R., and Williams, N.M. 2021. Integrated pest and pollinator management-expanding the concept. *Front. Ecol. Environ.* 19: 283–291.

Mao, W., Schuler, M.A., and Berenbaum, M.R. 2011. CYP9Q-mediated detoxification of acaricides in the honey bee (*Apis mellifera*). *Proceedings of the National Academy of Sciences USA* 108: 12657–12662.

Mineau, P., Harding, K.M., Whiteside, M., Fletcher, M.R., Garthwaite, D., and Knopper, L.D. 2008. Using reports of bee mortality in the field to calibrate laboratory-derived pesticide risk indices. *Environmental Entomology* 37(2): 546–554.

Nagaraja, N. 2016. Effect of insecticide poisoning on mortality of giant honeybee, *Apis dorsata* colonies. *PESQUISA-International Journal of Research* 2(1): 96–100.

Naidoo, S., Denby, K.J., and Berger, D.K. 2005. Microarray experiments: Considerations for experimental design. *S Afr J Sci* 101: 347–354.

Oldroyd, B.P., and Wongsiri, S. 2006. *Asian honey bees: Biology, conservation and human interactions*. Harvard University Press, Cambridge, MA.

Partap, U. 2011. The pollination role of honeybees. In: Hepburn, H.R., and Radloff, S.E. (Eds.), *Honeybees of Asia*. Springer-Verlag, Berlin.

Perveen, N., and Ahmad, M. 2017. Toxicity of some insecticides to the haemocytes of giant honeybee, Apis dorsata F. under laboratory conditions. *Saudi Journal of Biological Sciences* 24(5): 1016–1022.

Piccolomini, A.M., Whiten, S.R., Flenniken, M.L., O'Neill, K.M., and Peterson, R.K. 2018. Acute toxicity of permethrin, deltamethrin, and etofenprox to the alfalfa leafcutting bee. *Journal of Economic Entomology* 111(3): 1001–1005.

Raghunandan, K.S., and Basavarajappa, S. 2013. Analysis of multifloral honey of the giant honey bee, *Apis dorsata* F., for pesticide residues in southern Karnataka, India. *European Journal of Zoological Research* 2(3): 22–28.

Sadd, B.M., Barribeau, S.M., Bloch, G., de Graaf, D.C., Dearden, P., Elsik, C.G. et al. 2015. The genomes of two key bumblebee species with primitive eusocial organization. *Genome Biology* 16: 76.

Sakagami, S.F., Matsumura, T., and Ito, K. 1980. Apis laboriosa in Himalaya, the little known world largest honeybee (Hymenoptera: Apidae). *Insecta Matsumurana* 19: 47–77.

Sanchez-Bayo, F., and Goka, K. 2016. Impacts of pesticides on honey bees. In: *Beekeeping and bee conservation – advances in research*. InTech, Rijeka, 77–97.

Sanford, M.T. 2003. *Protecting honeybees from pesticides*. University of Florida Cooperative Extension Service Institute of Food and Agriculture Sciences, CIR534.

Sharma, A., Kumar, V., Shahzad, B., Tanveer, M., Sidhu, G.P.S., Handa, N., Kohli, S.K., Yadav, P., Bali, A.S., Parihar, R.D., and Dar, O.I. 2019. Worldwide pesticide usage and its impacts on ecosystem. *SN Applied Sciences* 1(11): 1–16.

Sihag, R.C., Kumar, S., and Gahlawat, S. 1998. Acute contact toxicity of endosulfan and carbaryl insecticides to two honeybee species, *Apis mellifera* L. and *Apis dorsata* F. *Journal of Apiculture* 13(2): 55–60.

Sihag, R.C., Kumar, S., and Gahlawat, S. 1999. Acute contact toxicity of five organophosphate insecticides to two honeybee species, *Apis mellifera* L. and *Apis dorsata* F. *Journal of Apiculture* 14(1): 1–4.

Sihag, R.C., Kumar, S., and Gahlawat, S.K. 2004. Acute oral toxicity of different insecticides to two honeybee Species, Apis mellifera L. and *Apis dorsata* F. *Journal of Apiculture* 19(2): 125–132.

Sparks, T.C., and Nauen, R. 2015. IRAC: Mode of action classification and insecticide resistance management. *Pesticide Biochemistry and Physiology* 121: 122–128.

Suwannapong, G., Benbow, M.E., and Nieh, J.C. 2012. Biology of Thai hoeybees: Natural history and threats. In: Florio, R.M. (Ed.), *Bees: Biology, threats and colonies*. Nova Science Publishers, Hauppauge, NY.

Suwannapong, G., Chaiwongwattanakul, S., and Benbow, M.E. 2010. Histochemical comparision of the hypopharyngeal gland in *Apiscerana* Fabricius, 1793 and *Apismellifera* Linneaus, 1758 workers. *Psyche: A Journal of Entomology* 7, Article ID 181025. https://doi.org/10.1155/2010/101025.

Suwannapong, G., and Wongsiri, S. 1999. Ultrastructure of the compound eyes of the giant honey bee queens, *Apisdorsata* Fabricius, 1793. *Journal STREC* 7(1–2): 60–68.

Thompson, H.M., Wilkins, S., Battersby, A.H., Waite, R.J., and Wilkinson, D. 2005. The effects of four insect growth-regulating (IGR) insecticides on honeybee (*Apis mellifera*) colony development, queen rearing and drone sperm production. *Ecotoxicology* 14: 757–769.

Wongsiri, S., Chanchao, C., Lekprayoon, C., Wattanasermkit, K., Deowanish, S., and Leepitakrat, S. 2000. *Honeybee diversity and management in the new millennium in Thailand*. Proceeding of the 7th International Conference on Tropical Bees, Chiang Mai, Thailand, 9–14, March 19–25.

Young, A.M., Kodabalagi, S., Brockmann, A., and Dyer, F.C. 2021. A hard day's night: Patterns in the diurnal and nocturnal foraging behavior of *Apisdorsata* across lunar cycles and seasons. *PLoS ONE* 16(10): e0258604.

14 Prospective Use of Giant Honeybees as Food and Feed
A Sustainable Underutilized Resource

Sampat Ghosh and Chuleui Jung

14.1 INTRODUCTION

Honeybees are not only considered for their incredible contribution to pollination and food production, but they also have been imbibed in human culture and traditions since early stages of civilization. There are plenty of depictions like the 8,000-year-old cave paintings of honey harvest in Arana cave, Spain, portrayals of beekeeping and related activities in Egyptian civilization, honey being offered by a monkey to Lord Buddha at Wat Jed Yod temple in Thailand, etc., which offer us insights into the rich human cultures and traditions of different parts of the world arisen out of human–ecological interaction. The use of honey as food and medicine has been mentioned in ages-old Buddhist, Hindu, and Islamic literature and its functional properties have already been demonstrated by a large amount of scientific evidence. Other hive products—i.e., bee pollen, royal jelly, propolis, and bee venom—have received attention, however, primarily from the commercial apiaries with managed honeybees *Apis mellifera*. What received the least appreciation as a hive product is the honeybee itself, although honeybees—especially broods—have been accepted as food among several communities around the world (reviewed by Ghosh et al. 2021a). To cite two examples, bee broods are cooked with chicken eggs and brood comb wrapped in banana leaves and steamed are included in the cuisine of Thailand (Chantawannakul 2018), and Hocking and Matsumura (1960) mentioned canned fried honeybee brood, especially pupae, with soy sauce in Japan for human consumption. In order to evaluate the nutritional potential of different species of honeybees at different developmental stages, we conducted a series of investigations (Ghosh et al. 2016, 2020a, 2020b, 2020c, 2021b). The objective of the present chapter is to represent the nutritional value of giant honeybee *Apis dorsata* and discuss prospective uses of this underutilized resource.

DOI: 10.1201/9781003294078-14

14.2 GIANT HONEYBEES AND BEEKEEPING

At 17–20 mm long, *Apis dorsata* is the largest among the common honeybee species. Their forewing length varies within the range of 12.5–14.5 mm. *A. dorsata* is distributed throughout South and Southeast Asia, ranging from the Indian subcontinent up to southern China and down throughout Indonesia and Malaysia. Compared to other honeybee hives, nests of the giant honeybee are large single-comb nests that can be up to 1.5 m in width and 1 m in depth and often are found in the forest canopy or cliff faces and coastal mangrove forests. Unlike colonies of dwarf honeybees (*A. florea*) or cavity-nesting honeybees (*A. cerana*), colonies of the giant honeybee *A. dorsata* are highly clustered in specific locations with some trees known as 'bee trees'. Giant honeybee nests are found in large aggregations (Oldroyd and Nanork 2009), and it therefore is common to find several colonies in a single tree (Figure 14.1A). Giant honeybee nests are usually built far off the ground, which may be 20–40 m high sometimes, on thick branches of 'bee trees', overhanging rocks and cliffs—sometimes even in buildings or other structures (Figure 14.1b). Being highly philopatric, giant honeybees often migrate a large distance and return to the same nesting site after certain time (Paar et al. 2004). Unlike dwarf honeybees (*A. florea*), instead of wrapping around the tree branches (see Ghosh et al. 2020b), giant honeybee nests hang underneath a branch or structure. They also have a wider foraging range than

FIGURE 14.1 Giant honeybee nests in a tree in Balaghat, India (A, photo credit: Dr. Sampat Ghosh); giant honeybee nest build in the building in Chiang Mai, Thailand (B, photo credit: Dr. Bajaree Chuttong); rafter beekeeping practice in southern Thailand (C–D, photo credit: Dr. Bajaree Chuttong); schematic diagram of rafter beekeeping (D).

that of European honeybees, and they play a crucial role in assisting the pollination of the forest ecosystem.

Honey harvesting from the giant honeybee colonies is a livelihood for many local or indigenous people, often augmenting their income, in several regions of South and Southeast Asia. Honey harvest from the wild, however, could not ensure the profit-making harvest all the time and also become a threat to the environment. Recently, giant honeybee *A. dorsata* beekeeping through artificial nesting substrate called rafter beekeeping has been developed and has been reported from Thailand (Figure 14.1C–E), Vietnam, Indonesia, Borneo, Sumatra, Malaysia, and Cambodia. However, it is of surprise—as Chuttong et al. (2019) reported—that instead of having demand of bee brood for human consumption, rafter beekeepers ignored the brood and brood combs.

14.3 GIANT HONEYBEES AS ETHNO-FOOD SOURCES

For the last few decades, edible insects have received attention as an alternative food resource for multiple reasons. First, the majority of edible insects are rich in protein and minerals like calcium, iron, zinc, etc. Very often, particularly the people underdeveloped and developing regions of the world cannot afford dietary protein sources like conventional animal meats, eggs, dairy products, etc., and therefore suffer from nutrition deficiency. In general, the protein content of edible insects is comparable—if not superior—to most conventional foods of animal origin (Ghosh et al. 2017). Besides, it contains essential amino acids required for the human diet, and the amino acid composition of insect protein satisfies the ideal protein pattern recommended by joint consultation of FAO (the Food and Agriculture Organization of United Nations), WHO (the World Health Organization), and UNU (United Nations University) (Ghosh et al. 2021a). Similarly, edible insects are a rich source of minerals—and if they are included in a planned diet, they could satisfy a substantial proportion of recommended dietary allowance of the minerals (Ghosh et al. 2021c). Second, multiple factors are ascribed to edible insects which make them sustainable. The requirements of natural resources like feed, water, and land to rear edible insects are far less than what is required to produce the same amount of any animal meat such as beef, pork, or chicken (Van Huis et al. 2013). The feed conversion ratio of edible insects is more efficient than that of cattle or swine. Moreover, insects emit a negligible amount of greenhouse gases (GHGs) like methane while the significant amount of GHGs emissions from conventional livestock negatively affects the environment (Oonincx and de Boer 2012). Finally, entomophagy in human history is not novel. Several communities of the world eat insects as a planned part of their diet. However, in the era of dietary transition due to several reasons, the decline of entomophagy is observed (Ghosh et al. 2021d). Honeybee, including giant honeybee *A. dorsata*, is not an exception in this context. In South and Southeast Asia, several communities eat giant honeybee *A. dorsata*. The consumption of *A. dorsata* has been mentioned in different scientific articles in northeast India (Kato and Gopi 2009; Megu et al. 2018; Mozhui et al. 2020), north-central Thailand (Chen et al. 1998), Sabah in Malaysia (Chung et al. 2002), east Kalimatan of Indonesia (Chung et al. 2003), Borneo (Chung et al. 2010), indigenous community of Palawan Forest in Philippines (Matias et al.

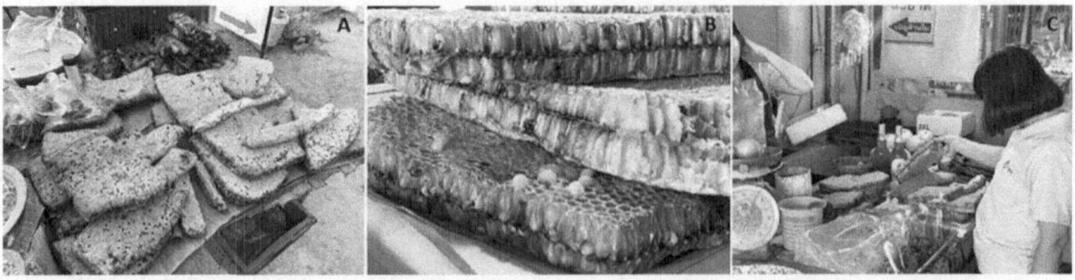

FIGURE 14.2 Selling of giant honeybee nests with bee brood in a market of Chiang Mai, Thailand (A–C, photos credit: Dr. Bajaree Chuttong).

2019), and Vedda tribal people in Sri Lanka (Nandasena et al. 2010). Although giant honeybee is a delicacy in several traditions, people frequently collect the nest from the wild and sell it at market (Figure 14.2A–C) or consume it themselves. Some of the delicacies and their recipes such as stir-fried queen bee larvae with fresh green peppercorns, queen bee larvae omelette, steamed bee brood with chili sauce, and scrambled brood have been mentioned by Chen et al. (1998). Some of these foods are also available in hotels and restaurants in Bangkok.

There is hardly any evidence of a systematic rearing system which has been developed for the purpose of utilizing giant honeybees as human food or animal feed. If there is no systemic approach for production, processing, and supplying resource like giant honeybees, their use will be limited.

14.4 NUTRITIONAL POTENTIAL OF GIANT HONEYBEES

Despite the use of *A. dorsata* as human food, only fragmentary information on the nutritional value of the species is available in the scientific literature. Ghosh et al. (2020a, 2021a) investigated the nutrient composition of pupae and adults of *A. dorsata* collected from Thailand. Chantawannakul (2020) presented the proximate composition and amino acid composition of *A. dorsata* collected from Thailand; however, the exact stage and caste have not been mentioned. Another study by Semuel et al. (2019) investigated the bioactive content and antioxidant activity of the honeybee nest of *A. dorsata* from Indonesia.

14.4.1 Protein

Protein receives the most attention in order to evaluate the quality of food, and the quality of protein depends on its amino acid composition. Considering that the total amino acid content as protein, the pupa and adult *A. dorsata* workers were found to contain 39 g and 55.7 g per 100 g of honeybees, respectively, on the basis of dry weight (Ghosh et al. 2020a, 2021a). Chantawannakul (2020) reported the protein content of *A. dorsata* at 41.26%, although the developmental stage was not mentioned. The protein content of giant honeybees is more than that of conventional protein sources such as pork, beef, chicken, soybeans, etc., as calculated on dry weight basis from the U.S. Department of Agriculture database (USDA, n.d.). Table 14.1 represents the amino

TABLE 14.1
Amino Acid Composition (g/100 g of Dry Matter and Percentage of Total Amino Acids) and Scoring Pattern of Essential Amino Acids of *Apis dorsata* Pupae and Adults

	Pupae				Adults			
	Content		Scoring Pattern		Content		Scoring Pattern	
	g/100 g	%	Tissue	Maintenance	g/100 g	%	Tissue	Maintenance
Val	2.2	5.6	115.1	144.6	3.3	5.9	120.9	151.9
Ile	1.7	4.4	124.5	145.3	2.6	4.7	133.4	155.6
Leu	3.3	8.5	112.8	143.4	4.5	8.1	107.7	136.9
Lys	2.2	5.6	77.3	125.4	3.3	5.9	81.2	131.7
Thr	1.7	4.4	103.8	189.5	1.9	3.4	81.2	148.31
His	1.0	2.6	95.0	170.9	1.2	2.2	79.8	143.6
SAA	1.9	4.9	139.2	221.4	5.1	9.2	261.6	416.2
AAA	2.8	7.2	98.3	188.9	4.3	7.7	105.8	203.2
Arg	1.9	4.9			2.5	4.5		
Asp	5.2	13.3			2.7	4.8		
Ser	1.9	4.9			2.7	4.8		
Glu	4.3	11.0			7.3	13.1		
Gly	2.9	7.4			4.6	8.3		
Ala	3.3	8.5			5.0	9.0		
Pro	2.7	6.9			4.7	8.43		
Total	**39.0**				**55.7**			

Notes: The values of content were obtained from Ghosh et al. 2020a, 2021a and the scoring pattern was calculated on the basis of ideal protein pattern for tissue amino acids and maintenance amino acids provided by WHO/FAO/UNU report (FAO 2013)

acid composition of pupa and adult *A. dorsata*, and scoring pattern of essential amino acids on the basis of ideal protein pattern for tissue amino acid and for maintenance amino acid, according to WHO/FAO/UNU consultation (FAO 2013). The proportion of essential amino acid was found to be 43.2% and 47.1% of the total amino acid for pupa and adult *A. dorsata*, respectively. Although the contents of individual amino acids, and therefore the total, were increased along with the developmental stages, the pattern of amino acid distribution remains the same. Leucine was found to be the most predominating among the essential amino acids, followed by valine and lysine. On the other hand, glutamic acid, aspartic acid, glycine, and alanine were abundant among non-essential amino acids. The essential amino acids of the protein (mg/g of protein) obtained from *A. dorsata* pupae and adults satisfied the requirement of amino acids for the maintenance of tissue (Table 14.1). However, lysine, threonine, and histidine remain as marginal amino acids for marginal to satisfy the requirement for tissue formation (Table 14.1). In this context, it is to be noted that tryptophan, cysteine, and methionine could not be recovered entirely or partially presumably because of the acid hydrolysis process (Ghosh et al. 2020a). Therefore, instead of consuming whole insects, nutrients such as protein can be extracted and be useful protein sources.

14.4.2 Fat

Fat content of *A. dorsata* was found 30.84%. The value was higher than that reported for *A. dorsata* pupa (6.18% and 6.13%) and adult (3.13%), considering the total fatty acids (Ghosh et al. 2020a). However, the values are comparable to the fatty acid content of honeybee species like *A. mellifera* and *A. florea* workers (Ghosh et al. 2016, 2020b). The quality of fat depends on the fatty acid composition. Based on the saturation point of view, fatty acids are of three types: saturated fatty acids (SFA), monounsaturated fatty acids (MUFA), and polyunsaturated fatty acids (PUFA). Table 14.2 represents the fatty acid composition of *A. dorsata* pupa and adult workers. Overall, the proportion of MUFA was found predominating, followed by SFA and the least amount of PUFA. The distribution of fatty acids obtained for *A. dorsata* pupae and adults was in agreement with the fatty acid composition for other worker and drone honeybee species and subspecies (Ghosh et al. 2020c, 2021a, 2021b). It is undesirable to have higher saturated fats in the diet because of its linkage to cardiovascular disorders. The total

TABLE 14.2
Fatty Acid Composition (mg/100 g of Dry Matter and Percentage of Total Fatty Acids) of *Apis dorsata* Pupae and Adults

	Pupae[1]		Pupae[2]		Adults	
	mg/100 g	%	mg/100 g	%	mg/100 g	%
C12:0	30	0.5	30	0.5	10	0.3
C13:0	0	0	10	0.2	10	0.3
C14:0	200	3.2	150	2.4	30	1.0
C16:0	2060	33.3	1830	29.9	450	14.4
C18:0	730	11.8	790	12.9	450	14.4
C20:0	30	0.5	40	0.7	30	1.0
SFA	**3050**	**49.4**	**2850**	**46.5**	**980**	**31.3**
C16:1	20	0.3	30	0.5	40	1.3
C18:1n-9, cis	2950	47.7	3090	50.4	1910	61.0
C20:1	110	1.8	110	1.8	130	4.2
MUFA	**3080**	**49.8**	**3230**	**52.7**	**2080**	**66.5**
C18:2n-6, cis	50	0.8	50	0.8	70	2.2
C18:3n-3	0	0	0	0	0	0
PUFA	**50**	**0.8**	**50**	**0.8**	**70**	**2.2**
Total	**6180**		**6130**		**3130**	

Notes: The values of content were obtained from Ghosh et al. (2020a); pupae used for the analyses were collected from Thaweephol Condo in metropolitan Chiang Mai[1] and from Mae Jo University campus of Thailand[2]

fatty acid was found to be lower along with the development stages. Oleic and palmitic acid was the most abundant among MUFA and SFA, respectively.

14.4.3 Minerals

Among micronutrients, minerals play crucial roles in human nutrition. Many of them such as zinc (Zn), copper (Cu), magnesium (Mg), etc., are required as co-factors for several biochemical reactions. Table 14.3 represents the minerals of nutritional importance and their content in worker *A. dorsata* pupae and adults. Almost all the minerals were present in higher amounts in adults than that of pupae. We further calculated whether the percentage of RDA (recommended dietary allowance) or AI (adequate intake) can be satisfied if one consumes 100 g of *A. dorsata* (dry weight basis). This amount of dried adult *A. dorsata* consumption could satisfy the complete requirement of zinc for adult females and more than 86% for males. On the other hand, it satisfies completely the requirement of iron for males and about 50% of RDA for females. Anemia is a major concern for women, especially for those of child-bearing and lactating age, and is highly prevalent in South and Southeast Asia (Sunuwar et al. 2020). Another concern is zinc deficiency. Wieringa et al. (2019) reported that zinc deficiency is highly prevalent in Southeast Asia, affecting more than 30% of the subjects. Assuming the high bioavailability, the consumption of giant honeybees could mitigate the problem of mineral deficiency.

TABLE 14.3
Mineral Content (mg/100 g of Dry Matter) of *Apis dorsata* Pupae and Adults and Percentage of RDA/AI Satisfied by Consumption of 100 g Dry Matter

	Pupae[1]			Pupae[2]			Adults		
		% of RDA Satisfied			% of RDA Satisfied			% of RDA Satisfied	
	mg/100 g	Male	Female	mg/100 g	Male	Female	mg/100 g	Male	Female
Ca	68.9	6.9	6.9	78.5	7.9	7.9	99.2	9.9	9.9
Mg	103.4	24.6	32.3	113.3	27.0	35.4	144.7	34.5	45.2
P	905	129.3	129.3	972.3	138.9	138.9	1187.9	169.7	169.7
Na	48.6	3.2	3.2	53.9	3.6	3.6	67.8	4.5	4.5
K	1136.6	33.4	43.7	1254.3	36.9	48.2	1573.9	46.3	60.5
Fe	5.8	72.5	32.2	7.6	95.0	42.2	8.9	111.3	49.4
Zn	6.4	58.2	80.0	7.4	67.3	92.5	9.5	86.4	118.8
Cu	1.1	122.2	122.2	1.2	133.3	133.3	1.6	177.8	177.8
Mn	0.1	4.3	5.6	0.1	4.3	5.6	0.2	8.7	11.1

Notes: The values of minerals content were obtained from Ghosh et al. 2020a and percentage of RDA/AI satisfied was calculated on the basis of values from Linus Pauling Institute of Oregon State University, USA; pupae used for the analyses were collected from Thaweephol Condo in metropolitan Chiang Mai[1] and from Mae Jo University campus of Thailand[2]

14.5 BIOFUNCTIONAL PROPERTIES AND FOOD PROCESSING

Currently, food processing is one important component to widen the scope of bioprospection of edible insects. The nutritional value of edible insects, including honeybees, can be utilized if they are consumed by people. However, the perception of consumers is very crucial in order to find acceptance for a prospective food. Several communities that already have the tradition of eating insects have developed different traditional processes like sun drying, boiling, smoking, roasting, etc., to increase the shelf life of the insect food as well as to avoid the undesired effects due to antinutrient contents of the insects—but members of the community which does not have a tradition of eating insects might have difficulty accepting the insects as food. Therefore, in order to exploit them as a rich nutritional resource, insects may be processed in the form of flour and used in food preparation of processed foods such as bread, biscuits, etc. In this way, the food made up of wheat flour can be fortified with insect flour to enhance its nutritional value. Also, the investigation of functional properties of the giant honeybees is a subject to be investigated. The drone of different subspecies of *A. mellifera* exhibited functional properties like antioxidants, containing polyphenols, etc. (Ghosh et al. 2020c). The results of the investigation of *A. dorsata* nest revealed that the nest extract contained alkaloids, at least 20 flavonoids, and their derivatives saponins, tannins, and steroids. In addition, ethanol and n-hexane—especially ethanol extract of the nest—exhibited high free radical scavenging activity (Semuel et al. 2019).

14.6 PROSPECTIVE USE OF GIANT HONEYBEES AS ANIMAL FEED

Honeybees are often prey for a wide variety of predators including insects, birds, reptiles, and mammals. Wasps and hornets (several species of *Vespa*) are serious predators of honeybees and cause significant damage of the colonies in managed apiaries, as well as colonies in the wild. Several birds such as Asian bee-eater, orange-rumped honey guide, Malaysian honey guide, oriental honey buzzard, woodpecker, drongo, etc., are enemies of honeybees. Most of them eat honeybees (Oldroyd and Nanork 2009). Arboreal reptiles like geckos, skinks, and lizards may damage honeybees, and a few of them have been regularly recorded as present in commercial apiaries, according to a report of National Bee Board of India (n.d.). Many mammals are dangerous to honeybee colonies. To mention a few, pine martins and raccoons often destroy honeybee colonies to drink honey. On the other hand, bears, skunks, honey badgers, and rodents eat honey as well as bees (Gupta et al. 2014). Therefore, in the food web, honeybees—including giant honeybees in the forest—occupy a critical place as many other animals get a significant proportion of their nourishment from honeybees. So not only as a human food, huge scopes also exist to utilize the giant honeybees—drones especially—as animal feed.

There are several scientific studies that exhibited the beneficial effect of honeybees—although not necessarily with *A. dorsata*—for animal feed. Yucel et al. (2011) investigated the effect of Apilarnil—i.e., 3–7-day-old drone larvae—administration on the growth performance and secondary sex characteristics of male broilers (Ross-308 genotype). The study showed that the administration of Apilarnil at the rate of 1600 mg/kg body weight/day, which was calculated as 4 g/broiler/day, increased significant body weight gain at 36–42 days, higher feed intake, higher feed conversation ratio at the beginning, and improved androgenic effects like comb length and wattle length and height. Another study by Muraviev and Kalatzinskaja (2014) showed an increase in egg production and enhance the quality of egg in hens fed with 15, 45, and 75 mg/bird/day; among them, the hens fed with 45 mg/bird/day performed the best (cf. Sawczuk et al. 2019). Yemets (2020) demonstrated that supplemental feeding with 0.5g of homogenate drone larvae improved the fertility of large white gilts (young female pigs). Although scientific study on the use of honeybee as fish feed is limited, a study by Linayati et al. (2021) demonstrated that the addition of honey to the fish feed increased the growth of white snapper seeds (*Latescalcarifer*). Another study conducted by Soror et al. (2021) revealed that the addition of bee pollen with ginger and moringa exhibited recuperative effect in Nile Tilapia (*Oreochromis niloticus* L.) when exposed to the sub-lethal dose of an organophosphate, dimethoate. Besides the use of honeybees as feed, they also exhibited as pharmaceutical properties in animal model study. Ingestion of raw drone larvae and pupae elicited a urotrophic effect in rat model, and E-dec-2-enedioc acid was found to be possible bioactive compound for the estrogenic activity (Seres et al. 2013). However, a systematic review of the scientific literature reveals that the majority of the work has been conducted on European honeybees *A. mellifera*. Scientific investigation on giant honeybees remains largely unexplored.

14.7 CONCLUSIONS

Giant honeybee workers are nutritious source, as they contain high protein with almost all essential amino acids and fewer fats, and they are rich in minerals like phosphorus, sodium, zinc and iron. Therefore, exploration of the resource might improve the nutritional status of the local population and broaden the bioindustrial prospects through the production of animal feed and enhance the economy. Only honey harvest from the giant honeybee hives is not often enough to support a human family, but if honey hunters could harvest the possible other products, including the brood, it could enhance the economic conditions of the honey collectors. Also, entering the core area of the forest for the purpose of honey collection could be a threat to the lives of honey collectors, as well as to biodiversity conservation. The introduction of rafter beekeeping in those areas which are suitable for giant bee rearing, such as the delta of the Sundarban region, can ensure the timely honey harvest and its scope to explore other valuable hive products, including bee brood. This, in turn, augments substantial income and enhances the local economy. As honeybees are primary pollinators in the forest ecosystem, it is essential to investigate the brood and drone production capacity of a colony without compromising the pollination service.

ACKNOWLEDGEMENT

This study was partly supported by the BSRP through the National Research Foundation of Korea (NRF), Ministry of Education (NRF-2018R1A6A1A03024862), and Korean Forest Service R&D grant (20211362A00–2123-BD01). The authors sincerely acknowledge Dr. Bajaree Chuttong of Chiang Mai University, Thailand, for providing us with some relevant photographs.

REFERENCES

Chantawannakul, P. 2018. Honey bees in modernized South East Asia: Adaptation or extinction? In: Lopez, M., and Suryomenggolo, J. (Eds.), *Environmental resources use and challenges in contemporary Southeast Asia. Asia in transition.* Vol. 7. Springer, Singapore. https://doi.org/10.1007/978-981-10-8881-0_8.

Chantawannakul, P. 2020. From entomophagy to entomotherapy. *Frontiers in Bioscience, Landmark* 25: 179–200.

Chen, P.P., Wongsiri, S., Jamyanya, T., Rinderer, T.E., Vongsamanode, S., Matsuka, M., Sylvester, H.A., and Oldroyd, B.P. 1998. Honey bees and other edible insects used as human food in Thailand. *American Entomologist* 44: 24–29. https://doi.org/10.1093/ae/44.1.24.

Chung, A.Y.C. 2010. Edible insects and entomophagy in Borneo. In: Durst, P.B., Johnson, D.V., Leslie, R.N., and Shono, K. (Eds.), *Forest insects as food: Human bite back, proceedings of a workshop on Asia-Pacific resources and their potential for Development.* Food and Agriculture Organization of the United Nations Regional Office for Asia and the Pacific, Bangkok, Thailand, 141–150.

Chung, A.Y.C., Binti, M., and Fadan, A. 2003. Ethnoentomological survey at the Northern Kayan Mentarang National Park, East Kalimatan, Indinesia. In: Mardiastuti, A., and Soehartono, T. (Eds.), *Joint biodiversity expedition in Kayan Mentarang National Park.* Ministry of Forestry-WWF Indinesia-ITTO, Jakarta, 343–356.

Chung, A.Y.C., Khen, C.V., Unchi, S., and Binti, M. 2002. Edible insects and entomophagy in Sabah, Malaysia. *Malayan Nature Journal* 56: 131–144.

Chuttong, B., Somana, W., and Burgett, M. 2019. Giant honey bee (*Apis dorsata* F.) Rafter beekeeping in southern Thailand. *Bee World* 96: 66–68. https://doi.org/10.1080/0005772X.2019.1596546.

FAO (Food and Agricultural Organisation of United Nations). 2013. *Dietary protein quality evaluation in human nutrition.* FAO, Rome, Italy.

Ghosh, S., Chuttong, B., Burgett, M., and Meyer-Rochow, V.B. 2020a. Nutritional value of brood and adult workers of the Asia honeybee species *Apis cerana* and *Apis dorsata.* In Mariod, A.A. (Ed.), *African edible insects as alternative source of food, oil, protein and bioactive componenets.* Springer Nature, Switzerland, 265–273. https://doi.org/10.1007/978-3-030-32952-5_19.

Ghosh, S., Herren, P., Meyer-Rochow, V.B., and Jung, C. 2021b. Nutritional composition of honey bee drones of two subspecies relative to their pupal developmental stages. *Insects* 12: 759. https://doi.org/10.3390/insects12080759.

Ghosh, S., Jung, C., Chuttong, B., and Burgett, M. 2020b. Nutritional aspects of the dwarf honeybee (*Apis florea*) for human consumption. In: Abrol, D.P. (Ed.), *The future role of dwarf honeybees in natural and agricultural systems.* CRC Press, Boca Raton, 137–145.

Ghosh, S., Jung, C., and Meyer-Rochow, V.B. 2016. Nutritional value and chemical composition of larvae, pupae, and adults of worker honey bee, *Apis mellifera ligustica* as a sustainable food source. *Journal of Asia-Pacific Entomology* 19: 487–495. https://doi.org/10.1016/j.aspen.2016.03.008.

Ghosh, S., Lee, S.-M., Jung, C., and Meyer-Rochow, V.B. 2017. Nutritional composition of five commercial edible insects in South Korea. *Journal of Asia-Pacific Entomology* 20: 686–694. https://doi.org/10.1016/j.aspen.2017.04.003.

Ghosh, S., Meyer-Rochow, V.B., and Jung, C. 2021a. Honey bees and their brood: A potentially valuable resurce of food, worthy of greater appreciation and scientific attention. *Journal of Ecology and Environment* 45: 31. https://doi.org/10.1186/s41610-021-00212-y.

Ghosh, S., Namin, S.M., Meyer-Rochow, V.B., and Jung, C. 2021c. Chemical composition and nutritional value of different species of *Vespa* hornets. *Foods* 10: 418. https://doi.org/10.3390/foods10020418.

Ghosh, S., Sohn, H.-Y., Pyo, S.-J., Jensen, A.B., Meyer-Rocow, V.B., and Jung, C. 2020c. Nutritional composition of *Apis mellifera* drones from Korea and Denmark as a potential sustainable alternative food source: Comparison between developmental stages. *Foods* 9: 389. https://doi.org/10.3390/foods9040389.

Ghosh, S., Tchibozo, S., Lanmantchion, E., Meyer-Rochow, V.B., and Jung, C. 2021d. Observations on how people in two locations of the Plateau Dėpartment of southeast Benin perceive entomophagy: A study from West Africa. *Frontiers in Nutrition* 8: 637385. https://doi.org/10.3389/fnut.2021.637385.

Gupta, R.K., Sharma, D., and Bali, K. 2014. Honeybee predators: Insects, reptiles and mammals. In: Gupta, R.K., Reybroeck, W., van Veen, J.W., and Gupta, A. (Eds.), *Beekeeping for poverty alleviation and livelihood security*. Springer, Dordrecht, 321–345. https://doi.org/10.1007/978-94-017-9199-1_11.

Hocking, B., and Matsumura, F. 1960. Bee brood as food. *Bee World* 41: 113–120. https://doi.org/10.1080/0005772X.1960.11096777.

Kato, D., and Gopi, G.V. 2009. Ethnozoology of *Galo* tribe with special reference to edible insects in Arunachal Pradesh. *Indian Journal of Traditional Knowledge* 8: 81–83.

Linayati, L., Rizkyansyah, B., Mardiana, T.Y., and Yahya, M.Z. 2021. The addition of honey bee to the feed to increase the growth of white snapper seed (*Latescalcarifer*). *Journal of Aquaculture and Fish Health* 10: 380–386. https://doi.org/10.20473/jafh.v10i3.26944.

Matias, D.M.S., Borgemeister, C., Sėmah, A.-M., and von Wehrden, H. 2019. The role of linked social-ecological system in a mobile agent-based ecosystem service from giant honey bees (*Apis dorsata*) in an indigenous community forst in Palawan, Philippines. *Human Ecology*. https://doi.org/10.1007/s10745-019-00114-7.

Megu, K., Chakravorty, J., and Meyer-Rochow, V.B. 2018. An ethnographic account of the role of edible insects in the Adi Tribe of Arunachal Pradesh, North-East India. In: Halloran, A., Flore, R., Vantomme, P., and Roos, N. (Eds.), *Edible insects in sustainable food systems*. Springer International Publishing AG, Cham, 35–54.

Mozhui, L., Kakati, L.N., Kiewhuo, P., and Changkija, S. 2020. Traditional knowledge of the utilization of edible insects in Nagaland, North-East India. *Foods* 9: 852. https://doi.org/10.3390/foods9070852.

Muraviev, D.W., and Kalatzinskaja, A.M. 2014. Drone homogenate and hens productivity. *ViestnikKazanskovo GAU* 1: 132–134.

Nandasena, M.R.M.P., Disanayake, D.M.S.K., and Weeratunga, L. 2010. Sri Lanka as a potential gene pool of edible insects. In: Durst, P.B., Johnson, D.V., Leslie, R.N., and Shono, K. (Eds.), *Forest insects as food: Human bite back, Proceedings of a workshop on Asia-Pacific resources and their potential for development*. Food and Agriculture Organization of the United Nations Regional Office for Asia and the Pacific, Bangkok, Thailand, 161–164.

National Bee Board of India. n.d. *Pests and diseases of honeybees and their management*. Department of Agriculture and Cooperation, Ministry of Agriculture, Government of India, New Delhi, India.

Oldroyd, B.P., and Nanork, P. 2009. Conservation of Asian honey bees. *Apidologie* 40: 296–312. https://doi.org/10.1051/apido/2009021.

Oonincx, D.G.A.B., and de Boer, I.J.M. 2012. Environemntal impact of the production of mealworms as a protein source for humans-A life cycle assessment. *PLoS One* 7: e51145. https://doi.org/10.1371/journal.pone.0051145.

Paar, J., Oldroyd, B.P., Huettinger, E., and Kastberger, G. 2004. Genetic structure of an *Apis dorsata* population: The significance of migration and colony aggregation. *Journal of Heredity* 95: 119–126.

Sawczuk, R., Karpinska, J., and Miltyk, W. 2019. What do we need to knowabout drone brood homogenate and what is known. *Journal of Ethnopharmacology* 245: 111581. https://doi.org/10.1061/j.jep.2018.10.042.

Semuel, M.Y., Kaunang, E.S.N., and Manopo, J.S. 2019. The bioactive contents and antioxidant activity of honey bee nest extract of *Apis dorsata. Binghami from the North Sulawesi. Molekul* 14: 92–102. https://doi.org/10.20884/1.jm.2019.14.2.502.

Seres, A.B., Duca, E., Báthori, M., Hunyadi, A., Béni, Z., Dékány, M., and Gáspár, R. 2013. Raw drone milk of honeybees elicits uterotrophic effect in rats: Evidence for estrogenic activity. *Journal of Medicinal Food* 16: 404–409. https://doi.org/10.1089/jmf.2012.0232.

Soror, E.I., Asely, M.E., Gawad, E.A.A., Radwan, H.A., and Abbass, A.A. 2021. Recuperative effects of honey bee pollen, ginger (*Zingiberofficinale*), and *Moringaoleifera* in Nile tilapia (*Orechromisniloticus* L.) after sub-lethal exposure to dimethoate. *Aquaculture* 530: 735886. https://doi.org/10.1016/j.aquaculture.2020.735886.

Sunuwar, D.R., Singh, D.R., Chaudhary, N.K., Pradhan, P.M.S., Rai, P., and Tiwari, K. 2020. Prevalence and factors associated with anemia among women of reproductive age in seven South and Southeast Asian countries: Evidence from nationally representative surveys. *PLoS One* 15: e0236449. https://doi.org/10.1371/journal.pone.0236449.

USDA Database on Nutrient Content of Food. n.d. www.data.nal.usda.gov (10.3.2022).

Van Huis, A., Itterbeeck, J.V., Klunder, H., Mertens, E., Halloran, A., Muir, G., and Vantomme, P. 2013. *Edible insects: Future prospects for food and feed security.* FAO Forestry Paper 171. Food and Agriculture Organization of the United Nations, Rome, Italy, 201.

Wieringa, F.T., Dikhuizen, M.A., and Berger, J. 2019. Micronutrient deficiencies and their public health implications for South-East Asia. *Current Opinion in Clinical Nutrition and Metabolic Care* 22: 479–482. https://doi.org/10.1097/MCO.0000000000000603.

Yemets, Y.M. 2020. Dietary effects of drone larvae homogenate on the homeostatic constants and the reproductive capacity of large white gilts. *Translational Research in Veterinary Science* 3: 27–39. https://doi.org/10.12775/TRVS.2020.008.

Yucel, B., Acikgoz, Z., Bayraktar, H., and Seremet, C. 2011. The effects of Apilarnil (drone bee larvae) administration on growth performance and secondary sex characteristics of male broilers. *Journal of Animal and Veterinary Advances* 10: 2263–2266.

15 Decline in Population of Giant Honeybees

Dharam P. Abrol

15.1 INTRODUCTION

Giant honeybees are the largest of the honeybee species and are widely distributed throughout Southeast Asia, ranging from the Indian subcontinent up to southern China and down throughout Indonesia and Malaysia. There are two closely related species of Giant honeybees *Apis dorsata* and the slightly larger Himalayan honeybee (*Apis laboriosa*) which is only present in the mountainous regions, particularly the Himalayas above 1,500 m. The other closely related species is the giant Philippine honeybee (*Apis breviligula*) which is restricted to a small cluster of islands in the Philippines.

The nests of giant honeybees are large single combs which can measure up to 1.5 m in width and 1 m in depth. This large single comb can contain upwards of 60,000 bees. Unlike dwarf honeybees or cavity-nesting honeybee species, colonies of giant honeybees can be highly clustered in a specific location, with some trees in Asia having more than 50 nests. Giant honeybee nests are mostly exposed built on thick branches of trees, rocks, or cliffs usually 20–40 m above the ground. The difference between giant honeybee nests and those of dwarf honeybees is not only the size of nests, but dwarf honeybee nests are mostly hidden in branches or foliage.

Giant honeybee colonies are quite aggressive, and more than three-quarters of the population is engaged in colony defense, forming a protective curtain of around 3–4 bees thick. Giant honeybees are mainly tropical and can migrate up to 200 km. Colonies will travel for many months, resting in trees along the way, building combs and honey reserves, and then moving on to new locations as the forage decreases, before setting up new nests for the mass flowering of the monsoon season. Interestingly, bees have been reported to return to the same site even if the original bees have been replaced by new ones. Such a mechanism of memory retention within the honeybee colony still remains unknown. Bees are strongly dependent upon the environment for food resources and nesting sites. They are vulnerable to transformation of their native habitat into large-scale agricultural landscapes. Most of the studies have predicted decline in their abundance and diversity. The data on pollinator decline is worse, as there is no quantification of existing populations of bees in India. There is no data available on economic estimates on the pollination of crops in India. However, Basu et al. (2011) reported that the economic value of the six most commonly cultivated pollinator-dependent crop species grown in India was US$3.72 billion. The total economic valuation of insect pollination (EVIP) of these six crops was calculated as US$760 million (around ₹5,400 crore now). Nearly 1 million animal

DOI: 10.1201/9781003294078-15

and plant species are threatened with extinction worldwide, but conservation in India continues to focus on big cats, elephants, and rhinos while no efforts have been made to conserve and protect the honeybees which are crucial for survival of ecosystems and ensuring food security. Loss of habitat and pesticide usage is severely affecting honeybee population.

Pollinator species are declining worldwide, which is a matter of great concern for sustaining pollination services (Gallai et al. 2009; Partap 2011; Dicks et al. 2016; Potts et al. 2016; Zattara and Aizen 2021). Ever-growing increases in human population requires increases in food production, which demands more insect pollinators. Decline in pollinators has not only threatened the food security but the existence of natural ecosystems. Insect pollinators are critical to the maintenance of biological diversity and stability of ecosystems (Potts et al. 2010, 2016).

Sihag (2014) reported that *A. dorsata* colonies have drastically declined during the last 30 years in northern India. He reported that this has mainly been due to decline of floral resources, attack of pests and diseases, and environmental degradation. Some earlier researchers (Koeniger and Koeniger 1980; Venkatesh and Reddy 1989; Woyke et al. 2012; Shrestha et al. 2002; Dyer and Seeley 1994; Pokhrel 2010) reported that seasonal migration depends upon existing weather and floral conditions. Long-distance migrations occurred due to floral dearth and inclement climatic conditions of the area (Venkatesh and Reddy 1989). Woyke et al. (2012) also reported that swarms of the giant honeybee did not migrate directly to the seasonal alternative nesting sites but established new colonies in surrounding areas.

Decline of wild honeybee populations has been attributed to crude methods of honey extraction, loss of habitat across southern Asia (Laurance 2007), and felling of tall trees that that are used as nesting sites (Oldroyd and Wongsiri 2006; Oldroyd and Nanork 2009; Corlett 2011; Sihag 2014). Entire colonies of *A. dorsata* migrate regularly depending upon seasonal changes between highland and lowland nesting sites. A single colony measuring 1–1.5 m wide may contain 50,000 individuals or more (Morse and Laigo 1969; Oldroyd et al. 2000; Oldroyd and Wongsiri 2006; Hawkeswood et al. 2019), which may produce number of swarms that may aggregate on single trees or other structures such as water towers.

Though there is general observation among the masses regarding decline of wild bee population, there is no comparative data on the importance of different honeybee species and their decline. *Apis dorsata* has the biggest colonies with large numbers of individuals which have the largest foraging range, making them unique for pollination of wide variety of plants. However, a recent survey on pollinators visiting plants in Bangkok showed that the smaller dwarf honeybee visited almost four times and stingless bees even 12 times the number of flowers as were visited by *Apis dorsata* (Stewart and Waitayachart 2020). Yet, pollination interactions are more complex. Mukherjee et al. (2019) reported that under normal rainy conditions, the fruit set was mainly dependent on stingless bees; however, in drought years when abundance was low, *Apis dorsata* foragers helped in fruit set. For the pollination of unevenly and irregularly flowering trees of tropical forests, the migrating honeybees—especially *Apis dorsata*—are probably indispensable (Oldroyd and Wongsiri 2006; Corlett 2011).

Giant honeybees often nest in aggregations, and more than 50 colonies can be found on a tree. In 2002, about 100 *Apis dorsata* colonies lived in a huge banyan tree on the campus of the University of Agricultural Sciences in Bengaluru (Woyke et al. 2005, 2012). They looked like big fruits. However, it was really surprising that after 15 years in 2017, only two colonies were left. Eucalyptus trees nearby providing nectar were cut and the bees were gone. Huge losses in abundance and diversity of bees and other pollinating insects are observed in different regions of the world (Petts et al. 2010; Hallmann et al. 2017; Sánchez-Bayo and Wyckhuys 2019). There is no clear data on the decline of honeybees and other pollinators in India, but there are clear indications that giant honeybees have declined (Oldroyd and Wongsiri 2006; Kastberger and Hötzl 2015).

Decline in honeybee and stingless bees has been reported from Karnatka, India (Basavarajappa 2010; Basavarajappa and Raghunandan 2013). Venkatesh (2014) reported that in a survey conducted in different parts of Bangalore, the number of *Apis dorsata* colonies declined by 50% between 1987 and 2014. He reported that the observed strong decline of *Apis dorsata* populations has probably many reasons which include the felling of trees which reduced important food sources, increasing air pollution, and the intentional killing of colonies nesting at buildings by means of chemicals. In particular, the giant honeybees have considerably declined due to the loss of food resources and natural nesting sites caused by deforestation and the fast urbanization and landscape fragmentation (Oldroyd and Wongsiri 2006; Kastberger and Hötzl 2015; Basavarajappa and Raghunandan 2013; Raghunandan and Basavarajappa 2015).

Ahmad et al. (2003) reported severe decline of *A. laboriosa* numbers in Nepal. Robinson (2021) in conversations with honey vendors and others acquainted with honey-hunting activity in Thailand reported that *A. dorsata* populations are threatened. He observed considerable decline in nesting populations of *Apis dorsata* in many areas of Thailand. However, there is no numerical data on the apparently dwindling numbers of colonies (Oldroyd and Wongsiri 2006; Oldroyd and Nanork 2009). *A. dorsata* colonies can travel 100–200 km during their annual migrations. In an alarming research, Oldroyd and Nanork (2009) warn about the decreasing densities of Asian honeybees. They predicted that the stress of honey hunting was affecting the bee population more than the threats of parasites and pathogens. They also stressed the urgent need for conservation efforts in the Asian subcontinent.

However, there are also positive reports of indigenous techniques developed to extract honey without killing the wild bees. Joshi and Gurung (2015) clarified that Nepalese honey hunters use non-violent honey gathering practices since they associate honey harvesting with spiritual, cultural, and social significance. Paliwal (2005) from Wardha (Maharashtra) devised new technique called Sevagram Nisarg technique of rock bee management (SNTRM). They taught the local Adivasis to use an indigenous kit containing bee suits, gloves, knives, torches, ropes, rope ladders, and containers. The men were advised to cut only some part of the upper portion of the comb which contains honey. The operated portions are generally refilled with honey by the bees within 25–30 days. This method allowed 2–3 harvests per season, without any destruction to the hive population. Rafter beekeeping is practiced in Vietnam

(Tan and Ha 2002) and Cambodia (Waring and Jump 2004). In Vietnam, *Apis dorsata* bees are 'received' upon these rafters every year. The honey is harvested 3–4 times in a single season through proper management (Mahindre 2004). Back home in India, the Siddhi tribes of Karnataka use long bamboo sticks with sickle tied to it so as to cut the honey portion from the comb. Fire is lit to drive away the bees (Sunil Kumar and Reddy 2014). Perumal et al. (2021) developed a new wireless technology to easily dispel the honeybees from their comb during the process of extraction of the honey. This process saves the bees from destruction, as well as human beings from being stung. Bees are dispelled by generating different frequencies using Arduino microcontrollers. The mobile application is used to send waves of different frequencies to the honeybee nest which are used to dispel honeybees from nest by causing irritation. These different frequencies will be irritating to the honeybees.

Migrating animals utilize stopover sites for resting and refueling energy reserves for further movement. However, destruction or degradation of a stopover sites could result in the death of exhausted migrating colonies (Wilcove 2008; Jesmer et al. 2018; Satterfield et al. 2020).

In a rainforest habitat in northeastern Thailand, migration of the giant honeybee allowed colonies to track seasonally varying floral resources in different regions (Dyer and Seeley 1994). The colonies returned back after the end of dry season and then again departed early in the rainy season.

Similarly, in the case of *Apis laboriosa*, Joshi et al. (2004) reported a sharp decline in the bee population during recent decades. They found that the number of cliffs having bees has declined and the number of nests per cliff decreased. They reported that in the Seti Valley of Nepal, there used to be 59 bee cliffs until 1980, but in 2003, only 33 cliffs with bee nests were found. Similarly, in the Mardi and Madi river valleys in Modi 20–30 years ago, there were 2–3 times more bee cliffs in the area than at present. Based on anecdotal evidence, Valli and Summers (1988) and Underwood (1992) also suggested that within the last 50 years, the number of colonies of *A. laboriosa* has greatly declined throughout much of Nepal.

In earlier studies, Underwood (1986) published a list of *A. laboriosa* cliffs observed during 1984. Joshi et al. (2004) revisited eight of these sites visited by Underwood and looked for signs of bee nests between February and May 2001 and found that four of the eight sites had no nests at all and one only site had two colonies. Information on nest aggression size reveals that the number of nests per cliff has decreased by at least 50% in the last 20–30 years (Joshi et al. 2004). Underwood (1986) reported up to 76 (averaging 14.53 ± 17.52) *A. laboriosa* nests aggregated on a single cliff, with 247 colonies of *A. laboriosa* recorded from 17 separate cliff sites. In 12 of the 17 cliffs (70.5%), he found more than ten nests (Underwood 1986).

Information gathered from honey hunters and general observations around the cliffs indicate certain features that provide clues about reasons for bee population decline. These include loss of forage, crude methods of honey harvesting, and loss of nesting sites (Ahmad et al. 2003). Lack of awareness about the importance of bees is responsible for their over-exploitation. Underwood (1992) reported that overgrazing by livestock and destruction of forests are main reasons for the decline of *A. laboriosa* populations. Introduction of improved crop varieties and indiscriminate use of pesticides are other factors responsible for the decline of bee populations.

In addition to these reasons, the presence of European foulbrood disease (caused by *Melissococcus plutonius*) and other pathogens of *A. mellifera* have accelerated the decline. *M. plutonius* has been identified in *A. laboriosa* colonies (Allen et al. 1990).

Gregory and Jack (2022) reported that in case of *Apis laboriosa*, honey hunters bring in more than 50% of the annual honey yield in the region (Ahmad et al. 2003). This destructive method of honey harvesting has led to rapid decline of the *Apis laboriosa* population in recent decades (Thapa et al. 2018; Joshi et al. 2004). Other possible reasons include land loss due to deforestation and landslides (Thapa et al. 2018; Aryal et al. 2015; Joshi et al. 2004; Ahmad et al. 2003; Anderson and Morgan 2007; Allen et al. 1990) or parasites and diseases (Chantawannakul et al. 2016).

15.2 THE ECOLOGICAL AND ECONOMIC IMPORTANCE OF HONEYBEES

Pollination is a key function in all terrestrial ecosystems, interlinking the fates of plants and animals. Pollinating animals are themselves a major part of the biodiversity worldwide; they are safeguarding plant biodiversity—and indirectly, all the animals depend on fruits or leaves of animal pollinated plants for their food. A world without pollination is beyond imagination and would have lost all its richness and most of its vegetation. Pollination services cannot just be replaced by humans; they are extremely valuable and are a precondition for or affecting an estimated 35% of the world's crop production, increasing outputs of 87 of the leading food crops (FAO 2008). The total economic value of pollination services worldwide is estimated to €153 billion, with vegetables and fruits (Gallai et al. 2009).

There is direct and indirect evidence of pollinators decline worldwide in different regions, which is likely to impact production of fruits and vegetables and to accelerate the loss of species diversity. This is particularly important in highly modified agricultural landscapes, where pollinators are most needed for food production and crop security. Many people assume that the only benefit of honeybees is that they produce honey for human consumption, but they are not aware of their in pollination of cultivated plants and stability of terrestrial ecosystems (Wenzel et al. 2020).

15.3 IMPACT ON PLANT–POLLINATOR INTERACTIONS

The International Union for the Conservation of Nature (IUCN 2022) predicts that 20,000 flowering plant species will have disappeared within decades. The decline in pollinators will have serious impact on both the agricultural and non-agricultural landscapes. It is not only the domesticated bees which are important—wild bees are equally important and need to be conserved for plant reproduction and food production.

Giant honeybee *Apis dorsata* workers continue to forage even after sunset. Different from all other honeybee species, giant honeybees are able to fly and forage at low light intensities (Somanathan et al. 2009). Most people are afraid of them. The bees behave in strange ways, react very sensitively to disturbances, and are considered to be rather aggressive, because when agitated, they start a mass attack

(Oldroyd and Wongsiri 2006; Koeniger and Koeniger 2015). This is one of the reasons provoking people for their elimination by destruction.

15.4 REMOVAL FROM GREEN SPACES

Urban green spaces are important refuges for wild bees, but they are also important for the recreation and health of urban citizens (Turo and Gardiner 2019). They are considered a threat to the visitors and are removed and killed.

15.5 PESTICIDAL KILLING OF BEE COLONIES

Giant honeybees are treated like a pest insect in many parts of India. Pest management companies are called to remove honeybee colonies nesting under roofs of houses or balconies. Investigations in rural and urban areas showed that especially in urban areas, a large number of bee colonies were destroyed with insecticides—"These brutal killings are responsible for the death of thousands of *Apis dorsata* colonies every year" (Nagaraja 2016)—and the problem will persist. Removing a colony is only a short-term solution. In search of better living conditions, Apis *dorsata* colonies can perform seasonal migrations over large distances (up to at least 200 km). These migrations help to overcome unfavorable weather conditions and to find good food resources. Due to frequent wandering of honeybee colonies, there is loss of honeybees that are available for pollination on farmland and in forests.

15.6 WHY DO GIANT HONEYBEES NEED TO BE PROTECTED?

Up to 30% of the world's food consumption is derived from plants that depend on pollination by bees and other insects (Oldroyd and Nanork 2009). Bees in particular play a crucial role in food production through the pollination of crops (Marshman et al. 2019). Since agricultural yields can be maximized, in terms of quantity and quality, by abundant and diverse populations of pollinators (Eardley et al. 2016), the presence of several native honeybees is an asset for agriculture. Most of the crops are pollinated by several species of honeybees, but there are some specific crops which are species-specific dependent. One of the most important features of the Asian giant honeybee *Apis dorsata* is its unique ability to forage at dusk, which is essential for the pollination of crops and trees that bloom at night. Therefore, preserving the abundance and diversity of bees is of great economic significance in maintaining and conserving ecosystems.

A collapse of native honeybee populations could result in serious adverse effects on the biodiversity of Southeast Asian ecosystems. Giant honeybees have socio-economic and socio-cultural values in Southeast Asian countries. honey hunting has been practiced for more than 40,000 years in Southeast Asia to generate income for thousands of families belonging to the poor communities. Wild honeybees *Apis dorsata* and *Apis laboriosa* are economically the most important due to large quantities of honey they produce: up to 50–80 kg per colony per year. *Apis dorsata* indirectly contribute to economic development and conservation of protected forests.

15.7 MAIN THREATS FOR SOUTHEAST ASIAN HONEYBEES AND THE NEED FOR CONSERVATION

Deforestation deprives bees of nesting sites, strong and high branches for *Apis dorsata*, shrubs for *Apis florea*, and dense forests for *Apis andreniformis* (Kremen et al. 2004; Freitas et al. 2009). Agricultural intensification, monocultural farming systems, and deforestation have contributed in the decline of honeybees. The monocultural agro-systems often rely on heavy use of pesticides. The impact of pesticides on Asian native honeybees is poorly documented. Bee colonies nesting downwind are affected by pesticide droplets or contaminated dust particles. Besides, systemic insecticides such as neonicotinoids pose a very serious risk to bees because of their higher toxicity and longer persistence in the environment than most pesticides. They are absorbed by a plant, and as it grows, the residues of the insecticide are present in all parts of the treated plants, including the flowers (pollen and nectar). Destructive honey hunting is a major problem in which villagers not only harvest the honey but cut down the whole nests, thereby destroying the whole colony. Night harvesting and indiscriminate use of insecticides are particularly destructive, as many colonies are killed (Deccan Herald 2015; Walter 2018). The high demand for bee brood in some countries is highly is responsible for destruction of bee colonies. Though some information is available on the impact of climate on the western honeybee, *Apis mellifera*, almost nothing is known about the likely impact of climate change on the native Asian honeybees. Forest fires, extended droughts, and high unusual temperatures severely affect honeybee populations. Climate change also may result in mismatch between blooming of flowers and honeybee migrations, leading to a deficit of food for honeybees and other pollinators.

Honey hunters are often receptive to ideas that may help conserve bees (Oldroyd and Nanork 2009). Non-destructive honey harvesting methods can help by preserving the bee brood, thereby allowing multiple harvests without damaging the colonies. Rafter beekeeping, a sustainable management of *Apis dorsata* developed by several communities, needs to be introduced to communities unfamiliar with the method. Protecting nesting sites by preserving forests and reducing the use of pesticides may help restore wild bee populations.

15.8 IMPACTS OF ARTIFICIAL LIGHT AT NIGHT

Apis dorsata is often found in cities (Saravanan et al. 2020) where it is likely to be affected by light pollution. *Apis dorsata* increases its activity into the night in the waxing or waning period of the moon between half-moon and full-moon cycles (Dyer 1985; Young et al. 2021), since that quantity of moonlight is above the illumination threshold required for the optimal functioning of its optical system. Still, the impact of artificial light at night (ALAN) on *A. dorsata* is presently unknown. It is possible that this aggressive bee with its large worker force could dominate night-blooming flowers under conditions of light pollution. The decline in honeybee population could be attributed to loss of bees to artificial light at night.

15.9 AIR POLLUTION KILLING GIANT ASIAN HONEYBEES

Air pollution is killing giant Asian honeybees. A quantitative analysis on the impacts of air pollution on insects by the National Centre for Biological Sciences (NCBS), Bengaluru indicates a decline in India's insect population, including giant Asian honeybees (Thimmegowda et al. 2020). They found that of the 1,800 *A. dorsata* nests collected from moderate to highly polluted areas, more than 80% of the bees died within 24 hours as compared to less polluted areas, where dead bees were less than 20%. The bees from polluted areas showed irregular heartbeats, lower blood cell count, and the expression of genes coding for stress, immunity, and metabolism. They had arrhythmic heartbeats and fewer immune cells, and they showed signs of stress. Ryalls (2022) studyied air pollutants from both urban and rural environments, and found that there were up to 70% fewer pollinators, up to 90% fewer flower visits, and an overall 31% decline in pollination in test plants.

15.10 BEE–HUMAN CONFLICT AND SUGGESTIONS TO OVERCOME IT

Bees help to sustain life on earth, but bee–human conflict is a matter of serious concern both for bees and survival of human beings. Hence, the significant reduction in bee population due to habitat loss and land use change is a matter of serious concern. Without insect pollination, many of our widely consumed and nutritious food crops, fruits, and vegetables would not exist today. According to the United Nations, 75% of the world's food crops depend on animal pollination, linking pollinators like bees directly to global food security (Potts et al. 2016).

Awareness is being spread by many organizations, yet there are many cases of bee–human conflict. For example, there are frequent cases of bee–human conflict in India. *Apis dorsata is* a bee species found in Asia that cannot be domesticated like *Apis mellifera* and *Apis cerana*. They attack only when provoked by humans or animals. Increased urbanization in India has taken a toll on many trees used as nesting sites for bees. They nest on roofs and balconies of the houses and offices, and when these bee colonies make nests, the residents end up calling pest management companies to exterminate them. Unfortunately, the general public is not yet aware of their importance in food security and treat them like cockroaches. They have not realized how important and necessary the honeybees are to the ecosystem.

The best way to overcome this problem is to not open the windows. Even opening of windows is no problem unless there is no artificial light in the room. In cases when the colonies need to be removed, smoking the colonies and removing the combs can make the bees move away rather than killing them.

Within the next decade, India is expected to become the world's most populous country (United Nations 2019) and to experience the world's second-largest growth of urban areas. This growth will mainly take place on former farmland (The Nature Conservancy 2018). Environmental degradation already causes major economic, social, and health problems (Thakur et al. 2014), and India's hunger situation is still serious (Global Hunger Index 2020). Therefore, it is time to act to stop further decline of important wild honeybee pollinators. Initiative is required to create awareness about the importance of wild honeybees. People must know nature to protect it. Apparently, insufficient knowledge and fear are the main reasons for destroying colonies of the giant honeybee, one of India's prime pollinators. People need to be provided better knowledge about wild honeybees and be taught how to interact with them in non-harmful ways. The history of biological evolution teaches us that protecting diversity is to our own advantage (Mayr 1997).

REFERENCES

Ahmad, F., Joshi, S.R., and Gurung, M.B. 2003. *Indigenous honeybees of the Himalayas: The Himalayancliff bee Apis laboriosa Smith and the honey hunters of Kaski*. Vol. 1. International Center for Integrated Mountain Development, Kathmandu, Nepal. https://lib.icimod.org/record/21832.

Allen, M.F., Ball, B.V., and Underwood, B.A. 1990. An isolate of *Melissococcus pluton* from *Apis laboriosa*. *Journal of Invertebrate Pathology* 55(3): 439–440. https://doi.org/10.1016/0022-2011(90)90090-S.

Anderson, D.L., and Morgan, M.J. 2007. Genetic and morphological variation of bee-parasitic *Tropilaelaps mites* (Acari: Laelapidae): New and re-defined species. *Experimental and Applied Acarology* 43: 1–24.

Aryal, S., Thapa, R., and Jung, C. 2015. An overview of beekeeping economy and its constraints in Nepal. *Journal of Apiculture* 30(3): 135–142.

Basavarajappa, S. 2010. Studies on the impact of anthropogenic interference on wild honeybees in Mysore District, Karnataka, India. *African Journal of Agricultural Research* 5(4): 298–305.

Basavarajappa, S., and Raghunandan, K.S. 2013. Colony status of Asian giant honeybee, *Apis dorsata* Fabricius in Southern Karnataka, India. *African Journal of Agricultural Research* 8(8): 680–689.

Basu, P., Bhattacharya, R., and Ianetta, P.A. 2011. Decline in pollinator dependent vegetable crop productivity in India indicates pollination limitation and consequent agro-economic crises. *Nat Prec.* https://doi.org/10.1038/npre.2011.6044.1.

Chantawannakul, P., de Guzman, L.I., Li, J., and Williams, G. 2016. Parasites, pathogens, and pests of honeybees in Asia. *Apidologie* 47(3): 301–324.

Corlett, R.T. 2011. Honeybees in natural ecosystems. In: Hepburn, H.R., and Radloff, S.E. (Eds.), *Honeybees of Asia*. Springer, Berlin, 215–225.

Deccan Herald. 2015. *Girl dies in bee attack during flower show at Lalbagh*. www.deccanherald.com/content/495995/girl-dies-bee-attack-during.html (16.10.2020); Bangalore Mirror 2016: Bee attack in Lalbagh. https://bangaloremirror.indiatimes.com/bangalore/others/Bee-attack-in-Lalbagh-2-suffer-severe-stings/articleshow/53714582.cms (16.10.2020).

Dicks, L.V., Viana, B., Bommarco, R., Brosi, B., Arizmendi, M.D., Cunningham, S.A., Galetto, L., Hill, R., Lopes, A.V., Pires, C. et al. 2016. Ten policies for pollinators. *Science* 354: 975–976.

Dyer, F.C. 1985. Nocturnal orientation by the Asian honey bee, *Apis dorsata*. *Animal Behaviour* 33: 769–774.

Dyer, F.C., and Seeley, T.D. 1994. Colony migration in the tropical honey bee *Apis dorsata* F. (Hymenoptera: Apidae). *Insectes Sociaux* 41(2): 129–140.

Eardley, C., Freitas, B.M., Kevan, P.G., Rader, R., Gikungu, M., Klein, A.M., and Hill, R. 2016. *Background to pollinators, pollination and food production in Annex to document* IPBES/4/INF/1/Rev.1 Thematic Assessment on Pollinators. https://www.ipbes.net/sites/default/files/downloads/pdf/individual_chapters_pollination_20170305.pdf

FAO. 2008. *A contribution to the international initiative for the conservation and sustainable use of pollinators – rapid assessment of pollinator's status.* FAO, Rome, January, 52.

Freitas, B.M., Imperatriz-Fonseca, V.L., Medina, L.M., de Matos, P., Kleinert, A., Galetto, L., Nates-Parra, G., and Quezada-Eúan, J.G. 2009. Diversity, threats and conservation of native bees in the neotropics. *Apidologie* 40: 332–346.

Gallai, N., Salles, J.M., Settele, J., and Vaissière, B. 2009. Economic valuation of the vulnerability of world agriculture confronted to pollinator decline. *Ecological Economics* 68: 810–821. https://doi.org/10.1016/j.ecolecon.2008.06.014.

Global Hunger Index. 2020. *Complete report 2020.* www.globalhungerindex.org (16.10.2020).

Gregory, M., and Jack, C. 2022. *IFAS extension university of Florida.* IFAS Extension at Hernando County, Brooksville. https://doi.org/10.32473/edis-IN1348-2022.

Hallmann, C.A., Sorg, M., Jongejans, E., Siepel, H., Hofland, N., Schwan, H., Stenmans, W., Müller, A., Sumser, H., Hörren, T., Goulson, D., and de Kroon, H. 2017: More than 75 percent decline over 27 years in total flying insect biomass in protected areas. *PLoS One* 12(10): e0185809.

Hawkeswood, T.J., Sommung, B., and Sommung, A. 2019. A record of the giant honey bee, *Apis dorsata* Fabricius, 1793 (Insecta: Hymenoptera: Apidae) from Sisaket Province, Thailand. *Calodema* 720: 1–4.

IUCN. 2022. *The IUCN red list of threatened species. Version 2022-2.* https://www.iucnredlist.org (09.05.2023).

Jesmer, B.R., Merkle, J.A., Goheen, J.R., Aikens, E.O., Beck, J.L., Courtemanch, A.B., Hurley, M.A., McWhirter, D.E., Miyasaki, H.M., Monteith, K.L. et al. 2018. Is ungulate migration culturally transmitted? Evidence of social learning from translocated animals. *Science* 361: 1023–1025.

Joshi, S.R., Ahmad, F., and Gurung, M.B. 2004. Status of *Apis laboriosa* populations in Kaski district, western Nepal. *Journal of Apicultural Research* 43(4): 176–180.

Joshi, S.R., and Gurung, M.B. 2015. Non-destructive method of honey hunting. *Bee World* 86(3): 63–64.

Kastberger, G., and Hötzl, T. 2015. Wie gefährdet sind die Riesenhonigbienen? In: Lorenz, S., and Stark, K. (Eds.), *Menschen und Bienen.* Oekom, München, 93–104.

Koeniger, N., and Koeniger, G. 1980. Observations and experiments on migration and dance communication of *Apis dorsata* in Sri Lanka. *Journal of Apicultural Research* 19: 21–34.

Koeniger, N., and Koeniger, G. 2015. *Apis dorsata* – champions of defence. *Bees for Development Journal* 115: 6–10.

Kremen, C., Williams, N.M., Bugg, R.L., Fay, J.P., and Thorp, R.W. 2004. The area requirements of an ecosystem service: Crop pollination by native bee communities in California. *Ecology Letters* 7: 1109–1119. https://doi.org/10.1111/j.1461-0248.2004.00662.x.

Laurance, W.F. 2007. Forest destruction in tropical Asia. *Curr. Sci.* 93: 1544–1550.

Mahindre, D.B.A. 2004. *'Apis dorsata' F.—the manageable bee.* Agrotec Publication, Udaipur Rajasthan, India, 160.

Marshman, J., Blay-Palmer, A., and Landman, K. 2019. Anthropocene crisis: Climate change, pollinators and food security. *Environments* 6–22. https://doi.org/10.3390/environments6020022.

Mayr, E. 1997. *This is biology.* Harvard University Press, Cambridge, MA.

Morse, R.A., and Laigo, F.M. 1969. *Apis dorsata in the Philippines.* Philippine Association of Entomologists, Laguana.

Mukherjee, R., Deb, R., and Devy, S.M. 2019. Diversity matters: Effects of density compensation in pollination service during rainfall shift. *Ecology and Evolution* 9(17): 9701–9711.

Nagaraja, N. 2016. Effect of insecticide poisoning on mortality of giant honeybee, *Apis dorsata* colonies. *PESQUISA- International Journal of Research* 2(1): 96–100.

The Nature Conservancy. 2018. *Nature in the urban century*. www.nature.org/content/dam/tnc/nature/en/documents/TNC_NatureintheUrbanCentury_FullReport.pdf (16.10.2020).

Oldroyd, B.P., and Nanork, P. 2009. Conservation of Asian honey bees. *Apidologie* 40(3): 296–312.

Oldroyd, B.P., Osborne, K.E., and Mardan, M. 2000. Colony relatedness in aggregations of *Apis dorsata* Fabricius (Hymenoptera, Apidae). *Insectes Sociaux* 47: 94–95.

Oldroyd, B.P., and Wongsiri, S. 2006. *Asian honey bees: Biology, conservation and human interactions*. Harvard University Press, Cambridge, MA, 340.

Paliwal, G. 2005. *Sustainable honey extraction shows way to self-reliance in Maharashtra's tribal district*. https://india.mongabay.com/2021/08/sustainable-honey-extraction-shows-way-to-self reliance-in-maharashtras-tribal-district/.

Partap, U. 2011. The pollination role of honeybees. In: Hepburn, R., and Radloff, S.E. (Eds.), *Honeybees of Asia*. Springer-Verlag, Berlin, Germany, 227–255.

Perumal, B., Muneeswaran, V., Pothirasan, N., Reddy, K.R.M., Sai Pranith, K.S., Chaitanya, K., and Krishna Kumar, R. 2021. Bee eloper: A novel perspective for emancipating honey bees from its comb using a contrivable technique. *AIP Conference Proceedings* 2378: 020003. https://doi.org/10.1063/5.0058827.

Pokhrel, S. 2010. Climato-cyclic immigrations with declining populations of wild honeybee, *Apis dorsata* F. in Chitwan valley, Nepal. *The Journal of Agriculture and Environment* 11(6): 51–58.

Potts, S.G., Biesmeijer, J.C., Kremen, C., Neumann, P., Schweiger, O., and Kunin, W.E. 2010. Global pollinator declines: Trends, impacts and drivers. *Trends in Ecology and Evolution* 25(6): 345–353. https://doi.org/10.1016/j.tree.2010.01.007.

Potts, S.G., Imperatriz-Fonseca, V., Ngo, H.T., Aizen, M.A., Biesmeijer, J.C., Breeze, T.D., Dicks, L.V., Garibaldi, L.A., Hill, R., Settele, J. et al. 2016. Safeguarding pollinators and their values to human well-being. *Nature* 540: 220–230.

Raghunandan, K.S., and Basavarajappa, S. 2015. Studies on the impact of man-made activities on *A. dorsata* population in South-Western Karnataka, India. *Research Journal of Social Science & Management* 5(1): 141–145.

Robinson, W.S. 2021. Surfing the sweet wave: Migrating giant honey bees (Hymenoptera: Apidae: *Apis dorsata*) display spatial and temporal fidelity to annual stopover site in Thailand. *Journal of Insect Science* 21(6), November. https://doi.org/10.1093/jisesa/ieab037.

Ryalls, J. 2022. Air pollution and pollination. *Bee Culture, the Magazine of American Beekeeping*, January 27.

Sánchez-Bayo, F., and Wyckhuys, K.A.G. 2019. Worldwide decline of the entomofauna: A review of its drivers. *Biological Conservation* 232: 8–27.

Saravanan, S., Kamaladhasan, N., Maruthupandian, J., Basu, M.J., and Chandrasekaran, S. 2020. Beehive removal practices in urban India: Bane of bee's life? *Materials Today: Proceedings*. https://doi.org/10.1016/j.matpr.2020.10.060.

Satterfield, D.A., Sillett, T.S., Chapman, J.W., Altizer, S., and Marra, P. 2020. Seasonal insect migrations: Massive, influential, and overlooked. *Front. Ecol. Environ.* 18: 335–344.

Shrestha, J.B., Mandal, C.K., Shrestha, S.M., and Ahmad, F. 2002. The trend of the giant honeybee, *Apis dorsata* Fabricius colony migration in Chitwan, Nepal. *The Wildlife* 7(10): 16–20.

Sihag, R.C. 2014. Phenology of migration and decline in colony numbers and crop hosts of giant honeybee (*Apis dorsata* F.) in semiarid environment of Northwest India. *Journal of Insects* 9, Article ID 639467. http://dx.doi.org/10.1155/2014/639467.

Somanathan, H., Warrant, E.J., Borges, R.M., Wallén, R., and Kelber, A. 2009. Resolution and sensitivity of the eyes of the Asian honeybees *Apis florea*, *Apis cerana* and *Apis dorsata*. *Journal of Experimental Biology* 212(15): 2448–2453.

Stewart, A.B., and Waitayachart, P. 2020. Year-round temporal stability of a tropical, urban plant-pollinator network. *PLoS One* 15(4): e0230490.

Sunil Kumar, S., and Reddy, M.S. 2014. Indegenous methods of honey harvesting from rock bee (Apis dorsata) colonies by Siddhi tribes of Uttar Kannada district, Karnataka, India. *Insect Environment* 19(4): 236–242.

Tan, N.Q., and Ha, D.H. 2002. Socio-economic factors in traditional rafter beekeeping with Apis dorsata inVietnam. *Bee World* 83: 165–170.

Thakur, B.K., Rout, H.S., and Chakraborty, T. 2014. Environmental degradation, sustainable development and human well-being: Evidence from India. *Manthan: Journal of Commerce and Management* 1(1): 101–120.

Thapa, R., Aryal, S., and Jung, C. 2018. Beekeeping and honey hunting in Nepal: Current status and future perspectives. In: *Asian beekeeping in the 21st century*. Springer, Singapore, 111–127.

Thimmegowda, G.G., Mullen, S., Sottilare, K., Sharma, A., Mohanta, S.S., Brockmann, A., Dhandapany, P.S., and Olsson, S.B. 2020. A field-based quantitative analysis of sublethal effects of air pollution on pollinators. *PNAS* 117(34): 20653–20661.

Turo, K.J., and Gardiner, M.M. 2019. From potential to practical: Conserving bees in urban public green spaces. *Frontiers in Ecology and the Environment* 17(3): 167–175.

Underwood, B.A. 1986. *The natural history of Apis laboriosa Smith in Nepal*. Master's thesis, Cornell University, Ithaca, 88.

Underwood, B.A. 1992. Impact of human activities on the Himalayan honeybee, *Apis laboriosa*. In: Verma, L.R. (Ed.), *Honeybees in mountain agriculture*. Oxford University Press, IBH Publishing Co. Pvt. Ltd., New Delhi, 51–57.

United Nations. 2019. *World population prospects 2019*. https://population.un.org/wpp/Download/Standard/Population/ (16.10.2020).

Valli, E., and Summers, D. 1988. Honey hunters of Nepal. *National Geographic* 174(5): 660–671.

Venkatesh, G. 2014. Study on population status in relation to urban development in few selected nesting site of rock bee colonies, *Apis dorsata* F. *International Journal of Scientific and Research Publications* 4(10): 22–23.

Venkatesh, G., and Reddy, C.C. 1989. Rates of swarming and absconding in the giant honey bee, *Apis dorsata* F. *Proceedings of the Indian Academy of Sciences (Animal Sciences)* 98(6): 425–430.

Walter, S.M. 2018. *Bee lab Bangalore*. 2nd edn. Goethe Institut & NCBS, Bengaluru. https://drive.google.com/open?id=1uaZ6KKf2FDlN7cHvUgMzCRlCHycuRw1I (16.10.2020).

Waring, C., and Jump, D.R. 2004. Rafter beekeeping in Cambodia with Apis dorsata. *Bee World* 84: 14–18.

Wenzel, A., Grassa, I., Belavadi, V.V., and Tscharntke, T. 2020. How urbanization is driving pollinator diversity and pollination – a systematic review. *Biological Conservation* 241: 108321.

Wilcove, D.S. 2008. Animal migration: An endangered phenomenon? *Issues Sci. Technol.* 24: 71–78.

Woyke, J., Wilde, J., Reddy, C.C., and Nagaraja, N. 2005. Periodic mass flights of the giant honey bee *Apis dorsata* in successive days at two nesting sites in different environmental conditions. *Journal of Apicultural Research* 44(4): 180–189.

Woyke, J., Wilde, J., and Wilde, M. 2012. Swarming and migration of *Apis dorsata* and *Apis laboriosa* honey bees in India, Nepal and Bhutan. *Journal of Apicultural Science* 56(1): 81–91.

Young, A.M., Kohl, P.L., Rutschmann, B., Steffan-Dewenter, I., Brockmann, A., and Dyer, F.C. 2021. Temporal and spatial foraging patterns of three Asian honeybee species in Bangalore, India. *Apidologie* 52: 503–523.

Zattara, E.E., and Aizen, M.A. 2021. Worldwide occurrence records suggest a global decline in bee species richness *One Earth* 4: 114–123.

16 Foraging in Giant Honeybees

Dharam P. Abrol and Parul Sharma

16.1 INTRODUCTION

Food is critical for survival of many organisms, ranging from bacteria to primates, for which they have to make movements known as search strategies. By studying the factors influencing how these organisms forage, we can gain insights into the interactions between organisms and their environments and make predictions about how organisms will react to changing environmental conditions. The habitat of most animals provides sufficient resources of different types that are essential for their survival; however, these may not be necessarily available in the close vicinity nearby. Accordingly, these animals should forage to bridge some temporal or spatial distance to locate the essential resources (Schoener 1971). Foraging can be defined as the process of searching food resources and their acquisition. It is one of the most important tasks in the life of an organism related to its survivability. The data on foraging ability and foraging behavior are therefore necessary to the understanding of population dynamics and community structure of bees, as well as to develop conservation strategies. The foraging behavior of social insects is especially interesting, as the individuals do not forage to meet their own requirements but rather gather food resources to meet the needs of the colony. Foraging force is divided in a way that some individuals forage for nectar and some for pollen, others return to the nest carrying both nectar and pollen, and a small proportion of the foraging force return with water. The foraging requires information of locations, distances between locations and the availability of food. The honeybees that consistently return to the same location are called central-place foragers in which every foraging trip consists of distinct 'outward' and 'homeward' paths.

Both flowering plants and honeybees have a special mutualistic relationship in which both benefit from each other. As honeybees get food, they in turn facilitate the pollination process. Honeybees visit flowers to collect pollen and nectar. Pollens are the principle source of protein, fats, vitamins and minerals, which are essential for growth and development, repairing of worn-out tissue and stimulating the development of hypopharyngeal glands in honeybees. Nectar containing carbohydrate compounds (sucrose, fructose and glucose) is food of the honeybees and is the raw material of honey (Jones and Yates 1991).

Foraging behavior of bees is a complex phenomenon which depends upon several factors. Physical features of flowers—such as color, shape, odor and the diversity of environmental factors such as temperature, humidity, light, solar radiation, time of the day and nectar flow—decisively shape the behavior of pollinating insects, thereby

DOI: 10.1201/9781003294078-16

influencing cross-pollination and the production of the crop (Visscher and Seeley 1982; Corbet et al. 1993). The activity spectra of both flowers and pollinators coincide on daily and yearly rhythms (Faegri and van der Pijl 1978). The bees and flowers form a unique mutually compatible system, the functions of which are affected by a set of common environmental factors in a manner that functional activity of bees and flowers synchronize. To achieve it, there are obvious interactions between biotic and abiotic factors such as temperature, light intensity, solar radiation, wind velocity, rains, dew drops, time of day, distance to food sources, availability of resources, abundance of nectar and concentration of sugar in nectar. Diel patterns of bee visits are related to the changes in the quality and quantity of floral rewards, as well as the direct influence of environmental factors. Hence, the complex interactions of biophysical parameters and food rewards determine foraging success (Heinrich 1975a). Bertsch (1983) stated that close dependence between pollinating insects and their flowers can be better understood if the functions of flowers, their adaptations and the needs of flower visitors are known, which would consequently lead to the management of a particular group of pollinators in a given set of ecological conditions. Floral patterns attracting pollinators and other specialized mechanisms resulting, therefore, seem to have evolved in relation to the sensory development of the bees which in the course of development gradually perfected their own sensory faculties and physical efficiency to the identify color, odor, iconic numerals and symmetry of flowers. Differentiation of nectaries accompanying floral evolution became responsible to trophic activities of progressively evolving contemporary pollinators which directed specific changes in their reward-getting capabilities (Leppik 1977). Thus, co-evolutionary adaptations by the vast variety of angiosperms passing through several niches of time have chiefly been attributed to the mutually interrelated modifications among flowers and their pollinators. Obviously, the rewards system offered by the flowers plays a determinant role besides the interplay of other factors which participate in attracting floral visitors. Evidently, the flower visitors relationship depends upon the energy requirements of flower visitors and the amount of quantity they can harvest from flowers (Hocking 1953; Churchill and Christenenson 1970; Hainsworth and Wolf 1972a, 1972b, 1972c; Heinrich and Raven 1972; Stiles 1971), which influences the frequency of visits to flowers (Heinrich 1973, 1975a, 1975b). The social setup of honeybees requires a large quantity of food supply for their brood—many times more than their own energy requirements is secured by them by repeated visits to flowers (Reddy and Reddy 1984); this can keep happening only if a flower provides sufficient reward to attract the foragers on one hand and limits the reward on the other hand, compelling the visitors to frequent other flowers of the same species (Heinrich 1975a, 1975b) and thus enabling development of an optimum strategy both by flowers and pollinators for maximum cross-pollination (Emlen 1966; Schoener 1971; Cody 1974).

16.2 DIVERSITY OF HONEYBEES IN ASIA

The Asian continent is the richest in the world in honeybee species diversity. The only exotic honeybee is *Apis mellifera*, which has been widely used for honey production (Ruttner 1988). Among the nine species, eight species are distributed in Asian countries under different agro-climatic conditions as detailed.

TABLE 16.1
Diversity of Different Honeybee Species

Scientific name of species	Common name	Distribution
A. dorsata Fab.	Rock Bee	Asia
Apis florea Fab.	Little Bee	Asia
Apis laboriosa Smith.	Largest Bee	Asia
Apis cerana Fab.	Asiatic Hive Bee	Asia
Apis andreniformis Smith.	Smallest Bee	Asia
Apis mellifera Lin.	European Bee	Europe
Apis koschevnikovi Enderlein	Red Bee	Malaysia
Apis nuluensis Lin.	Malaysian Bee	Malaysia
Apis nigrocinta Smith	Black Bee	Indonesia

Of the nine species of honeybees, *Apis mellifera*, *Apis cerana*, *Apis koschevnikovi*, *Apis nuluensis* and *Apis nigrocinta* all build nests containing a series of parallel combs. These species usually nest in cavities. The nests of honeybee species *Apis florea*, *Apis andreniformis*, *A. dorsata* and *Apis laboriosa* consist of single combs.

16.3 FACTORS IMPACTING FORAGING ACTIVITY

There are many factors which influence foraging activity. They include presence of the queen, whether mated or virgin. Higher foraging activity with less pollen collection was found in colonies headed by virgin queens than it was in colonies headed by mated queens, while lower foraging activity and pollen collection were found in queenless colonies than in colonies with a mated or virgin queen (Free et al. 1985). Also, foraging activity is influenced by colony strength and brood-rearing activity (Amdam et al. 2009; Abou-Shaara et al. 2013), and by the degree of pollen need (Weidenmuller and Tautz 2002). Foraging activity is also influenced by the type of bee hive (Abou-Shaara et al. 2013). The infection of honeybee foragers with diseases and parasites such as *Nosema* sp. or *Varroa destructor* may result in the inability of foragers to return to their colonies or increase their time to return (Kralj and Fuchs 2006, 2010). The genotype of honeybee strains (high and low pollen-hoarding bees) strongly affected foraging behavior for nectar or for pollen (Pankiw et al. 2002). The bees have preferences for some plant resources over others. Moreover, Fulop and Menzel (2000) found that the reward volume has an impact on foraging activity and that bees can perceive the amount of reward from the feeding source.

With respect to environmental factors, foraging activity is influenced passively by elevated temperature (Cooper and Schaffer 1985). In contrast, relative humidity had less of an effect on flight activity (Joshi and Joshi 2010). Other environmental factors that impact foraging activity include solar ultraviolet-B (UV-B) rays (Collins et al. 1997), altitude (Mattu et al. 2012), electromotive fields (Sharma and Kumar 2010) and exhaust fuels, which reduce the ability of worker bees to recognize floral odors (Girling et al. 2013).

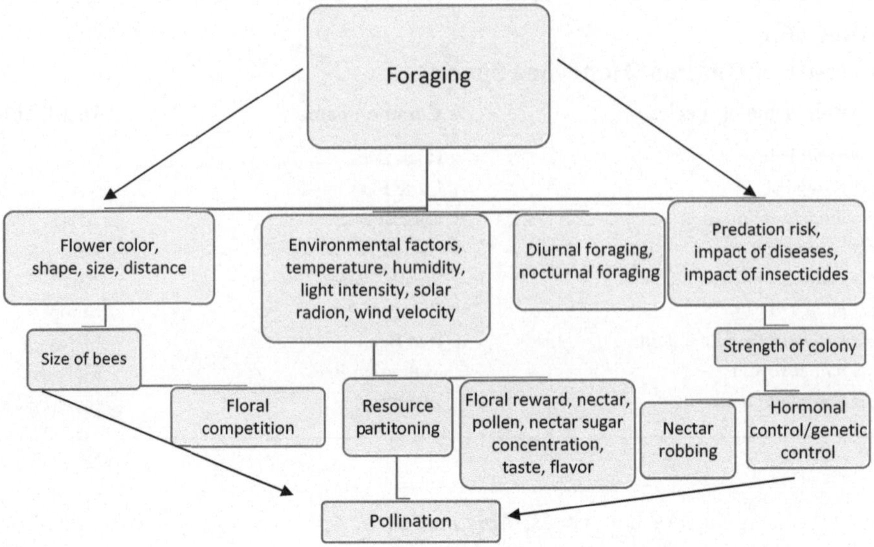

FIGURE 16.1 Schematic model exhibiting factors influencing foraging behavior of honeybees.

Foraging behavior can also be influenced by natural enemies of honeybees in the field. Tan et al. (2013) reported a reduction in foraging activity of honeybees by 55%–79% by the presence of predators such as hornets. Similarly, the presence of bee-eaters strongly deterred foraging activity of honeybees (Ali and Taha 2012). Insecticides also influence foraging behavior. Schneider et al. (2012) found a significant reduction in foraging activity, as well as longer foraging flights at doses of two neonicotinoid insecticides: 0.5 ng/bee or more for clothianidin and 1.5 ng/bee or more for imidacloprid during the first three hours after treatment. In contrast, the presence of residues in the nectar and pollen of oilseed rape and maize due to seed treatment with thiamethoxam was reported to represent a low risk to honeybees (Pilling et al. 2013). More investigations of these factors are urgently required, especially since neonictinoids are so widely used. Other factors may also have an impact on foraging behavior are foraging distance and time of year (Steffan-Dewenter and Kuhn 2003; Beekman et al. 2004). Pearce et al. (2013) found no considerable effects of moving beehives from their location to another location as far as 26 km from their original site of honeybee foraging activity (Figure 16.1).

16.3.1 Foraging as a Function of Energetic Relationships

16.3.1.1 Foraging in Relation to Reward from Flowers

Pollinators differ in their energy requirements, from low-energy groups such as ants and flies to higher energy–requiring endothermic groups of mammals and birds (Heinrich 1983). Besides color, shapes and odor of flowers, energy requirements and caloric rewards offered by the flowers also determine whether or not an animal can be a dependable flower visitor (Heinrich and Raven 1972; Heinrich 1975a; Abrol 1991, 1992a, 1993, 1998a, 1998b). Rewarding system developed by flowers enable bees to make distinction between them and closely related species or ecotypes. This has resulted in a mutualistic development of a co-evolutionary process and have evolved co-partnership between the two (Leppik 1977). The interdependence of pollinating bees with flowers depends much upon their energy requirement and the balance sheet they share with flowers (Heinrich and Raven 1972). The pollinators with high energy requirements may not forage at the flowers, which provide low caloric rewards (Heinrich 1983).

Energetics of pollinator plant interaction has been investigated in only few instances. Many studies of foraging economics have focused on birds—especially nectarivores (Stiles 1971; Wolf and Hainsworth 1971; Gill and Wolf 1975), as they are usually easy to observe and their activities are easily classifiable into discrete categories (e.g. flight, rest and torpor). Evidently, their energy intake is rather easily quantified. Some interesting information on bumblebees and honeybees has also been generated (Heinrich 1975a, 1975b; Waddington et al. 1981; Schaffer et al. 1979). Because of their high energy requirements, both bumblebees and honeybees have served as ideal subjects for studying foraging energetics. Their foraging patterns are largely determined by the relationship of energy requirements to flight to nectar reward in flowers (Heinrich 1975a, 1975b, 1983; Wolf et al. 1972, 1975). The high energy requirements of bumblebees and honeybees result in part from their social organization and ability to thermoregulate (Heinrich 1983)

Most bee–plant interactions depend upon the energy needs of the pollinators and the energy available from plants (Heinrich and Raven 1972; Heinrich 1975a, 1975b, 1979a, 1979b, 1983; Abrol 1986a, 1992a). The foraging strategies of bees and allocation of time to various daily activities is probably related to their level of sociality. Honeybees, which are active throughout the year, need continuous influxes of energy for thermoregulation and brood development. The solitary bees, in contrast to honeybees, have less energy demands due to their individual nesting habits, as they need not incubate brood or support nest mates. Foraging efficiency of solitary bees differs from eusocial bees, which have longer-lived colonies (Wcislo and Cane 1996). Evidently, foraging strategies of honeybees could be entirely different from earlier studied wild solitary bees.

Abrol (2007) studied the energy used by a bee *Apis cerana* and *Apis mellifera* in relation to various rapeseed/mustard (*Brassica juncea*, *B. carinata*, *Brassica campestris var toria*) and umbelliferous crops (*Foeniculum vulgare* and *Coriandrum sativum*). The energy balance varied from one day to another and crop to crop. In general, the foraging profitability per bee was in the order: *B. carinata* > *Brassica juncea* > *Brassica campestris var toria* > *Foeniculum vulgare* > *Coriandrum sativum* for both the species. *B. carinata* appears to be competitively superior as a forage crop over *Brassica juncea* > *Brassica campestris var toria* > *Foeniculum vulgare* > *Coriandrum sativum*. He further found that the flight activities of *A. cerana* commenced much earlier and ceased later than did those of *A. mellifera*. Evidently, *A. cerana* worked for longer durations in the field as compared to *A. mellifera*. *Apis cerana*, on average, spent 300 minutes in foraging, 336.78 minutes in nest activity and 803.21 minutes resting as compared to *A. mellifera*, which under similar conditions spent less time in foraging (284.28 minutes) and nest activity (222.50 minutes) and more time in resting (933.21 minutes). *Apis cerana*, on average, made 10–15 foraging trips/day for nectar collection and each trip averaged 23.0–25.0 minutes, while *A. mellifera* made 10–12 trips/day and on an average spent 26.5 minutes/trip. Abrol (2016) reported that of the two honeybee species, *Apis dorsata* preferred relatively large sucrose-dominated flowers, whereas *Apis florea* preferred small flowers with glucose- + fructose-dominated sugars. However, energy cost and reward provided a clear picture which indicates that *A. dorsata*—bigger in size and tongue length—forage flowers providing high energy rewards compared to *A. florea*, which forage low energy–rewarding flowers. Such preferences are associated with the foraging profitability of the pollinators.

Apis florea and *A. dorsata* have been recorded as versatile pollinators of wide variety of field and fruit crops which varied in time and space. Abrol (1992a) reported that bees with higher energy requirements do not forage on the flowers providing low caloric rewards. The flowers of *F. vulgare*, *C. sativum*, *D. carota*, *A. cepa*. *Trigonella foenumgraecum* var Kasuri and *Mangifera indica* were not attractive to *A. dorsata* because the bee could make no profit from these crops. However, foraging population of *A. florea* was observed in large numbers. The bee with its smaller size and body weight was morphologically better adapted to extract maximum caloric reward from the flowers. Because of relatively low energy requirements, the bees were still able to maintain an energy balance and visited the crop in large numbers. Sihag and Kapil (1983) reported that *Apis florea* visited *Brassica* crops in larger numbers than *A. dorsata*. They attributed this difference to the different energy demands of the bees. In case of sunflower, a wide variety of pollinators were equally attracted, though caloric reward per floret was low, yet each bee species was able to maintain the energy balance. Since the flower head with platform provides no barrier for the landing of foragers, the energetic cost is reduced due to temporary suspension of hovering flight and large number of flowers can be visited in rapid succession. Similarly, *Luffa cylindrica* flowers are large with open nectaries and are easily accessible to each type of pollinator irrespective of their body size. An interesting situation was observed in the case of *M. sativa* and *T. alexandrium* which bloom simultaneously and compete for pollinators. *Medicago sativa* has comparatively higher caloric reward than *T. alexandrium*. The honeybees (*Apis* sp.) were attracted to *M. sativa* during the early hours of the day at peak periods of nectar production, but after 11:00 hours onward, foraging populations shifted to *T. alexandrium* due to reduction in quantity of available rewards from *M. sativa*. Evidently, pollinators even prefer low-rewarding flowers, when high floral rewards are not available or the nearby resources are depleted. *A. florea* visited flowers with low caloric rewards, whereas *A. dorsata* preferred those with high rewards. This behavior was largely guided by their energy demands. Generally, the smaller flowers with little nectar are unattractive to large hovering animals such as hummingbirds and sphinax moth, which probably cannot suffice their energetic demand. However, insects such as bumblebees which land on inflorescences of *Spiraea latifolia* L. (Rosaceae) and *Solidago canadensis* (Asteraceae) are still able to maintain an energy balance despite minute amounts of nectar in flowers (Heinrich 1972) because the energy expended in walking from one flower to another is 100 times less than an equivalent period of flight and because the flowers can be visited in rapid succession (Heinrich and Raven 1972).

Abrol (1998b) studied the foraging behavior of *Apis florea* and *A. dorsata* in relation to the energy production pattern of four plant species. The energy produced ranged from 0.35 joules per flower per day in *Lantana indica* to 1,031 joules in *Adhatoda vasica*. However, flowers of *A. vasica* which are structurally more complex were less preferred than those of the other three species. The bees preferred to forage on *Brassica juncea*, followed by *B. campestris var. toria*, from which they could harvest more energy per unit time. In a similar study, Abrol (2006) observed the foraging behavior of four honeybee species (*Apis mellifera*, *A. cerana*, *A. dorsata* and *A. florea*) in relation to energy production rates of *Prunus persica* and a simultaneously blooming weed, *Lepidagathus incurva*. Energy produced ranged from a minimum of 0.642 ± 0.01 joules (*L. incurva*)/flower/day to a maximum of 1.49 ± 0.14 joules/day (*P. persica*). The weed having higher nectar sugar concentration and high flowering density attracted more number of bees as compared to peach. The foraging rates of bees were much higher on the weed and they could harvest more energy per unit time from the weed flowers. The results indicate that food acquisition efficiency and quality of food determines the foraging decision of bees. Bees are generally attracted to flowers that are available in large numbers and are more rewarding in terms of energy (Abrol 1993). Sihag and Rathi (1992) reported that *Apis florea* populations were more attracted to *Brassica compestris* flowers providing less caloric reward than *Cajanas cajan* having high caloric reward. This species could easily harvest more energy in a unit time from simple flowers of *Brassica compestris* compared to more complex papilionaceous *flowers of Cajanas cajan*, making the visitor approach much more complex. Earlier, the same author Sihag (1990) observed similar behavior of two honeybees species, *Apis florea* and *Apis dorsata*, on *Brassica chinensis* and *Eruca sativa* wherein the flowers of *Brassica chinensis* were more preferred. Verma (1995) also observed in his study that when alternate sources were available, *A. mellifera* did not prefer to forage on the apple flowers. These investigations revealed that the bees prefer to forage on more attractive plants and neglect the others when they flower, at the same time and location affecting the pollination requirements of the less attractive plant species to a greater extent.

The pollinators with high energy requirements may not forage at the flowers which provide insufficient caloric reward (Heinrich 1983). Smaller flowers with low caloric rewards are unattractive to larger hovering animals which cannot meet their energetic requirements through them. In general, size of the flower and caloric rewards in relation to size of visitor and energy requirements seem to be determinants for resource partitioning among the various bees and thus permitting co-existence under similar ecological conditions. The population of certain species of pollinating bees was found to be the function of their body size, as well as the size of the flowers, since the feeding pattern of many animals suffer as a function of their triplicate structure (Heinrich 1979a, 1979b; Harder 1985, 1986).

16.3.1.2 Foraging Preference of Bees in Relation to Nectar Sugar Compenents

Sihag and Kapil (1983) found that *Apis dorsata* with higher energy requirements visited sucrose-dominated flowers more frequently than did *Apis florea*, which mostly relied on glucose-dominated ones. Wyke (1952) reported that honeybees preferred equi-proportioned sugars. Waller (1972) and Abrol (1985), on the contrary, stated that bees prefer nectar with one dominant sugar rather than the equi-proportioned sugars. Abrol and Kapil (1991) found that honeybee *Apis dorsata* and solitary bees *Megachile lanata* Lepel, *M. cephalotes* Smith and *Xylocopa fenstrata* F. visited sucrose-dominated nectars more frequently than did *Apis florea* F. and *Pithitis smargdula* F., which heavily relied on glucose-dominant nectars. Preference of sucrose- or glucose-dominant nectar indicates that total caloric reward is an important factor determining the attractiveness of foraging population on a particular crop (Abrol and Kapil 1991).

Abrol (1991) reported that *A. dorsata* bees were generally most active from 11:00–14:00 hours when temperature ranged from 24.5–34.5°C, relative humidity 22.5–43%, light intensity 5,100–5,500 lux, solar radiation 60–85 mW/cm^2, soil temperature 22.5°–32°C, nectar sugar concentration 32%–45% and wind velocity 2.9–4.2 km/hr. Increases in temperature above this range resulted in decline of bee activity. Of the seven factors studied, the direct effect of light intensity, solar radiation and soil temperature were positive and other factors negative.

16.3.1.3 Nectar Productivity and Bee Activity

One source of variation in foraging cues between genotypes is the nectar, which may provide variety of stimuli. Nectar is the potential energy rewards provided by the flowers to their visitors. It has been found to be very significant parameter that decisively shapes the behavior and physiology of pollinators in relation to their energy demands (Heinrich 1975e; Abrol 1986a, 1990, 1995a, 2007). Abrol (1990) studied the energetics of nectar production in 54 cultivars of apple in relation to foraging behavior of *Apis cerana* and *A. mellifera*, finding that cultivars varied greatly in pattern of nectar production characteristics and pollinator attractiveness. Cultivars providing higher amounts of sugar, nectar concentration and energy were highly attractive to bees, which was reflected in their population dynamics. In a similar study, (Abrol 1992b) studied foraging preferences of honeybee *Apis cerana* and *Apis mellifera* in relation to nectar productivity of 13 cultivars of strawberry (*Fragaria* sp.). Foraging of both *Apis cerana* and *Apis mellifera* correlated with energy yields. The results suggest that cultivars with higher caloric rewards had a competitive edge over others in attracting foraging populations of both species. Similar results were obtained in relation to the foraging preferences of honeybees *A. mellifera* and *A. cerana* on almond cultivars (Abrol 1995a).

16.3.1.4 Floral Constancy

Floral constancy is one of the striking features of honeybees wherein single trip, a honeybee almost invariably collects reward from just one type of flower (Free 1970; Grant 1950). The examination of pollen types can help determine floral constancy of bees. For example, 99% of the pollen grains in pollen pellets carried by A. *mellijera* have been shown to be from the same species of plant (Stimec et al. 1997). Raj et al. (1993) found that *A. mellifera* and *A. cerana* foraging on rapeseed *Brassica campestris* exhibited pollen consistency, with 66.78% of the pollen loads collected by *A. mellif*era and 65.45% by *Apis cerana* found to be from mustard crop. Priti and Sihag

(1997) also observed similar phenomenon in case of *A. mellifera* and *A. cerana* visiting cauliflower (*Brassica olearacea* L. var. *botrytis* cv. Hazipur Local) blossoms at Hisar, India.

Choudary (1978) reported that *Apis cerana indica* showed higher floral fidelity as compared to other foragers, i.e. *A. florea*, *A. dorsata* and *A. mellifera* foraging on alfalfa (*Medicago sativa*).

Honeybees' floral constancy makes them well-suited for pollination purposes. At any one time, a hive of bees may be collecting pollen from dozens of different plant species. Floral constancy and local foraging has an adaptive advantage leading to sympatric and allotropic speciation in orchids, which are pollinated only by male euglossine bees (Dressler 1968; Dodson 1975). Nataraj et al. (2000) also reported flower constancy in *Apis cerana* under Tamilnadu conditions, whereas Thiyagesan et al. (2001a, 2001b) reported risk-sensitive and central foraging in honeybee *A. cerana*. Devkota and Thapa (2005) found that the pollen loads of *Apis cerana* bees in March and June were mostly from coconut (90%), gramineae (6%) and jackfruit (4%). In *Apis mellifera*, 67% of pollen loads were from coconut, 25% from gramineae and 10% from Rubiaceae, thereby indicating floral constancy in collection of pollen.

16.3.1.5 Species-Specific Variations in Foraging Activity

Abrol (1992a) studied the foraging activity of honeybee *Apis cerana indica* and *A. dorsata* on peach flowers in relation to weather conditions, and found that initiations of flight was a dual threshold of temperature and light intensity and cessation controlled by reduction in light intensity for both the bee species, though the temperature was still over the levels required for initiation of the activity. Foraging population correlated significantly and positively with ambient temperature, light, solar radiations and nectar sugar concentration, and negatively with relative humidity. Only one factor—namely light intensity—directly influenced the flight activity of *Apis cerana* indica, while the flight of *Apis dorsata* was influenced both by temperature and light intensity. This demonstrates that different—yet closely related—bee species differ in their responses to environmental conditions. The differences in response of the bees are species specific and indicative of their physiological adaptations. For instance, Sihag and Abrol (1986) found that for *Apis florea*, relative humidity and solar radiations were the important factors influencing flight activity. Burill and Dietz (1981) found solar radiations important for *Apis mellifera*.

Abrol (1992a) studied the foraging strategies of six bee species in relation to 25 crops. *Apis dorsata* with its larger size and higher energetic demands preferred *Cajanus cajan*, *Parkinsonia aculeata*, *Pongamia glabra* and *Luffa cylindrica* rather than flowers of *F. vulgare*, *C. sativum D. carota A. cepa*, *T. foengraceum* and *M. indica*. The latter group of flowers was visited by *A. florea* throughout the day. The latter bee species with its small size and body weight is psychologically better adapted to extract maximum reward from these flowers. In the case of Brassica crops, *Apis dorsata* visited flowers during early hours of the day at peak periods of nectar production. But from 11:00 hours onward, foraging population shifted to *T. alexandrium* due to reduction in quantity of reward available from *M. sativa* and *Brassica* crops. *A. florea*, which commenced activity between 10:00–11:00 hours, dominated throughout the day. In case of sunflower, all types of pollinators continued foraging throughout the day. The caloric reward per flower is low, yet each bee species was able to maintain energy balance; since flower heads with platforms provide no barrier for landing of the foragers, the energetic cost is reduced due to temporarily suspension of hovering flight and large number of flowers can be visited in rapid succession.

16.3.1.6 Genetic Control of Foraging

Genetics of a particular individual determines the foraging behavior. Hunt et al. (2007) reported that two genes have been found associated with the preference of the collection of sugar concentration in nectar in honeybee *A. mellifera*. In addition to the number of genes associated with foraging behavior of honeybees with reference to the onset of foraging behavior, task division between foragers and workers, and bias in foraging for either pollen or nectar have been studied (Dugatkin 2020; Hunt et al. 2007).

16.3.1.7 Hormonal Control of Foraging

Honeybee foraging behavior is accompanied by large changes in hemolymph sugar concentrations—concentrations of glucose, fructose and trehalose in hemolymph. The concentration of trehalose is high during departure from the hive and it decreases during the hive return. The concentrations of glucose and fructose are high in the follower bees and in the foraging bees. Biogenic amines and juvenile hormones also control some aspects of the foraging behavior. Concentrations of biogenic amines change during the waggle dance activity. The follower bees have higher activity of dopamine than the inactive bees. Increased octopamine concentrations are linked to feeding arousal of honeybees. Serotonin concentrations increase during the decline of foraging activity. The queen pheromone is transferred passively from the queen to its attending bees, and further from the queen attending bees to other bees in the hive (Božič and Woodring 1997).

16.3.1.8 Speed of Flight

In honeybees, the speed of flight has been found to vary from 20.9–25.7 kph (average 24 km per hour) for loaded bees and from 10.9–29.0 kph (average 20 kph) for empty bees. This is because of the fact that a bee on outward flight does not make a beeline in all the cases for the source of supply. Bees do not like to work in a wind blowing over 24 kph.

16.3.1.9 Floral Rewards—Nectar Foraging

Strength of a bee colony is important for surplus honey gathering. Sharma and Sharma (1950) found that in the case of *Apis cerana indica*, colonies' nectar gathering capacity per 100 bees increased steadily as the number of bees in the colonies increased. Kapil and Kumar (1975) reported nectar robbing by *A. dorsata* on *Brassica juncea*. Sihag (1983) observed *A. dorsata* and *A. florea* bees foraging on the juice of ruptured/damaged grapes (*Vitis vinifera* L.) during dearth period. The grape juice contained 14%–20% dissolved sugars and was comparable with nectars of cruciferous crops in having glucose dominated sugars. Around 10,000–45,000 bees/ha of *A. dorsata* and 20,000–60,000 bees/ha of *A. florea* were attracted on damaged grapes. The author observed several species of honeybees and nectar feeding birds collecting juice from grapes. The honeybees *A. dorsata*, *A. florea*, *A. cerana* and *A. mellifera* were most common visitors throughout the day. *A. dorsata* commenced foraging activity much earlier than all the honeybee species. *A. dorsata* started foraging at 04:52 hours, much before sunrise (05:23 hours) and continued foraging even after the sun set (19:42 hours) at 2035 hours. Foraging activity of other honeybees commenced in the order of *A. cerana* followed by *A. mellifera* and *A. florea*. All the honeybees continued foraging throughout the day. The nectar-feeding hummingbird was observed throughout the day from morning to evening.

16.3.1.10 Floral Rewards—Pollen Foraging

Foraging patterns showed inverse relationship of pollen-gathering activity with nectar gathering and non-foraging activity (Reddy 1980a). Pollen-gathering activity of *A. cerana indica* in Delhi was reported by Bisht and Pant (1968). The bees collected pollen throughout the year and they maximum was from January–March. But in Shimla hills of north Himalayas, this bee collected pollen and nectar throughout the year (Mattu and Verma 1985). Foraging activity was greater in summer and autumn months than in winter and monsoon periods. Time of greater flight activity varied from season to season. The correlations between pollen foraging activity and temperature and relative humidity were also given. Similar observations were reported in Saharanpur by Singh (1981), who also found that foraging hours varied with season of the year. Varma and Joshi (1983) found that peak pollen collected by *A. cerana indica* in Jeolikote was between 08:00 and 11:00 hours in Feburary–March. Proportions of bees in a colony that foraged for pollen, the pollen stores in hive and amount of brood reared followed a similar pattern (Reddy 1980b). Thakur et al. (1982) on comparative foraging behavior reported that *A. c. indica* on mustard had considerable activity in the morning and *Apis mellifera* picked up by 10:00–10:30 hours, and remained active until late in the evening. At Pune, *A. cerana* and *A florea* commenced foraging activity on onion between 09:00 and 09:30 hours and 08:15–08:40 hours, respectively, and continued until 18:30 hours (Rao and Lazar 1980). The peak foraging of *A. cerana* was observed in the afternoon from 15:00–16:00 hours. *Apis cerana indica* foragers started foraging on mustard under mid-hill conditions between 09:00 and 10:00 hours and stopped foraging beyond 15:00 hours (Bhalla et al. 1983a; Mattu et al. 1994), but its activity on stone fruits continued up to 16:00 hours (Bhalla et al. 1983b). Choudary (1978) analyzed 5,200 pollen loads and only 56 contained pollen from more than one plant species, which showed high floral fidelity of *A. cerana indica*. Similar observations were reported by Chaturvedi (1973, 1977), Sharma (1970a, 1970b) and Jhajj and Goyal (1979b). Dhaliwal and Atwal (1986) reported that the foragers of *A. florea*, *A. dorsata*, and *A. mellifera* foraging on alfalfa (*Medicago sativa* L.) carried 60.0%, 60.0% and 66.6% pure pollen loads, respectively.

16.3.1.11 Foraging Speed and Foraging Rate

Atwal et al. (1970) reported that *A. cerana indica*, *A. dorsata* and *A. florea* visited 12.7, 16.6 and 17.2 flowers of sarson per minute, respectively, but *A. cerana indica* was a faster worker than the other two species on *Eranthenum*, radish, *Eucalyptus* and berseem. In general, *A. florea* was the slowest forager; average number of *Brassica juncea* flowers visited by *A. dorsata* was 12.3 per min. Under mid hill conditions, visiting mustard blooms, *A. c. indica* foraged 10 flowers per minute (Bhalla et al. 1983a; Raj and Rana 1993, 1994), but on stone fruit flowers, the rates were 6.9 on peach and 4.9 on almond (Bhalla et al. 1983b). Dhaliwal and Atwal (1986) observed that on alfalfa (*M. sativa*) foraging behavior of honeybees was related with their body weights. But foraging rates were in the order of *A. mellifera* < *A. florea* < *A. dorsata*. On onion blooms, *A. cerana* indica was faster flier between the umbels than *A. florea*, visiting 1.93 and 1.34 umbels per minute, respectively (Rao and Lazar 1980). But on an umbel, *A. florea* showed better foraging rate than *A. c. Indica*, visiting 6.53 and 5.93 flowers per minute, respectively. Dhaliwal and Atwal (1986) compared the tripping efficiency of different bees visiting alfalfa (*M. sativa*), where order of efficiency was *A. dorsata* > *A. mellifera* > *A. florea*. Kapil and Kumar (1975) reported the temporal change in foraging rate of *A. dorsata* which was 9.23 flowers per minute in the morning but 16 flowers per minute in the evening *A. c. Indica* was found to be slower forager than *A. mellifera* on *Plectranthus rugosus*; these species foraged 17.5–25.8 flowers per minute in the morning and 25.0–33.6 flowers per minute in the evening, respectively (Gupta et al. 1984; Shah and Shah 1989). Amounts and nectar concentrations were responsible for morning and evening differences in the rate of foraging.

Collection of full pollen load from sarson flowers took longer for a forager than for a load of nectar. Diwan and Salvi (1965) and Sihag (1984) have confirmed nocturnal foraging by *A. dorsata*, and night flowering plants were generally visited. Jadhav (1982) found that *A. florea* and *A. dorsata* stopped foraging when solar eclipse was at its peak and did not resume even after the eclipse was over, whereas Diwan (1980) found that returning foragers of *Tetragonula iridipennis* had difficulty in locating the hive.

Chaudhary and Sihag (2004) reported that *A. dorsata F.* was the most abundant on onion blossoms, followed by *A. florea* F. and then the dipterous insects, whereas the number of *A. florea* F. was significantly higher on the blossoms of carrot and fennel than *A. dorsata* F. Chandel et al. (2004) found that in case of onion (*Allium cepa* L.) flowers, A. *dorsata* proved to be the dominant visitor (7.4 bees/m^2/2 minutes) and most efficient pollinator visiting on an average seven flowers/umbel/visit during peak hours of their foraging activity (12:00–14:00 hours) compared with *A. cerana* (5.4 flowers/umbel/visit). Kapil and Kumar (1975) found that *A. dorsata* on an average visited 12.30 and 10.79 *B. juncea* flowers per minute. They found that *A. dorsata* on an average visited fewer flowers of *B. juncea* during morning between 09:30 and 11:00 hours, when the temperature was low, and the number visited increased with the advance of day and their maximum visits occurred between 14:30 and 15:45 hours, when the temperature was relatively higher.

16.3.1.12 Foraging in Relation to Weather Factors

The physical environment influences flower visiting, foraging behavior and effectiveness of pollination in complex ways. Foraging activities of pollinating insects are under the key control of different environmental variables.

16.3.1.13 Commencement and Cessation of Foraging Activity

Each bee species was found to have its specific ecological threshold below which activity does not occur (Osgood 1974; Szabo and Smith 1972; Reddy 1979; Lerer et al. 1982; Burill and Dietz 1981). The time of commencement of bee activity varies from one day to another, which depends upon attainment of minimum threshold conditions for their foraging activity. Abrol (2007) studied the foraging behavior of four honeybee species—*A. dorsata F, A. mellifera* L, *A. cerana* F. and *A. florea* F.—on litchi flowers (*Litchi chinensis* Sonn.) in relation to weather factors and found that 16°C temperature, 74% relative humidity (RH), 600× light intensity and 10 mW/cm² solar radiation for *A. dorsata*; 16°C temperature, 75% relative humidity, 800 1× light intensity, 10 mW/cm² solar radiation for *A. mellifera*; 15.5°C temperature, 76% relative humidity, 600 1× light intensity, 9 mW/cm² solar radiation for *A. cerana* and 18.5°C temperature, 64% RH, 1700 1x light intensity and 20 mW/cm² solar radiation for *A. florea* appeared to be minimum threshold condition for initiation of flight activity. Cessation of activities in all the honeybee species was governed mainly by decline in values of light intensity and solar radiation, irrespective of other factors. Earlier, Kapil and Kumar (1975) also reported 15–18°C as the minimum threshold temperature for commencement of field activities in honeybees.

Sihag (1984) studied the limiting effect of light intensity, ambient temperature and humidity on the commencement and cessation of flight activity of *A. dorsata* and *A. florea. A. dorsata* maintained foraging activity when illumination was as low as 0.5–1.0 lux during nights of May–June, but foraging activity was limited if the temperature was below 16.5°C or above 38°C and if humidity was below 32% or there was rain. On the other hand, in *A. florea*, illumination below 200 lux, temperature below 18.5°C or above 40°C and humidity below 25% or if there was rain caused the cessation of flight activity. The highest limits of light intensity for *A. florea* were 6,500 lux and 7,000 lux, respectively. Before the approaching of these limits, temperature had also crossed the upper limiting values; however, once commenced, the flight activity in *A. florea* was maintained by the direct effect of solar radiation and relative humidity. Temperature, relative humidity and nectar sugar concentration had no direct effect on the flight activity (Sihag and Abrol 1986). Dhaliwal and Atwal (1986) reported 33°C as the optimum field temperature for peak activity of *A. mellifera* on alfalfa (*M. sativa*). Severe robbing during dearth period was also reported between *A. c. indica* and *A. mellifera* (Atwal and Dhaliwal 1970; Atwal and Sharma 1971; Adlakha and Sharma 1974, 1975). Abrol (1985) found that *A. dorsata* was physiologically more tolerant to large fluctuations in ambient temperature than *A. florea*, which restricts its activity during unfavourableweather conditions.

Abrol (1985) found that *Apis florea* were found to be more sensitive to variations in ambient temperature than *A. dorsata*. This implies that *A. florea* is adapted to work at relatively higher temperature (43°C) and is limited by fluctuations in lower ambient temperature. It is probably advantageous for the small bees to work at extremely higher temperatures, since they can efficiently radiate enough heat to the environment by their larger surface area. Similar observations were made by Willmer (1983), who reported that small lightly colored species visited fully isolated flowers and were highly correlated with solar radiation. Kapil and Kumar (1975) found foraging population of *A. dorsata* on *B. juncea* peaked when the temperature ranged from 21–23°C.

In the case of *A. laboriosa*, the bee's large size is a sign of its adaptation to the cold and low-oxygen climates in which it lives, where there is a need for increased muscle mass among other characteristics (Summers 1990). The bees' foraging area is larger than most of the other honeybee species. It can fly up to a radius of 3 m in thin air. *Apis laboriosa* bees live in the Himalayas under harsh conditions at altitudes of 1,200–3,600 m where a temperature range of 4.5°–23.4°C is encountered. The lower range of temperatures recorded in *A. laboriosa* territory does not occur in *A. dorsata* territory (6°–27°C), and the upper temperature range recorded in *A. dorsata* territory does not occur in *A. laboriosa* territory. However, a large temperature range is the same for the active life of both species mentioned. According to Dyer and Seeley (1987), *A. dorsata* workers are not able to fly at temperatures lower than 17°C. Woyke et al. (2003) in a similar study found that closely related *A. dorsata* workers started foraging at a temperature of 18°C, and *A. laboriosa* at a lower temperature of 10°C. The increase in temperature to 12°C resulted in a tenfold increase in the number of foragers leaving the nests of both species, and their flight activity reaction to temperature changes was similar. Usha et al. (2016) reported that maximum temperature exerted a significant positive correlation with the foraging activity of *A. dorsata*, *A. mellifera* and *A. cerana indica* within the optimal range (21.0°–33.5°C). However, minimum temperature correlated significantly with the foraging activity of *A. dorsata* and *A. cerana indica* only. It reveals that the *A. dorsata* and *A. cerana indica* are able to forage with optimal range of maximum and minimum temperatures.

16.3.1.14 Diurnal Trends in Foraging Activity in Relation to Environmental Factors

Foraging activity occurs only when ecological conditions within which foraging occurs are attained. Szabo and Smith (1972) reported that greatest foraging activity of *M. rotundata* in Hungary occurred at 30°C in bright sunshine and declined at higher temperature. Kapil and Brar (1971) recorded peak activity of *A. florea* on *Brassica campestirs* L. var. toria between 21–25°C temperature and 50%–57% relative humidity during November. Similar results were obtained by Cirudarescu (1971), who found that the number of insect visitors on lucerne varied directly with temperature and inversely with relative humidity. Nunez (1977) reported that morning activity of *A. mellifera* was related to nectar flow, while the afternoon, with the photoperiod (Cirudarescu 1971; Szabo and Smith 1972; Nunez 1977; Corbet 1978a, 1978b; Bailey et al. 1982; Burill and Dietz 1981; Lerer et al. 1982). Abrol (2006, 2007) found that in the case of *A. dorsata*, temperature, light intensity and solar radiation; in case of *A. mellifera*, temperature, light intensity, solar radiation and relative humidity; in case of *A. cerana*, light intensity; and in case of *A. florea*, solar radiation and relative humidity were the important factors. In between the commencement and cessation, *Apis florea* activity was highest on flowers when temperature ranged from 25°–38°C and declined at higher temperatures (Abrol 2006). Similar observations were made by Free (1993), who found that metabolic activity of insects increases as the temperature increase and they visit many flowers at that time. Similar results have been reported by several earlier investigators (Lerer et al. 1983; Burill and Dietz 1981; Nunez 1977). In general, suitability of optimum bee activity varies from season to season, depending upon the geographical regions, time of the year, melliferous crops or species of bees. Neupane et al. (2006) found that the activity of *A. dorsata* foragers was more pronounced during the morning hours when air temperature was comparatively lower (19.3°C at 07:30 hours) than in the afternoon (33.1°C at 17:30 hours). The pronounced activity of the bees during morning hours was due to the early release of fresh pollen and nectar, and suitable foraging temperatures. The *A. dorsata* foragers forage for long distances by showing maximum foraging activity at temperatures from 25°–35°C, the distance and range being higher than the *Apis mellifera* (Sihag 1998).

Sihag (2000a) found that *Apis florea* started foraging when ambient temperatures surpass 18°C and continue until temperature approaches 43°C. Maximum foraging activity is shown at 30°–40°C. These ranges are highest than those shown by *A. dorsata*, *A. mellifera* and *A. cerana*. In the case of *A. dorsata* (Sihag 2000b), foraging commenced when temperatures surpass 16°C and continue to around 40°C. Maximum foraging is shown at 25°–35°C. These ranges are lower than those shown by *A. mellifera*. Abrol (1992a) found that during winter, temperature acts as a stimulus for initiation of flight activity in honeybees *Apis florea*, *A. dorsata*, *A. cerana* and *A. mellifera*, whereas during summer, light intensity provided the minimum threshold. Woyke et al. (2003) found that workers of *A. laboriosa* did not initiate flight activities below 10°C and the flights were not performed during overcast days.

The large size of *A. laboriosa* workers has enabled this species to survive in temperate climates while the other open-nesting honeybees are confined to the tropics and subtropics. *A. laboriosa* workers are able to forage at ambient temperatures at least 5–6°C lower than the minimum ambient *at* which *A. dorsata* workers can fly (Dyer and Seeley 1987; Underwood 1991) (Figure 16.2). This has apparently been accomplished largely through an increase in body size, without resorting to creating a higher-powered bee and without a disproportionate increase in thoracic mass, such as seems to have been the case with *A. cerana* in Nepal.

FIGURE 16.2 Foraging on mustard flowers: a) *Apis laboriosa*; b) *A. dorsata*.

16.3.1.15 Ecological Threshold of Nectar Secretion and Foraging Activity of Honeybees

Co-evolutionary adaptations have brought a close correlation between plants and their pollen vectors in a manner that functional activity of both more often synchronizes. Each bee has its ecological threshold for initiation of field activity until surpassed activity is limited (Abrol 1986c). In plants, also, a characteristic rhythm of nectar secretion occurs, with peak periods of nectar production usually synchronizing when the conditions are optimum for bee activity (Corbet 1978). Abrol (1985) reported that in *Apis* species such as *A. dorsata* and *A. florea*, onset of flight activities occurred at 15°C under subtropical conditions of Hisar during winter months. Abrol (1986a) found that in flowers of *Brassica juncea* nectar secretion occurred from 12°–15°C, thereby indicating 12°–15°C as a minimum threshold temperature for activation of enzymatic machinery to secrete nectar. This overlap of activity threshold of two different groups make them interdependent on each other—one for energy reward and another for reproduction and perpetuation of its race. Abrol et al. (1988) reported that under temperate conditions of Kashmir, no nectar secretion was observed in flowers of *Eriobotrya japonica* below 8°C but began at 10°C, coinciding with initiation of flight activity of *Apis cerana*. Similarly, threshold temperature below which no nectar is produced has been reported for many plants: Boss wood (*Tilia Americana*) begins to secrete at 18°C (Desmuth 1933), bird cherry (*Prunus laurocerasus*) at 18°–20°C, cucumber (*Cucumis sativa*) at 17°–21°C (Collision 1973) and *Brassica juncea* at 12°–15°C (Abrol 1986a). Evidently, the physiology of plants and their pollen vectors have synchronized in a manner than functional activity of both most often synchronizes for the benefit of both.

16.3.1.16 Sequence and Timing of Bee Visits

Visits of bees to flowers depend largely upon the energy requirement reward system and occur in a very predictable and ordered manner (Schlising 1970; Heinrich 1975e; Murrell and Nash 1981). Abrol (1987) found that larger bees like *A. dorsata*, *A. mellifera* and *A. cerana* with higher energetic demands and ssuperior thermoregulatory capabilities visited flowers early in the morning when the caloric reward is more and competition with other pollinators is minimal, whereas the smaller bees such as *A. florea* with relatively low energy requirements and poor thermoregulatory capabilities visited flowers in the late morning hours, when energy expenditure due to prevailing higher temperatures was minimized.

16.3.1.17 Foraging as a Function of Species-Specific Adaptation

Several species of highly social *Trigona* (Johnson and Hubbell 1974, 1975) and *Apis mellifera* (Sakagami 1959; Schaffer et al. 1979) specialize on large productive flower clusters. The large size of their colonies (Michener 1974) results in large pollen and nectar requirements for colony maintenance and growth, which favor the efficient exploitation of large resource patches. Abrol (2006) studied the foraging behavior of four honeybee species *A. dorsata F*, *A. mellifera* L, *A. cerana* F. and *A. florea* F. visiting litchi flowers *(Litchi chinensis* Sonn.) and found that all the four honeybee species differed in their responses to environmental factors prevailing under similar sets of conditions depending upon physiological adaptation of each honeybee species. Of all the factors, temperature, light intensity and solar radiation had strong influence on foraging population of bees.

16.3.1.18 Competition for Floral Resources

A. cerana is sympatric in distribution and can co-exist with the two other species of Asiatic honeybees, *A. dorsata* and *A. florea*, without any adverse ecological consequences. Chahal et al. (1986) found that *Apis florea* showed aggressiveness toward *A. mellifera* both at hive and on flowers. Koeniger and Vorwoh (1979) in their experiments at an artificial feeding dish reported interactions among the species according to body size: *Trigona* attacked mainly *A. fiorea* and *A. cerana*; *A. florea* reacted to *A. cerana*; and *A. cerana* attacked *A. dorsata.* They reported that smaller species proved more successful than larger ones. They reported that the four competing bee species are not equal partners. They differ in body size. The smallest species were found to be the most aggressive. According to Lindauer (1956), *T. iridipennis* has a range of about 100 m, so it must make full use of the flowers in its territory and defense of them is profitable. On the other hand, *A. dorsata*, with a flight range of more than 5 km, can afford to retreat in a similar situation because unoccupied food sources are likely to be available at an accessible distance.

16.3.1.19 Resource Partitioning

In a mutualistic relationship with flowering plants, honeybees gather nectar and pollen from blossoms resources may be partitioned based on size. Sihag (2000a) reported that *A. florea* starts foraging when the ambient temperature surpasses 18°C and continues foraging until ambient temperature approaches 43°C. Maximum foraging activity is shown at 30°–40°C. These ranges are higher than those shown by *A. mellifera* and *A. dorsata.* In the case of *A. dorsata* (Sihag 2000b), foraging commenced when temperature surpasses 16°C and continues to around 40°c. Maximum foraging activity is shown at 25°–40°C. These ranges are lower than *A. florea* but higher than *A. mellifera.* Body size and tongue length play predominant roles in determining the resource partitioning of the bees. For example, Oldroyd et al. (1992a, 1992b) found evidence suggesting that pollen resources were partitioned among A. *dorsata*, *A. cerana*, *A. florea* and *A. andreniformis* based on bee size, with the larger two species using the richest pollen resources. They found that larger *A. dorsata* and *A. cerana* foraged earliest but in low numbers, probably exploiting the resource at its most productive time, followed by smaller *A. andreniformes* and *A. florea* in large numbers.

Neupane et al. (2006) found that A. *dorsata* preferentially concentrated their visits to litchi flowers in the early morning hours and then declined toward the later part of the day significantly during the entire periods of flowering. The bees more preferentially collected nectar from bottlebrush and litchi flowers and pollen from bottlebrush, citrus and summer squash flowers, showing distinct resources partitioning. Rao and Lazar (1983) in a study of bee behavior and pollination in onion reported that A. *cerana* and *A. florea* were observed collecting nectar, whereas *A. dorsata* was not observed. Finally, *A. dorsata* has a unique ability to forage on moonlit nights surrounding the full moon; this ability may provide an additional release from competition by enabling this species to forage when other honeybee competitors are unable to fly (Dyer 1985). Sometimes resources are partitioned when separate species forage at different times of day. In Bhubaneswar, India, *A. cerana* and *A. dorsata* were most active on niger plants at 11:00 hours. *A. florea*, in contrast, was more active in the afternoon (Panda et al. 1993).

16.3.1.20 Foraging in Relation to Flavor, Taste and Color of Flowers

The actual recognition of source of food or nectar is influenced by various factors such as color, flavor and taste. For example, bees can distinguish green, blue, white, yellow and violet flowers, but are insensitive to red flowers. The bees can also discriminate between different concentrations of sweetness and varying salt content in flowers. If flowers of different concentrations are blooming simultaneously, the bees will prefer those which have higher sugar content. Once the bees have started to work on a certain source for nectar or pollen, they continue to visit it and seldom visit more than one or two species on a single trip. Bees also drink water and they like stagnant water rather than fresh. If a pinch of salt is added to the fresh water, it also becomes attractive. It is again with the sense of smell that bees can recognize their nest mates. If by mistake a bee of other colony intrudes, it is recognized and killed at once—but if the stranger is laden with nectar or pollen, it is not driven away or killed. However, this rule does not apply to drones. Drones of all hives are allowed free entry.

Foraging behavior of *A. florea* was studied on *Portulaca grandiflora* flowers having different parches of straw yellow, dark yellow, purple and red flowers (Abrol 2003). Maximum foraging populations was observed on straw yellow flowers followed by dark yellow, purple and red flowers, respectively. Earlier, Faegri and van der Pijl (1971) reported that *A. mellifera* preferred yellow blue and purple flowers. Sharma et al. (1999) found that pink, red and blue attracted more honeybee and bumblebee pollinators; black and green were less effective than the control (white).

16.3.1.21 Nocturnal Foraging by *Apis dorsata*

Diurnal foraging activity is most common in almost all of the bees; however, some of the bees have been reported foraging in dim light intensities in the evening or moonlit nights. Dyer (1985) observed that *A. dorsata* were found to collect pollen at night. Nocturnal flights occurred only when the moon was half full or larger. On dark nights, diurnal activity ceased about one hour after sunset. On the night of a full moon, bees would fly from dusk to dawn at a nearly constant rate of about 20 outward flights per minute. Somanathan et al. (2009) reported that of the four species, the giant honeybee *A. dorsata* (which forages during moonlit nights) has the lowest spatial resolution and the most sensitive eyes, followed by *A. mellifera*, *A. cerana* and the dwarf honeybee *A. florea* (which has the smallest acceptance angles and the least sensitive eyes).

The major advantage of nocturnal foraging is that the bees which are able to fly in the evening or during night can collect large pollen loads very quickly and without competition with other bees or moths, and this is probably of major adaptive value (Bohart and Youssef 1976). In addition, there might be fewer predators and nest parasites such as satellite flies, ants, sphecidae and parasitic bees active during and after sunset (Smith et al. 2003; Wcislo et al. 2004).

16.3.1.22 Impact of Neonicotinoids on Foraging

Pesticide application may also change the physiology of nectar- and pollen-producing plants, change the attraction of bees to flowers, affect pollen viability and reduce pollen germination on contaminated stigma. It is not only the foragers visiting the crop that are exposed, but hive bees and larvae which feed on stored nectar and pollen are also exposed. This means that different life stages of honeybees (larvae and adults of various ages or involved in different duties) can be exposed. Pollinator foraging behavior is negatively influenced by pesticides at sublethal doses (Desneux et al. 2007; Tan et al. 2014). Impaired foraging reduces colony fitness (Whitehorn et al. 2012) and results in population declines (Cresswell et al. 2012). Neonicotinoids bind to insect cholinergic receptors, causing death at sufficient concentrations (Jeschke and Nauen 2008).

A. dorsata covers larger areas (>10 km) to gather good amount of nectar and pollen to supplement its huge colony population. During foraging, various flowering plant species which are partly or completely exposed to pesticide sprays during their bloom by farmers are visited. As a result, the nectar and pollen gets contaminated with residues of various pesticides (Bright et al. 1998). During foraging, several hundreds or thousands of forager bees become victims to pesticide poisoning. Moreover, hive bees and developing brood also become victim to pesticide poisoning. Ultimately, this could cause colony decline (Bright et al. 1998).

16.3.1.23 Honeybee Foragers as Indicators of Environmental Contamination

The great mobility and flying range of honeybees allows them to monitor vast areas. They make numerous inspections per day. A variety of materials are brought by honeybees into the hive through foraging (nectar, pollen, honeydew, propolis and water) and stored. Porrini et al. (2003) at the University of Bologna have been studying the use of honeybees as bioindicators to detect pesticides, heavy metals and radionuclides in many areas. Heavy metals present in the atmosphere can be deposited on the hairy bodies of bees and be brought back to the hive with pollen, or they may be absorbed together with the nectar of flowers or through the water or honeydew. In the urban environment, lead was found in higher quantities inside the bees than on the bees' surfaces. Honeybees have also been proposed to detect and locate land mines. Trained bees can act as biological detectors of these explosive devices, as they momentarily hover over and thereby mark a land mine.

16.3.1.24 General Flight Activity and Foraging Range

Flight activity of honeybees is an index of their foraging potential for pollination and honey production. Bees recognize their hive and surroundings, and they can return to their hive after completing field activities. Bees generally avoid longer flights to minimize the cost of foraging and maximize net energy gain. Many bees mostly prefer to forage on crops within their daily travel distance (Ricketts 2004). Foraging distance therefore determines the spatial scale at which bees can provide pollination services to crops (Kremen 2005). Foraging range of pollinators provides information regarding their intrinsic capabilities in gathering nectar and pollen resources (Cherian and Mahadeven 1945a, 1945b; Cherian et al. 1947; Eckert 1955; Dhaliwal and Sharma 1972, 1973, 1974; Naim and Phadke 1972; Kapyla 1978; Naim 1984; Abrol 1986a; Dyer and Seeley 1991). The foraging range of bees depends largely on their body size and energy needs (Abrol 1986a, 1986b), yet are determined also by the quality of reserve fuel carried when bees are in flight. The smaller bees with low energy requirements and little reserve fuel may hardly risk foraging large distances (Hocking 1953). Abrol (1988) reported *Apis florea* workers to forage up to 150 m from their nesting sites compared to *Megachile flavipes* and *Megachile nana*, which foraged up to 250 m and 150 m, respectively. Bees generally prefer to visit flowers within a radius of 1–3 km. *Apis cerana* normally forages 1–1.5 km, while *Apis mellifera* may forage up to 3 km, but if the food sources are limited and the competition is greater, the foragers may go long distances. The foragers of *Apis cerana* have been reported to fly 600–1,040 m, while those of *A. mellifera* from 3–14 km.

16.3.1.25 Flight Intensity of Foraging Bees

Wing beat frequency is an expression of flight intensity and is an important parameter, not only from the aerodynamics and bioenergetics points of view, but also helps in assessing the pollinating efficiency of bees (Puranik et al. 1977; Heinrich 1975e). The wing beat frequency is generally species specific. In general, smaller insects were found to have higher wing beat frequency than the larger ones (Sotavalta 1952). Generally, the insects with smaller wings have higher wing beat frequency than those with long or heavy wings. The smaller the surface area of the wings, the more rapidly the wings must beat to keep the animal in flight (Kammer and Heinrich 1978). Wing beat frequency is a function of body parameters. Besides the body parameters, wing beat frequency is also influenced by environmental temperature in all the insects except Hymenoptera (Reed et al. 1942).

Abrol (1985) found that wing beat frequency of *Apis florea* weighing 12–19 mg ranged from 92–106 cps, with an average of 101, and that of *A. dorsata* weighing 106–129 mg was 111–116 with an average of 113.95. Interestingly, smaller bees on the basis of per-gram body weight were found to have higher values of wing beat frequency than the larger ones. It was found that *Apis florea* weighing $12–19 \times 10^{-3}$ g had a wing beat frequency of 101 cps, in contrast to *A. dorsata* weighing $106–127 \times 10^{-3}$ g which had a wing beat frequency of 113.95 + 0.60. *Apis florea* per-gram body weight exhibited 6,688.79 cps compared to *A. dorsata*, for which the number was found to be 1,162.16 cps/g/body weight—a difference of more than fivefold. These results suggested that smaller bees exhibited higher wing beat frequency than the smaller ones.

Goyal and Atwal (1977) reported that *A. cerana* had higher wing beat frequencies than *A. mellifera* both for workers and drones. 306 wing beats per second for workers compared with 235 bps for drones in case of *A. cerana* and 283 bps for workers compared with 225 bps for drone in case of *A. mellifera*. These frequency differences are attributed to different wing sizes.

16.3.1.26 Influence of Pheromones and Bee Attractants on Foraging Activities of *Apis dorsata*

Achievement of desired pollination lies in the planned and efficient use of honeybees to increase the yield, as well as improving qualitative and quantitative parameters of the crop. Any material to increase visit of honeybees to specific crop could be of great practical value to harness the benefits of cross-pollination. Attracting honeybees in sufficient numbers for efficient pollination requires evaluation of bee attractants and their impact on the pollinating effectiveness of bees. Dorjay et al. (2022) found that in case of *Apis dorsata* visiting bitter gourd flowers, the population increased following first, second and third sprays of different attractants. All the treatments were effective in enhancing population visits as compared to control. Of all the different treatments, Bee-Q + Jaggery was most effective, whereas sugar solution of 100 g/l was the least attractive (1.00–2.40 bees/five plants/five minutes) and the other treatments were in between the two. Lingappa et al. (1999) reported that the number of fruits formed and total yield, respectively, when Bee-Q was sprayed was twice on watermelon. Application of Bee-Q of 12.50 g/l and 15 g/l resulted in higher yield (19.56 and 19.45 t/ha, respectively), maximum good fruits, minimum malformed fruits and higher size and weight in watermelon (Sattigi et al. 2001).

16.3.1.27 Nectar Robbing

Nectar robbing specifically refers to the phenomenon of animals, insects or birds getting nectar rewards in an illegitimate way by drilling holes on the corolla tubes of flowers (Inouye 1983; Maloof 2000). This phenomenon commonly exists in nature. The insects or birds that perform nectar-robbing behavior are called nectar robbers. According to the earlier research, there are two types of nectar robbers found in nature: primary nectar robbers (which get nectar by biting holes on the corolla) and secondary nectar robbers (which do not bite holes, but get nectar through the holes created by primary nectar robbers). The most common nectar robbers are bees, mostly bumblebees and carpenter bees. Some species of hummingbirds are also the frequently found nectar robbers in the North and South America.

Kapil and Kumar (1975) reported that few foragers of *A. dorsata* collected nectar from *Brassica juncea* flowers without actually entering the flowers. Sharma et al. (2001) found that all the bees—*A. florea, A. dorsata, A. cerana and A. mellifera*—foraged as top workers on flowers of *Brassica campestris var. sarson*, *Allium cepa*, *Daucus carota*, *Trifolium alexandrinum* and *Helianthus annuus*, except that *A. florea* foraged as side worker on *Brassica* flowers.

16.3.1.28 Effect of Pathogens and Diseases on Foraging Activity of Bees

Anderson and Giacon (1992) reported that honeybee, *Apis mellifera* colonies fed prepations of 50% of sucrose containg sac brood virus particle and *Nosema apis* Zander spores collected significantly less pollen than colonies fed only on the sucrose solution. Abrol (1995b) found that *Apis cerana* and *A. mellifera* infested with ectoparasitic mites had reduced pollen collection activity as compared to healthy ones. Schmid-Hempel and Schmid-Hempel (1990) were the first to document an association between parasite infections and bee foraging behavior in the field. Similar information on giant honeybees is not available.

16.3.1.29 Predation during Foraging

Pollinators have mutualistic relationships with flowering plants which coevolved over centuries of natural selection. During the process of foraging, pollinators are vulnerable to several predators, eventually causing the ecosystem to suffer from low pollination services, loss of plant diversity and poor or reduced seed set in flowering plants. The impact of predators on foraging behavior of pollinators has been little studied, with the assumption that predation is too rare phenomenon of general or ecological significance (Schmid-Hempel 1991; Dukas 2001; Pyke 1979; Morse 1986; Schmalhofer 2001; Reader et al. 2006). However, there is strong evidence that predation has important consequences for fitness and the nature of interactions between species. Crab spiders (Araneae: Thomisidae) (Dukas and Morse 2003, 2005; Suttle 2003; Dukas 2005; Robertson and Maguire 2005) and lizards (Muñoz and Arroyo 2004) are the important predators on bees foraging in the field. Their presence on or near flowers has been reported to reduce visitation rates and also seed production because of reduced population of bees. (Suttle 2003; Muñoz and Arroyo 2004; Dukas 2005).

16.3.1.30 Foraging in Risky Conditions

Woyciechowski (2007) studied the behavior of *A. mellifera* foragers and observed that collecting water is more risky than collecting nectar. Foraging is further more risky during inclement weather than it is during fair weather (Woyciechowski and Kozlowski 1998). Robbing from the nests of other bees is extremely risky. Individuals undertaking robbery are vulnerable to direct attack by nest defenders and can have their wings damaged or even can be killed during their mission (Free 1954; Winston 1987). Robbing is observed during dearth periods when resources are not readily available in the field (Free 1954; Winston 1987). Robbing occurs when pollen and nectar are in short supply. However, no information is available on foraging risks encountered by giant honeybees *Apis dorsata* and *A. laboriosa*.

16.4 CONCLUSIONS

Bees are excellent subjects for testing foraging theories because energy gained from nectar can be accurately measured. Most energetics research has been conducted on bumblebees and honeybees because they have high energy demands and can be easily monitored while foraging. More research on energetics of foraging in other taxa of bees is required to understand the complexity of pollinator–plant interaction. Studies are also required to understand the importance of factors other than energy, such as nutritive values of different pollens, in determining bee foraging patterns. No bee species is an absolute generalist forager. All bees specialize, to some extent, through innate or learned preferences for particular plant taxa, foraging times, spatial distributions of flowers, floral structures or floral products. Although the importance of foraging activity is generally recognized, it is not yet well elucidated in some honeybee species. The impacts of climatic changes, magnetic radiation, pollution and depleting natural resources, and degradation of habitat need to be investigated. Further studies are required on both the in-colony and out-colony factors that affect foraging activity. Most studies of competition for floral resources among bee species, especially between honeybees and wild bees, have been inconclusive because competition is difficult to prove in the field. Future studies should include manipulation of forager and resource levels, and also behavioral observations of the interactions between bees and flowers.

REFERENCES

Abou-Shaara, H.F., Al-Ghamdi, A.A., and Mohamed, A.A. 2013. Honey bee colonies performance enhance by newly modified beehives. *Journal of Apicultural Science* 57: 45–57.

Abrol, D.P. 1985. *Analysis of biophysical interactions in causing foraging behaviour of some bees-A study in bioenergetics*. Ph.D thesis, CCS Haryana Agricultural University Hisar, 286.

Abrol, D.P. 1986a. Flight range and significance of wing-hooks in Megachile femorata Smith (Hymenoptera, Megachilidae). *J. Anim. Morphol. Physiol.* 3: 107–112.

Abrol, D.P. 1986b. Time and energy budgets of alfalfa pollinating bee, *Megachile nana* Bingh and *M. flavipes* Spinola (Hymenoptera: Megachilidae). *Proc. Indian Acad. Sci. (Anim. Sci.)* 95: 579–586.

Abrol, D.P. 1986c. Ecophysiological adaptations between pollinating bees and their flowers. *Environ. Ecol.* 4: 161.

Abrol, D.P. 1987. Activity time budgets and pollination potential of *Apis cerana indica* workers. *Environ. Ecol.* 5(4): 707–709.

Abrol, D.P. 1988. Foraging range of subtropical bees, *Megachile flavipes*, *Megachile nana* (Hymenoptera: Megachilidae) and *Apis florea* (Hymenoptera: Apidae). *Journal of the Indian Institute of Science* 68(1–2): 43–48.

Abrol, D.P. 1990. Energetics of nectar production in some apple cultivars as a predictor of floral choice by honeybees. *Trop. Ecol.* 31(1): 116–122.

Abrol, D.P. 1991. Foraging strategies of honeybees in pollinating apple flowers. *Journal of Animal Morphology and Physiology* 38(1–2): 109–113.

Abrol, D.P. 1992a. Bioenergetics in beeflower interrelationship and analysis of foraging behaviour. *Korean Journal of Apiculture* 7(1): 39–66.

Abrol, D.P. 1992b. Energetics of nectar production in some strawberry cultivars as a predictor of floral choice by honeybees. *J. Biosci.* 17(1): 41–44.

Abrol, D.P. 1993. Energy intake and expenditure in *Megachile femorata* Smith (Hymenoptera: Megachilidae). *Trop. Ecol.* 34(2): 173–180.

Abrol, D.P. 1995a. Energetics of nectar producion in some almond cultivars as a predictor of floral choice by honeybees *Apis cerana indica* F. and *A. mellifera* L. (Hymenoptera: Apidae). *Proc. Indian Natn. Sci. Acad. B* 57(2): 127–132.

Abrol, D.P. 1995b. Haemocytes of honeybees *Apis mellifera* L. and *Apis cerana* F. (Hymenoptera: Apidae) and their alteration by mite parasitosis. *Indian Bee. J.* 57(1): 6–7.

Abrol, D.P. 1998a. Environmental factors influencing flight activity in honey bees, *Apis cerana* Fab. and *Apis mellifera* L. (Hymenoptera: Apidae). *Indian Bee Journal* 60(2): 71–75.

Abrol, D.P. 1998b. Foraging in honeybees *Apis cerana indica* F. and *A. dorsata* F. (Hymenoptera: Apidae) activity and weather conditions. *Journal of the Indian Institute of Science* 72(5): 395–401, 1992.

Abrol, D.P. 2003. Foraging behaviour of *Apis florea* F. on *Portulaca grandiflora* L. *Sci. & Cult.* 69: 289–290.

Abrol, D.P. 2006. Foraging behaviour of bees as influenced by quality and quantity of rewards from flowers. *Journal of Asia-Pacific Entomology* 9(2): 1–4.

Abrol, D.P. 2007. Honeybee and rapeseed: A pollinator plant interaction. In: Gupta, S.K. (Ed.), *Advances in botanical research: Rapeseed breeding*. Vol. 45. Elsevier, London, UK, 339–367.

Abrol, D.P. 2016. Foraging strategies in honeybees, *Apis dorsata* F. and *Apis florea* F. in relation to availability of energy rewards. *Journal of Apiculture* 31(1): 9–18.

Abrol, D.P., and Kapil, R.P. 1991. Foraging strategies of honeybees and solitary bees as determined by nectar sugar components. *Proc. Indian Natn. Sci. Acad. B* 57(2): 127–132.

Abrol, D.P., Masoodi, M.A., and Bhagat, K.C. 1988. Bee flower interaction, parallel evoluation and adaptations with reference to *Apis cerana indica* F. pollinating loquat (*Eriobotrya japonica* Lindley) flowers in Kashmir. *Environment and Ecology* 6(3): 734–736.

Adlakha, R.L., and Sharma, O.P. 1974. Intraspecific robbing. *Amer. Bee J.* 114(7): 248–249.

Adlakha, R.L., and Sharma, O.P. 1975. *Apis mellifera* Vs *Apis indica. Glean. Bee Cult.* 103(5): 160.

Ali, M.A.M., and Taha, E.A. 2012. Bee-Eating birds (Coraciiformes: Meropidae) reduce virgin honey bee queen survival during mating flights and foraging activity of honey bees (Apis mellifera L.). *International Journal of Scientific and Engineering Research* 3: 1–8.

Amdam, G.V., Ihle, K.E., and Page, R.E. 2009. Regulation of honey bee (*Apis mellifera*) life histories by vitellogenin. In: Pfaff, D., Arnold, A., Etgen, A., Fahrbach, S., and Rubin, R. (Eds.), *Hormones, brain and behavior.* Elsevier Academic Press, San Diego.

Anderson, D.L., and Giacon, H. 1992. Reduced pollen collection by honeybee (Hymenoptera: Apidae) colonies infected with *Nosema apis* and sacbrood virus. *J. Econ. Entomol.* 85(1): 47–51.

Atwal, A.S., and Dhaliwal, G.S. 1970. Interspecific relation between *Apis Indica* F. and *A.mellifera* L. *J. Apic Research* 2: 53–59.

Atwal, A.S., and Sharma, O.P. 1971. The dominance of *A.mellifera* L. over *Apis Indica* F. *American Bee J.* 111(9): 343.

Bailey, W.G., Lerer, H., and Mills, P.F. 1982. Humidity and the pollination activity of *Megachile rotundata. Environ. Entomol.* 11: 1063–1066.

Beekman, M., Sumpter, D.J.T., Seraphides, N., and Ratnieks, F.L.W. 2004. Comparing foraging behaviour of small and large honey-bee colonies by decoding waggle dances made by foragers. *Functional Ecology* 18: 829–835.

Bertsch, A. 1983. Nectar production of *Epilobium angustifolium* L. at different air humidities: Nectar sugar in individual flowers and the optimal—foraging theory. *Oecologia* 91: 187–193.

Bhalla, O.P., Verma, A.K., and Dhaliwal, H.S. 1983a. Foraging activity of insect pollinators visiting stonefruits: Anonymous. *Journal of Entomological Research* 7(2): 91–94.

Bhalla, O.P,. Verma, A.K., and Dhaliwal, H.S. 1983b. Insect visitors of mustard bloom (*Brassica campestris* var. *sarson*), their number and foraging behaviour under midhill conditions. *Journal of Entomological Research* 7(1): 15–17.

Bisht, D.S., and Pant, N.C. 1968. Studies on the pollen gathering activity of the Indian honeybee, *Apis cerana indica* F. under Delhi conditions. *Indian Journal of Entomology* 44: 368–372.

Bohart, G.E., and Youssef, N.N. 1976. The biology and behaviour of *Evylaeus galpinsiae* Cockerell (Hymenoptera: Halictidae). *Wasmann J Biol* 34: 185–234.

Božič, J., and Woodring, J. 1997. Effect of activity on the haemolymph sugar titres in honey bees. *J. Apic. Res.* 36: 33–39.

Bright, A.A., Chandrasekaran, M., and Muthuswami, M. 1998. Bee pollinators importance and preservation. *Kissan World* 25(4): 61–63.

Burill, R.M., and Dietz, A. 1981. The response of honeybees to variations in solar radiation. *Apidologie* 12: 319–328.

Chahal, B.S., Brar, H.S., Gatoria, G.S., and Jhajj, H.S. 1986. Aggressive behaviour of *Apis florea* towards *Apis mellifera* in hive robbing and in foraging. *Journal of Apicultural Research* 25(3): 134–138.

Chandel, R.S., Thakur, R.K., Bhardwaj, N.R., and Pathania, N. 2004. Onion seed crop pollination: A missing dimension in mountain horticulture. *Acta Hort (ISHS)* 631: 79–86.

Chaturvedi, M. 1973. An analysis of honeybee pollen loads from Banthra, Lucknow, India. *Grana* 13(3): 139–144.

Chaturvedi, M. 1977. Further investigation on the pollen analysis of bee loads from Banthra, India. *New Botanist* 4(1–4): 41–47.

Chaudhary, N., and Sihag, R.C. 2004. *Relative abundance, diversity and dominance concentration of two honeybee species on three concurrently flowering crops, onion (Allium cepa L.), carrot (Daucus carota L.) and fennel (Foeniculum vulgare L.).* First European Conference of Apidology, Udine, September 19–23.

Cherian, M.C., and Mahadevan, V. 1945a. The longevity of Indian bee (*Apis indica*). *Indian Bee Journal* 8(3–4): 41–43.

Cherian, M.C., and Mahadevan, V. 1945b. Preliminary studies on the pollen carrying capacity o the Indian bee (*Apis indica* F). *Indian Bee Journal* 8(7–8): 109–111.

Cherian, M.C., Ramachandran, S., and Mahadevan, V. 1947. Studies in bee behaviour. *Indian Bee Journal* 9: 116–124.

Choudary, R.K. 1978. Floral fidelity in the Indian honey bee (*Apis cerana indica*). *Indian Bee J.* 40(2): 33–35.

Churchill, D.M., and Christenenson, P. 1970. Observations on pollen harvesting by brush tongued lorikeets. *Austr J. Zool.* 18: 427–437.

Cirudarescu, G. 1971. Pollinators of lucerne and factors influencing their effectiveness in South Eastern parts of Birse depression. *Ann. Univ. Bucuresti Biol. Animalia* 20: 77–81.

Cody, M. 1974. Optimization in ecology. *Science* 183: 1156–1164.

Collins, S.A., Conner, J.K., and Robinson, G.E. 1997. Foraging behavior of honey bees (Hymenoptera: Apidae) on Brassica nigra and B. rapa grown under simulated ambient and enhanced UV-B radiation. *Annals of the Entomological Society of America* 90: 102–106.

Collision, C.H. 1973. *Nectar secretion and how it affects the bee activity of honeybees in the pollination of hybrid pickling cucumbers, Cucumis melo L.* M.Sc. thesis. Michigan State University, Ann Arbor.

Cooper, P.D., and Schaffer, W.M. 1985. Temperature regulation of honey bees (Apis mellifera) foraging in the Sonoran desert. *Journal of Experimental Biology* 114: 1–15.

Corbet, S.A. 1978a. A bee's view of nectar. *Bee World* 59: 25–32.

Corbet, S.A. 1978b. Bees and nectar of *Echium vulgare.* In: Richards, A.J. (Ed.), *The pollination of flowers by insects.* Academic Press, London, 21–30.

Corbet, S.A., Fussel, M., Ake, R., Fraser, A., Gunsen, E., Salvage, A., and Smith, K. 1993. Temperature and the pollinating activity of social bees. *Ecol. Entomol.* 18: 17–30.

Cresswell, J.E., Desneux, N., and van Engelsdorp, D. 2012. Dietary traces of neonicotinoid pesticides as a cause of population declines in honey bees: An evaluation by Hill's epidemiological criteria. *Pest Manag Sci* 68: 819–827. https://doi.org/10.1002/ps.3290.

Desmuth, G. 1933. Temperature and netar secretion. *Glean Bee Cult.* 51: 582–584.

Desneux, N., Decourtye, A., and Delpuech, J.M. 2007. The sublethal effects of pesticides on beneficial arthropods. *Annu Rev Entomol* 52: 81–106. https://doi.org/10.1146/annurev.ento.52.110405.091440.

Devkota, F.R., and Thapa, R.B. 2005. Foraging Preference of *Apis cerana* F. and *Apis mellifera* L. to Broccoli under Caged and Open Conditions in Chitwan J. *Inst. Agric. Anim. Sci.* 26: 167–168.

Dhaliwal, H.S., and Atwal, A.S. 1986. Factors influencing tripping efficiency of bees on alfalfa (*Medicago sativa* L.). In: Kapil, R.P. (Ed.), *Pollination biology: An analysis.* Inter-Indian Publication, New Delhi, 79–90.

Dhaliwal, H.S., and Sharma, P.L. 1972. Homing ability and the foraging range of the Indian honeybee, *Apis indica* F. *Indian J. Ent.* 34(3): 225–227.

Dhaliwal, H.S., and Sharma, P.L. 1973. The foraging range of Indian honeybee on two crops *J. Apic Res.* 12(2): 131–134.

Dhaliwal, H.S., and Sharma, P.L. 1974. Foraging range of the Indian honeybee. *J. Apic Res.* 13(2): 137–141.

Diwan, V.V. 1980. Honey bee behaviour during total solar eclipse: Anonymous. *Indian Bee Journal* 42(4): 97–99.

Diwan, V.V., and Salvi, S.R. 1965. Some interesting behavioural features of Apis dorsata Fab. *Indian Bee Journal* 27(1): 52.

Dodson, C.H. 1975. Coevolution of orchids and bees. In: Gilbert, L.E., and Raven, P.H. (Eds.), *Coevolution of animals and plants*. University of Texas Press, Austin, 91–99.

Dorjay, N., Abrol, D.P., and Vikram, S. 2022. Effect of bee attractants on foraging activities of honeybees *Apis mellifera, A. dorsata* and *A. cerana* on *Cucumis sativus* L. and *Memordica charantia* L. flowers. *Journal of Apiculture* 37(2): 123–134.

Dressler, R.L. 1968. Observations on orchids and euglossine bees in Panama and Costa Rica. *Rev. Biol. Trop.* 15: 143–183.

Dugatkin, L.A. 2020. *Principles of animal behavior.* 4th edn. The University of Chicago Press, Chicago and London.

Dukas, R. 2001. Effects of predation risk on pollinators and plants. In: Chittka, L., and Thompson, J.D. (Eds.), *Cognitive ecology of pollination*. Cambridge University Press, Cambridge, 214–236.

Dukas, R. 2005. Bumble bee predators reduce pollinator density and plant fitness. *Ecology* 86: 1401–1406.

Dukas, R., and Morse, D.H. 2003. Crab spiders affect flower visitation by bees. *Oikos* 101: 157–163.

Dukas, R., and Morse, D.H. 2005. Crab spiders show mixed effects on flower visiting bees and no effect on plant fitness components. *Ecoscience* 12: 244–247.

Dyer, F.C. 1985. Nocturnal orientation by the Asian honeybee, *Apis dorsata*. *Anim Behav* 33: 769–774.

Dyer, F.C., and Seeley, T.D. 1987. Interspecific comparisons of endothermy in honey-bees (*Apis*): Deviations from the expected size-related patterns. *J. Exp. Biol.* 127: 1–26.

Dyer, F.C., and Seeley, T.D. 1991. Dance dialects and foraging range in three Asian honey bee species. *Behavioral Ecology and Sociobiology* 28(4): 227–233.

Eckert, J.E. 1955. The flight of the honeybee. *Am. Bee J.* 95: 395–401.

Emlen, J.M. 1966. The role of time and energy in food preferences. *Am. Natur.* 100: 611–617.

Faegri, K., and van der Pijl, L. 1971. *The principles of pollination ecology.* 2nd edn. Pergamon, Oxford, 291.

Faegri, K., and van der Pijil, L. 1978. *The principles of pollination ecology.* 3rd edn. Oxford University Press, Oxford, New York and Toronto, 244.

Free, J.B. 1954. The behaviour of robber honeybees. *Behaviour* 7: 223–240.

Free, J.B. 1970. *Insect pollination of craps.* Academic Press, London.

Free, J.B. 1993. *Insect pollination of crops.* 2nd enlarged edn. Academic Press, London, 684.

Free, J.B., Ferguson, A.W., and Simpkins, J.R. 1985. Influence of virgin honeybees (*Apis mellifera*) on queen rearing and foraging. *Physiological Entomology* 10: 271–274.

Fulop, A., and Menzel, R. 2000. Risk-indifferent foraging behaviour in honeybees. *Animal Behaviour* 60: 657–666.

Gill, F.B., and Wolf, L.L. 1975. Economics of feeding territoriality in the golden winged sunbird. *Ecology* 56: 333–345.

Girling, R.D., Lusebrink, I., Farthing, E., Newman, T.A., and Poppy, G.M. 2013. Diesel exhaust rapidly degrades floral odours used by honeybees. *Scientific Reports* 3: 2779. https://doi.org/10.1038/srep02779.

Goyal, N.P., and Atwal, A.S. 1977. Wing beat frequency of *Apis Indica* indica F. and *A.mellifera* L. *J. Apic Research* 16(1): 47–48.

Grant, V. 1950. The flower constancy of bees. *Bot. Rev.* 16: 379–398.

Gupta, J.K., Kumar, J., and Mishra, R.C. 1984. Nectar sugar production and honeybee foraging activity in different cultivars of cauliflower. *Brassica oleracea* var. *botrytis*. *Indian Bee Journal* 46: 21–22.

Hainsworth, F.R., and Wolf, L.L. 1972a. Energetics of nectar extraction in a small, high altitude, tropical hummingbird, *Selasphorus flammula*. *J. Comp. Physiol.* 80: 377–387.
Hainsworth, F.R., and Wolf, L.L. 1972b. Power for hovering flight in relation to body size in hummingbirds. *Amer. Naturalist.* 106: 589–596.
Hainsworth, F.R., and Wolf, L.L. 1972c. Crop volume, nectar concentration, and hummingbird energetics. *Comparative Biochemistry and Physiology* 42: 359–366.
Harder, L.D. 1985. Morphology as a predictor of floral choice by bumblebees. *Ecology* 66: 190–210.
Harder, L.D. 1986. Effects of nectar concentration and flower depth on flower handling efficiency of bumblebees. *Oecologia* 69: 309–325.
Heinrich, B. 1972. Energetics of temperature regulation and foraging in a bumble bee, *Bornbus terricola* Kirby. *J. Comp. Physiol.* 77: 49–65.
Heinrich, B. 1973. The energetics of bumblebees. *Sci. Amer.* 288(4): 96–102.
Heinrieh, B. 1975a. Energeties of pollination. *Ann. Rev. Ecol. Syst.* 6: 139–170.
Heinrich, B. 1975b. Bee flowers: A hypothesis on flower variety and blooming times. *Evolution* 29: 325–334.
Heinrich, B. 1975e. The role of energetics in bumblebee-flower interrelationships. In: Gilbert, L.E., and Raven, P.H. (Eds.), *Coevolution of animals and plants*. University of Texas Press, Austin, 141–158.
Heinrich, B. 1979a. *Bumblebee economics*. Harvard University Press, Cambridge, MA.
Heinrich, B. 1979b. Keeping a cool head: Honeybee (*Apis mellifera*) thermal regulation. *Science* 205: 1269–1271.
Heinrich, B. 1983. Insect foraging energetics. In: Jones, C.E., and Little, R.J. (Eds.), *Handbook of experimental pollination biology*. Academic Press, New York, 187–215.
Heinrich, B., and Raven, P.H. 1972. Energetics and pollination ecology. *Science* 176: 597–602.
Hocking, B. 1953. The intrinsic range and speed of flight of insects. *Transactions of the Royal Entomological Society of London* 104: 223–345.
Hunt, G.J., Amdam, G.V., Schlipalius, D., Emore, C., Sardesai, N., Williams, C.E., Rueppell, O., Guzmán-Novoa, E., Arechavaleta-Velasco, M., Chandra, S., Fondrk, M.K., Beye, M., and Page Jr, R.E. 2007. Behavioral genomics of honeybee foraging and nest defens e. *Naturwissenschaften* 94(4): 247–267.
Inouye, D.W. 1983. The ecology of nectar robbing. In: Bentley, B., and Elias, T. (Eds.), *The biology of nectaries*. Columbia University Press, New York, 153–173.
Jadhav, L.D. 1982. Effect of solar eclipse on the foraging activity of honey bees. *Biovigyanam* 8(2): 157–161.
Jeschke, P., and Nauen, R. 2008. Neonicotinoids–from zero to hero in insecticide chemistry. *Pest Manag Sci* 64: 1084–1098.
Jhajj, H.S., and Goyal, N.P. 1979b. *Comparative studies on the flower constancy of Apis cerana indica F. and Apis mellifera L.* Proceedings of the Fourth International Symposium on Pollination, Maryland, 333–339.
Johnson, L.K., and Hubbell, S.P. 1974. Aggression and competition among stingless bees: Field studies. *Ecology* 55: 120–127.
Johnson, L.K., and Hubbell, S.P. 1975. Contrasting foraging strategies and co-existence of two bee species on a single resource. *Ecology* 56(6): 1398–1406.
Jones, D., and Yates, B.D. 1991. *Beekeeping study notes*. 2nd edn. Bee Books New and Old Tapping Wall Farm, Burrowbridge, Bridgwater TA7 ORY, Somerset.
Joshi, N.C., and Joshi, P.C. 2010. Foraging behaviour of Apis Spp. on apple flowers in a subtropical environment. *New York Science Journal* 3: 71–76.
Kammer, A.E., and Heinrich, B. 1978. Insect flight metabolism. *Adv. Insect Physiol.* 13: 133–228.
Kapil, R.P., and Brar, H.S. 1971. *Foraging behaviour of Apis florea F. on Brassica campestris var toria*. Proc. XXIIth Intl. Cong. Apic., Moscow, 335–339.

Kapil, R.P., and Kumar, S. 1975. Foraging activity of *Apis dorsata* F. on *Brassica juncea* Czern and Coss. *J. Bombay Nat. Hist. Soc.* 71(2): 327–331.

Kapyla, M. 1978. Foraging distance of a small solitary bee, *Chelostoma maflillosum* (Hymenoptera, Megachilidae). *Ann Entomologici Fenn* 35: 63–64.

Koeniger, N., and Vorwohl, G. 1979. Competition for food among four sympatric species of Apini in Sri Lanka *(Apis dorsata, Apis cerana, Apis florea* and *Trigona iridipennis*). *Journal of Apicultural Research* 18(2): 95–109.

Kralj, J., and Fuchs, S. 2006. Parasitic Varroa destructor mites influence flight duration and homing ability of infested Apis mellifera foragers. *Apidologie* 37: 577–587.

Kralj, J., and Fuchs, S. 2010. *Nosema* sp. influences flight behavior of infected honey bee (*Apis mellifera*) foragers. *Apidologie* 41: 21–28.

Kremen, C. 2005. Managing ecosystem services: What do we need to know about their ecology? *Ecol Lett* 8: 468–479.

Leppik, E.E. 1977. *Floral evolution in relation pollination ecology.* Today and Tomorrow Prizters and Publishers, New Delhi, 104–108.

Lerer, H., Bailey, W.G., Mills, P.F., and Pankiw, M. 1982. Pollination activity of *megachile rotundata* (Hymenoptera: Apidae). *Environ. Entomol.* 11: 997–1000.

Lindauer, M. 1956. Uber die Verstandigung bei indischen Bienen. *Z. vergl. Physiol.* 38: 521–557.

Lingappa, S., Viraktamath, S.A., Vastrad, A.S., and Williams, R. 1999. *Utilization of honey bee Apis cerana Fab. for pollination of watermelon and safflower.* Proceedings of XXXVI International Apicultural Congress (Apimondia) Vancouver, Canada, Book of Abstracts, 235.

Maloof, J.E. 2000. Reproductive biology of a North American subalpine plant: *Corydalis caseana* A.Gray *ssp.brandegei* (S.Watson) G.B. Ownbey. *Plant Species Biology* 15: 281–288.

Mattu, V.K., Chaudhary, D.K., and Kumar, L. 1994. Foraging ecology of *Apis cerana* F. and *A. mellifera* L. in pollinating stone fruit crops. *Pest Management and Economic Zoology* 2(1): 35–39.

Mattu, V.K., Raj, H., and Thakur, M.L. 2012. Foraging behavior of honeybees on apple crop and its variation with altitude in Shimla hills of western Himalaya. *International Journal of Science and Nature* 3: 296–301.

Mattu, V.K., and Verma, L.R. 1985. Studies on the annual foraging cycle of *Apis cerana indica* F. in Simla hills of Northwest Himalayas. *Apidologie* 16(1): 1–18.

Michener, C.D. 1974. *The social behavior of the bees.* Harvard University Press, Cambridge, 404.

Morse, D.H. 1986. Predatory risk to insects foraging at flowers. *Oikos* 46: 223–228.

Muñoz, A.A., and Arroyo, A.T.K. 2004. Negative impacts of a vertebrate predator on insect pollinator visitation and seed output in *Chuquiraga oppositifolia*, a high Andean shrub. *Oecologia* 138: 66–73.

Murrell, D.C., and Nash, W.T. 1981. Nectar secretion by toria brassica campestris var toria and foraging behavior of 3 apis species on toria in Bangladesh. *Journal of Apicultural Research* 20(1): 34–38.

Naim, M. 1984. Field behaviour of honeybee foragers. *Indian Journal of Entomology* 46(3): 335–339.

Naim, M., and Phadke, K.G. 1972. Flight range of *Apis cerana indica* when enticed to a feeding dish. *Journal of Apicultural Research* 11(3): 167–169.

Nataraj, S., Varadharajan, M., Thiyagesan, K., and Sankari, A. 2000. *Flower constancy in the Indian honeybee Apis cerana indica.* Proc. of 8th National Tamil Science Congress (in Tamil) (in press).

Neupane, K.R., Dhakal, D.D., Thapa, R.B., and Gautam, D.M. 2006. Foraging preference of giant honeybee, *Apis dorsata* F., to selected horticultural crops. *Inst. Agric. Anim. Sci.* 27: 87–92.

Nunez, J.A. 1977. Nectar flow by melliferous flora and gathering flow by *Apis mellifera ligustica*. *J. Insect Physiol.* 23: 265–276.

Oldroyd, B.P., Rinderer, T.E., and Buco, S.M. 1992a. Intra-colonial foraging specialism by honey bees *(Apis mellifera)* (Hymenoptera: Apidae). *Behavioral Ecology and Sociobiology* 30: 291–295.

Oldroyd, B.P., Rinderer, T.E., and Wongsiri, S. 1992b. Pollen resource partitioning by *Apis dorsata, A. cerana, A. andreniformis* and *A. florea* in Thailand. *Journal of Apicultural Research* 31(1): 3–7.

Osgood, C.E. 1974. Relocation of nesting populations of *Megichile rotundata* as an important pollinator of alfalfa. *J. Apic. Res.* 13: 63–73.

Panda, P., Sontakke, B.K., and Panda, B. 1993. Foraging behaviour of honey bee species on different varieties of niger, *Guizotia abyssinica* Cass. in *Orissa. Journal of lnsect Science* 6(1): 104–106.

Pankiw, T., Tarpy, D.R., and Page Jr, R.E. 2002. Genotype and rearing environment affect honey bee perception and foraging behaviour. *Anim Behav.* 64: 663–672.

Pearce, F.C.R., Couvillon, M.J., and Ratnieks, F.L.W. 2013. Hive relocation does not adversely affect honey bee (Hymenoptera: Apidae) foraging. *Psyche* 1–8.

Pilling, E., Campbell, P., Coulson, M., Ruddle, N., and Tornier, I. 2013. A four-year field program investigating long-term effects of repeated exposure of honey bee colonies to flowering crops treated with thiamethoxam. *PLoS One* 8(10): e77193.

Porrini, C., Sabatini Anna, G., Stefano, G., Severino, G., Piotr, M., Francesca, G., Laura, B., Enrico, G., and Giorgio, C. 2003. Honey bees and bee products as monitors of the environmental contamination. *Apiacta* 38: 63–70.

Priti, M.R.C., and Sihag, R.C. 1997. Diversity, visitation frequency, foraging behaviour and pollinating efficiency of insect pollinators visiting cauliflower (*Brassica olearacea* L. var. *botrytis* cv. Hazipur Local) blossoms. *Indian Bee Journal* 59(4): 230–237.

Puranik, P.G., Krishna, G.G., Ahmad, A., and Chari, N.C.1977.Wing beat frequency of a flier – mass flow theory. *Proc. Indian Acad. Sci.* 85: 327–339.

Pyke, G.H. 1979. Optimal foraging in bumblebees: Rules of movement between flowers within inflorescences. *Anim Behav* 27: 1167–1181.

Raj, D., and Rana, V.K. 1993. Foraging behaviour of *Apis mellifera* L. and *Apis cerana indica* F. at the hive entrance during rapeseed bloom. *Journal of Entomological Research* 17(2): 105–110.

Raj, D., and Rana, V.K. 1994. Time spent by *Apis mellifera* L. and *Apis cerana indica* F. foragers on rapeseed bloom. *Journal of Entomological Research* 18(4): 335–339.

Raj, D., Rana, V.K., and Devi, N. 1993.Comparative studies on pollen loads and pollen constancy of of *Apis mellifera* L. and *Apis cerana indica* F. on rapeseed. *Journal of Entomological Research* 17(1): 43–48.

Rao, G.M., and Lazar, M. 1980. *Studies on the behaviour and pollination on onion (Allium cepa L.).* Proceedings of the 2nd International Conference on Apiculture in Tropical Climates, New Delhi, 580–589.

Rao, G.M., and Lazar, M. 1983. *Studies on bee behavior and pollination in onion (Allium cepa L.).* Proceedings of the 2nd International Conference on Apiculture in Tropical Climates, New Delhi, 3, 580–589.

Reader, T., Higginson, A.D., Barnard, C.J., Gilbert, F.S., and The Behavioural Ecology Field Course. 2006. The effects of predation risk from crab spiders on bee foraging behavior. *Behavioral Ecology* 17(6): 933–939.

Reddy, C.C. 1979. *The influence of various meteorological factors on the foraging behavior of Apis cerana indica.* Proceedings of the Fourth International Symposium on Pollination, Maryland, 341–345.

Reddy, C.C. 1980a. Observations on the annual cycle of foraging and brood reanng by *Apis cerana indica* colonies. *Journal of Apicultural Research* 19(1): 17–20.

Reddy, C.C. 1980b. *Studies on trhe nesting behaviour of Apis dorsata F.* Proceedings of the 2nd International Conference on Apiculture in Tropical Climates, Indian Agricultural Research Institute, New Delhi, 2, 391–397.

Reddy, C.S., and Reddy, E.U.S. 1984. Bee flower interaction and pollination potential. *Proc. Indian Acad. Sci.* 93: 373–390.

Reed, S.C., William, C.M., and Chadwick, L.E. 1942. Frequency of wing beat as a character as a character for separating species, races and geographic varieties of *Drosophila. Genetics* 27: 349–361.

Ricketts, T.H. 2004. Tropical forest fragments enhance pollinator activity in nearby coffee crops. *Conserv Biol* 18: 1262–1271.

Robertson, I.C., and Maguire, D.K. 2005. Crab spiders deter insect visitations to slickspot peppergrass flowers. *Oikos* 109: 577–582.

Ruttner, F. 1988. *Biogeography and taronomy of honeybees.* Springer-Verlag, Berlin.

Sakagami, S.F. 1959. Some interspecific relations between Japanese and European honey bees. *J. Anim. Ecol.* 28: 51–68.

Sattigi, H.N., Rajasekhar, D.W., and Kulkarni, K.A. 2001. *Effect of attractants in enhancing the productivity of watermelon.* Paper presented at National Symposium on Environment and Evolutionary Biology, Dharwad, 24, March 1–3.

Schaffer, W.M., Jensen, D.B., Hobbs, D.E., Gurevitch, I., Todd, J.R., and Schaffer, M.V. 1979. Cornpetition, foraging energetics, and the cost of sociality in three species of bees. *Ecology* 60(5): 976–987.

Schlising, R.A. 1970. Sequence and timing of bee foraging on flowers of *Ipomoea* and *Aniseia* (convolvalacae). *Ecology* 51: 1061–1067.

Schmalhofer, V.R. 2001. Tritrophic interactions in a pollination system: Impacts of species composition and size of flower patches on the hunting success of a flower-dwelling spider. *Oecologia* 129: 292–303.

Schmid-Hempel, P. 1991. The ergonomics of worker behavior in social hymenoptera. *Adv Study Behav* 20: 87–134.

Schmid-Hempel, P., and Schmid-Hempel, R. 1990. Endoparasitic larvae of conopid flies alter pollination behavior of bumblebees. *Naturwissenschaften* 77: 450–452.

Schneider, C.W., Tautz, J.R., Grünewald, B., and Fuchs, S. 2012. RFID tracking of sublethal effects of two neonicotinoid insecticides on the foraging behavior of Apis mellifera. *PLoS One* 7: e30023. https://doi.org/10.1371/journal.pone.0030023.g005.

Schoener, T.W. 1971. Theory of feeding strategies. *Ann. Rev. Ecol. Syst.* 2: 369–404.

Shah, F.A., and Shah, T.A. 1989. Comparative foraging behaviour of *Apis cerana* Fabr. and *Apis mellifera* L. during *Isodon rugosus* (Wall.) Codd bloom in Kashmir Valley. *Indian Bee Journal* 51(4): 138–139.

Sharma, M.K. 1970a. An analysis of pollen loads of honeybees from Kangra, India. *Grana* 10: 35–42.

Sharma, M.K. 1970b. Studies on the pollen loads of honeybees from Kangra, India. *J. Palynol.* 6: 104–110.

Sharma, M.K., Shashikala, K., and Kapil, A. 1999. Studies on the effect of different colour attractants on cross pollination in *Boswellia serrata* Roxb. ex Colebr. *Indian Forester* 125(12): 1244–1247.

Sharma, P.L., and Sharma, A.C. 1950. Influence of numbers in a colony on the honey gathering capacity of bees (*Apis indica* hill variety). *Indian Bee Journal* 12(7–8): 106–107.

Sharma, S.K., Singh, J.R., and Mahla, J.C. 2001. Foraging behaviour of *Apis* spp. in semi-arid sub-tropical climate. 3. On flowers of mustard, onion, carrot, berseem and sunflower. *Crop Research (Hisar)* 21(3): 332–334.

Sharma, V.P., and Kumar, N.R. 2010. Changes in honeybee behaviour and biology under the influence of cellphone radiations. *Current Science* 98: 1376–1378.

Sihag, R.C. 1983. Honeybees forage on grape (*Vitis vinifera* L.) during dearth period. *Indian Bee Journal* 45: 111–112.

Sihag, R.C. 1984. Influence of environmental factors on pollination activity of bees. *Environ & Ecol.* 2: 149–152.

Sihag, R.C. 1990. *Differential attractiveness of two cruciferous species to honeybees.* Proceedings of the 11th International Congress of the International Union for the Study of Social Insects, India, 434–436.

Sihag, R.C. 1998. Ecology of the giant honeybee (*Apis dorsata*) in semi-arid subtropical climate of India. In: Matsuka, M., Verma, L.R., Wongsiri, S., Shrestha, K.K., and Pratap, U. (Eds.), *Asian bees and beekeeping progress of research and development.* IBH Pub. Co. Pvt. Ltd., New Delhi and Oxford, 51–52.

Sihag, R.C. 2000a. *Ecobiology of the little honeybee (Apis florea) in semi and subtropical climates of India.* Proceedings of the Fourth Asian Apicultural Association Conference, Asian Apicultural Association, Kathmandu, 46–49.

Sihag, R.C. 2000b. *Eco-biology of the giant honeybee (Apis dorsata) in semi-arid subtropical climates of India.* Proceedings of the Fourth Asian Apicultural Association Conference, Asian Apicultural Association, Kathmandu, 49–52.

Sihag, R.C., and Abrol, D.P. 1986. Correlation and path-coefficient analysis of environmental factors influencing flight activity of *Apis florea* F. *Journal of Apicultural Research* 25(4): 202–208.

Sihag, R.C., and Kapil, R.P. 1983. *Foraging strategies of honeybees as determined by quality and quantity of nectar.* Proc. 5th Int Symo. Polln. Versailles, Hissar, 51–59.

Sihag, R.C., and Rathi, A. 1992. Foraging modes and foraging rates of different bee pollination of pigeon pea (Cajanus cajan (L.) MILLSP. *Pollination in Tropics: Proc. Intl Symp Pollin. Trop.* 93–95.

Singh, Y. 1981. Studies on pollen gathering activity of Indian honey bees (Apis cerana indica Fabr.) under Saharanpur conditions. *Progress in Horticulture* 12: 31–38.

Smith, A.R., Wcislo, W.T., and O'Donnell, S. 2003. Assured fitness returns favor sociality in a mass-provisioning sweat bee, *Megalopta genalis* (Hymenoptera, Halictidae). *Behav Ecol Sociobiol* 54: 14–21.

Somanathan, H., Warrant, E.J., Borges, R.M., Wallén, R., and Kelber, A. 2009. Resolution and sensitivity of the eyes of the Asian honeybees Apis florea, Apis cerana and Apis dorsata. *The Journal of Experimental Biology* 212: 2448–2453.

Sotavalta, O. 1952. On the difference and variation of the wing stroke frequency of insects in wing mutation and loading experiments and in experiments at sub-atmospheric pressure. *Ann. Zool. Soc., Vanamo* 15: 1–67.

Steffan-Dewenter, I., and Kuhn, A. 2003. Honeybee foraging in differentially structured landscapes. *Proc R Soc Lond Ser B Biol Sci* 270: 569–575.

Stiles, F.G. 1971. Time, energy, and territorially of the Anna Hummingbird (*Calypte anna*). *Science* 173: 818–821.

Stimec, J., Scott-Dupree, C.D., and McAndrews, J.H. 1997. Honey bee, *Apis* mellifera, pollen foraging in southern Ontario. *The Canadian Field Naturalist* 111(3): 454–456.

Summers, S.D. 1990. Facing the bees. *International Wildlife* 20: 5.

Suttle, K.B. 2003. Predators as mediators of top-down effects on plants. *Ecol Lett* 6: 688–694.

Szabo, T.I., and Smith, M.V. 1972. The influence of light intensity and temperature on the activity of alfalfa leaf cutter bee. *Megachile rotundata* under field conditions. *J. Apic. Res.* 11: 157–165.

Tan, K., Chen, W., Dong, S., Liu, X., Wang, Y. et al. 2014. Imidacloprid alters foraging and decreases bee avoidance of predators. *PLoS One* 9(7): e102725. https://doi.org/10.1371/journal.pone.0102725.

Tan, K., Hu, Z., Chen, W., Wang, Z., Wang, Y., and Nieh, J.C. 2013. Fearful foragers: Honey bees tune colony and individual foraging to multi-predator presence and food quality. *PLoS One* 8(9): e75841. https://doi.org/10.1371/journal. pone.0075841.

Thakur, A.K., Sharma, O.P., Garg, R., and Dogra, G.S. 1982. Comparative studies on foraging behaviour of *Apis mellifera* and *A. cerana indica* on mustard. *Indian Bee Journal* 44(4): 91–92.

Thiyagesan, K., Varadharajan, M., and Abirami, S. 2001a. Studies on risk-sensitive foraging in the Indian honey bee (*Apis cerana indica*). *Journal of Apicultural Research* 40(3–4): 16–20.

Thiyagesan, K., Varadharajan, M., and Abirami, S. 2001b. Studies on central place foraging in the Indian honey bee, *Apis cerana indica. Journal of Apicultural Research* 40(3–4): 21–26.

Underwood, B.A. 1991.Thermoregulation and energetic decision-making by the honeybees *Apis cerana, Apis dorsata* and *Apis laboriosa. Journal of Experimental Biology* 157: 19–34.

Usha, S.P., Upadhyay, R.G., and Yadav, M.K. 2016. Impact of abiotic factors on the foraging activity of the major insect pollinators on mango in Tarai Region of Uttarakhand. *International Journal of Tropical Agriculture* 34: 767–737.

Varma, S.K., and Joshi, N.K. 1983. Studies on honeybees in the pollination of cauliflower (Brassica olercea L. var. Botryitis). *Indian Bee Journal* 45(3): 45–55.

Verma, L.R. 1995. Apis cerana: Biometric genetic and behavioural aspects. In: *The Asiatic hive bee*. Enviroquest, Cambridge, 41–53.

Visscher, P.K., and Seeley, T.D. 1982. Foraging strategy of honey bee colonies in a temperate deciduous forest. *Ecology* 63: 1790–1801.

Waddington, K.D., Allen, T., and Heinrich, B. 1981. Floral preferences of bumblebees *(Bombus edwardsii)* in relation to intermittent vs. continuous rewards. *Anim. Behav.* 29: 779–784.

Waller, G.D. 1972. Evaluating responses of honeybees to sugar solutions using an artificial flower feeder. *Ann. Ent. Soc. Am.* 65: 857–862.

Wcislo, W.T., Arneson, L., Roesch, K., Gonzalez, V., Smith, A., and Fernandez, H. 2004. The evolution of nocturnal behaviour in sweat bees, *Megalopta genalis* and *M. ecuadoria* (Hymenoptera: Halictidae): An escape from competitors and enemies? *Biol J Linn Soc* 83: 377–387.

Wcislo, W.T., and Cane, J.H. 1996. Resource utilization by solitary bees (Hymenoptera: Apoidea), and exploitation by their natural enemies. *Annu Rev Entomol* 41: 257–286.

Weidenmuller, A., and Tautz, J. 2002. In-hive behavior of pollen foragers (*Apis mellifera*) in honey bee colonies under conditions of high and low pollen need. *Ethology* 108(3): 205–221.

Whitehorn, P.R., O'Connor, S., Wackers, F.L., and Goulson, D. 2012. Neonicotinoid pesticide reduces bumble bee colony growth and queen production. *Science* 336: 351–352. https://doi.org/10.1126/science.1215025.

Willmer, P.G. 1983. Thermal constraints on activity pattern of nectar feeding insects. *Ecological Entomology* 8: 455–469.

Winston, M.L. 1987. *The biology of the honey bee*. Harvard University Press, Cambridge.

Wolf, L.L., and Hainsworth, F.R. 1971. Time and energy budgets of territorial humming birds. *Ecology* 52: 980–988.

Wolf, L.L., Hainsworth, F.R., and Gill, F.B. 1975. Foraging efficiencies and time budgets in nectar feeding birds. *Ecology* 56: 117–128.

Wolf, L.L., Hainsworth, F.R., and Stiles, F.G. 1972. Energetics of foraging: Rate and efficiency of nectar extraction by hummingbirds. *Science* 176: 1351–1352.

Woyciechowski, M. 2007. Risk of water collecting in honeybee (Apis mellifera) workers (Hymenoptera: Apidae). *Sociobiology* 50(3): 1059–1068.

Woyciechowski, M., and Kozlowski, J. 1998. Division of labour by division of risk according to worker life expectancy in the honey bee (Apis mellifera L.). *Apidologie* 29(1–2): 191–205.

Woyke, J., Wilde, J., and Wilde, M. 2003. Flight activity reaction to temperature changes in *Apis dorsata*, *Apis laboriosa* and *Apis mellifera*. *Journal of Apicultural Science* 47(2): 73–80.

Wyke, G.R. 1952. The preferences of honeybees for honeybee for solutions of various sugars which occur in nectar. *J. Exp. Biol.* 29: 511–528.

17 Management and Conservation of *Apis dorsata*

M.R. Srinivasan and S. Pradeep

17.1 INTRODUCTION

Apis species are found in a wide range of agroecosystems across India. *Apis dorsata* is the largest honeybee among the described bee species, and so these bees are called Asian giant honeybees and rock bees. These bees are about 17–20 mm long. Rock bees are found throughout India in sub-mountainous areas up to an altitude of 2,700 m. They create a massive single comb that is up to 2 metres long and 1 metre deep. This comb contains large amounts of honey and wax. Rock bees frequently change locations. A single rock bee colony can produce roughly 50–80 kg of honey per year (Mishra 1995). They frequently move from one location to another in colonies. These bees prefer to build comb in open conditions. Numerous colonies move around seasonally, residing in two or more forage-rich areas throughout the course of the year.

17.2 STATUS OF *APIS DORSATA*

The geographic distribution of *A. dorsata* has been examined, and the results show that they are found in Pakistan (possibly also in some areas of southern Afghanistan) in the west, through the Indian subcontinent and Sri Lanka to Indonesia and some areas of the Philippines in the east (Crane 2004). Its distribution runs from southern China to Indonesia in a north–south direction; neither New Guinea nor Australia include the species. Recently, the giant honeybees of Nepal and the Himalayas were reclassified as *A. dorsata laboriosa*, a different species of *Apis* (Akratanakul 1990). Another subspecies of *A. dorsata* known as *A. dorsata binghami* is found in the Khasi Hills, Sikkim, and Meghalaya, which are all restricted areas of Northeast India (Allen 1995; Otis 1996). The majority of the time, these bees build their combs more than 6 metres above ground, but occasionally, we can see the colonies hanging on branches over 0.5 metres from the ground. *A. dorsata* can grow in single or multiple colonies. The foraging and scout bees will land and take off in the active lower portion of the comb.

They are difficult to rear due to their aggressive nature. As these bees are aggressive, they will attack intruders and occasionally chase them for up to 100 m (Ramchandra et al. 2012). Bee stings can occasionally be fatal to people. Because of

DOI: 10.1201/9781003294078-17

the risks involved in collecting rock bee honey, this honey is typically very expensive in the area. Some experienced bee hunters favour working at night. Bees are calmed by smoke, and in some locations, experts add chicken feathers to the smoke made by burning coal. This irritates rock bees and makes honey extraction easier because the colony was driven out due to the smell produced by the addition of the chicken feathers. There is widespread concern that the overall number of *A. dorsata* nesting sites throughout Asia may be about to decrease, partially as a result of declining forested areas, the use of pesticides on foraging farmlands, and bee hunting.

17.3 WHY THEY SHOULD BE PROTECTED

Honeybees from Southeast Asia have socio-cultural and economic values. Several thousands of honey hunters, most of whom come from the poorest communities, still make a living by gathering wild honey, also known as "honey hunting," which has been done in Southeast Asia for over than 40,000 years. In comparison to farmed honey, wild honey frequently has a better reputation and is frequently used in herbal medicine. Due to their ability to produce large amounts of honey (up to 80 kg per colony per harvest), the two giant honeybee species, *Apis dorsata* and *Apis laboriosa*, are the most frequently hunted species. Various animal species rely on plants pollinated by bees for their survival, and fruits and seeds produced by pollination are essential to the regeneration of the forest. Native honeybees play a crucial role in the overall food chain as they are also prey to other insects, amphibians, reptiles, birds, and mammals (Oldroyd and Nanork 2009). Maintaining the diversity and abundance of bees is essential for preserving and retaining balanced ecosystems in Southeast Asia because forests there developed in the existence of two or more species of honeybees. The biodiversity of Southeast Asian ecosystems could be negatively impacted by the extinction of native honeybee populations (Corlett 2004).

Up to 30% of the food consumed worldwide comes from plants that rely on bees and other insects for pollination (Oldroyd and Nanork 2009). Through the pollination of crops, bees in particular are essential to the production of food (Marshman et al. 2019). The existence of several native honeybee species is a benefit for agriculture in Southeast Asia because pollinator populations that are diverse and abundant can increase agricultural yields in terms of quantity and quality. Some crops can be pollinated by a variety of species of bees, whereas other crops are reliant on a single species. Dragonfruit (*Hylocereus undatus*) and other plants and trees that bloom at night require the special ability of the Asian giant honeybee *Apis dorsata* to forage at dusk. Honey hunters frequently ignore *Apis dorsata*, which are thought to be more defensive, and prefer collecting the honey of *Apis laboriosa* where both species of giant honeybees are present. In many parts of Southeast Asia, villagers are encouraged to collect wild honey from *Apis dorsata* colonies in exchange for agreeing to a logging ban in compliance with non-timber forest product programmes. As a result, *Apis dorsata* indirectly contribute to the economic prosperity and environmental conservation of protected forests and non-timber forest products (NTFP).

17.4 MAJOR THREATS FOR *APIS DORSATA*

17.4.1 Deforestation

In the next 100 years, the Southeast Asian region—which currently has the highest rate of tropical deforestation in the world—is expected to lose 42% of its biodiversity and 75% of its original forest cover (Sodhi et al. 2004). Oil palm plantations had a very low proportion of Apidae (honeybees and stingless bees), whereas undisturbed sites had a very high proportion. This strongly implies that honeybees are not favoured in oil palm plantations. This makes sense, given that nesting sites are scarce within the palm plantations and that nectar is only produced by ground flora. There are no hollows available for cavity-nesting bees, and *A. dorsata* cannot build a nest on the palm fronds due to their dense leaves. Loss of the habitat leads *Apis dorsata* to build nests in the window sun shades of apartments, buildings, and other anthropogenic structures.

Venkatesh (2014) reported that different nesting locations are used by *Apis dorsata* in urban, rural, and wild areas. The number of colonies in the study area, which is in Bangalore district (six perennial sites in Bangalore CBD), decreased from 72 during the study period of 1987–1989 to just seven during the period of 2013–2014 as a result of anthropocentric development activities.

17.4.2 Habitat Loss of *Apis dorsata*

The removal of giant honeybee nesting trees should raise more concerns because these colonies are highly philopatric and frequently migrate over great distances but return to the same nesting site each year (Butani 1950; Koeniger and Koeniger 1980). Additionally, *Apis dorsata* frequently nest in large groups, with as many as 100 colonies on a single tree (Oldroyd et al. 2000). The colonies attempt to construct their nests on the building after a large bee tree is cut down to make room for a building or other structure. This could result in unpleasant interactions with people and the eradication of the colonies, depending on the structure.

17.4.3 Hunting

Honeybee hunting has been a tradition among Asian people for more than 40,000 years, and it is still very common on the continent (Crane 1999). *A. dorsata* hunting is much more devastating, and it frequently involves setting the bees ablaze with a smouldering torch made of tightly bound brush. Some destroyed colonies might be able to reassemble, particularly if the hunt is conducted during the day, but the hunt is frequently done at night. During these harvests, many queens perish and their colonies also perish. Many hunters prefer night-time hunting because it results in fewer bee stings. The majority of colonies are destroyed by this method of hunting. For instance, an ecologist saw a tree being cut down in the Nilgiris Hills of Tamil Nadu, India, where hunters killed more than 100 colonies in a single night. In recent times, there have even been reports of "hunters" using pesticides to kill bees before collecting honey.

17.4.4 Threats by Pathogens and Parasites

Numerous parasitic mite species are natural hosts of Asian honeybees. In Asia, different species of honeybees and the parasitic mites that live on them coexist—which may encourage the exchange of parasites, as well as concurrent infestations by different mite species at the colony or individual levels (Anderson 1994).

Tropilaelaps mites are native parasites of the giant honeybee species *A. dorsata*, *A. laboriosa*, and *A. breviligula* (Laigo and Morse 1968). *Tropilaelaps koenigerum* was reported in *Apis dorsata* (Anderson and Morgan 2007; Delfinado-Baker and Baker 1982). Contrary to *Varroa* and *Euvarroa* in the native host of mites, *A. dorsata*, *Tropilaelaps* does not appear to exhibit a preference for worker or drone brood (Buawangpong et al. 2013; Koeniger et al. 2002). More recently, *Tropilaelaps mercedesae* and *Tropilaelaps thaii*, which parasitize *A. dorsata*, were described by Anderson and Morgan (2007).

Chinese sac brood virus is also referred to as Thailand sac brood virus (TSBV). In Kashmir, TSBV killed more than 90% of populations of domesticated *A. cerana* (Abrol and Bhat 1990); it was also discovered in India's *A. dorsata* and *A. florea* (Allen and Ball 1996).

Apis dorsata is also affected by *Nosema ceranae*. Different degrees of intensity were seen, however. *A. dorsata* had an infection rate comparable to that of *A. mellifera*, but *A. florea* had a lower infection rate and *A. ceranae* the lowest.

17.4.5 Effect of Insecticides on *Apis dorsata*

Giant honeybees (*Apis dorsata* F.), which are excellent pollinators of a variety of entomophilous crops, have been severely harmed by greater use of pesticides in crop protection efforts. In terms of acute contact toxicity, Treban was the most toxic insecticide. In the event of acute stomach toxicity, methyl-parathion was thought to be the most dangerous substance for honeybees. Due to their non-toxicity in both toxicity modes, Carbosulfan, Diflubenzuron, Flufenoxuron, and Azadirachtin can be used in the field at the usual recommended concentrations (Nirmala and Nehra 2015). At the start of their nesting period, more colonies were poisoned with insecticides than at later times. In comparison to Malathion and Sevin, Lindane was the insecticide most frequently used to eradicate *A. dorsata* colonies (Nagaraja 2016). Toxicity of some insecticides—namely Ethofenprox, Diafenthiuron and Imidacloprid—are found to affect haemocytes of *Apis dorsata* (Perveen and Ahmad 2017).

17.4.6 Effect of Street Lights on Honeybees

Foragers are drawn to lights at night when open-nesting species such as *A. dorsata* and *A. andreniformis* nest close to sources of light. This results in the deaths of numerous bees. Although the impact on colony survival may be minimal, this cannot help a nest that is already under stress. If *A. dorsata* queens are drawn to lights during mating flights, which we do not know, then queens may also be lost in this manner (Oldroyd and Nanork 2009).

17.5 CONSERVATION AND MANAGEMENT OF *APIS DORSARTA*

Being an open-nesting species, *Apis dorsata* is not managed in the same manner as other honeybee species. Since *Apis dorsata* bees have not evolved to live in dark cavities like *Apis mellifera* and *Apis cerana* have, all attempts to house *Apis dorsata* colonies through wooden Langstroth hives used for those species have failed (Koeniger et al. 2010). Honey collecting from feral *Apis dorsata* colonies is a common practice in many areas of the bee's distribution range rather than beekeeping to obtain honey, wax, and brood.

People in South Asia have put themselves in danger for many years by collecting honey from colonies of *A. dorsata*. Colonies of *Apis dorsata* typically cluster on tree trunks or cliffs, making access extremely difficult. To reduce the number of flying bees that will emerge once the colony is disturbed, traditional honey collection is typically done on a moonless night. Ropes or improvised ladders are used to scale cliffs or trees, and a flame is used to scare the bees off the comb. In order to get both the honey and the brood, honey hunters frequently harvest the entire nest. However, recent conservation efforts are urging honey gatherers to just remove the honeycomb sections rather than destroying the entire colony. In areas of Vietnam, Cambodia, and Indonesia with submerged forests, rafter beekeeping continues to be practiced.

The cause of honeybee decline is evident, and it can only add to the chorus of other plants and animals that are also suffering from the same problem. Old growth forests should no longer be cut down on this planet. However, practicality and sound science must be a part of conservation strategies, so we should concentrate on problems that can be usefully and realistically solved in the near future.

17.5.1 Prevention of Honey Hunting in *A. dorsata* Honey

The effect of hunting on the viability of a species depends on the size and growth rate of the population, the percentage of colonies that survive a typical harvest, the percentage of colonies that are harvested, rates of migration from nearby regions, the lifespan and reproductive rate of colonies, and other factors (Caughley and Sinclair 1994). It is challenging to determine the effect of hunting on the longevity of honeybee populations because so little information exists on any of these factors. Oldroyd and Wongsiri (2009) suggested that hunting of *A. florea* is unusual to endanger populations because the level of harvesting is likely to be much less than the potential for population growth. Their reasoning was based on the hypotheses of undefined survival of established colonies, production of 2.5 swarms per year, and 100 colonies per square kilometre (i.e., a potential growth rate of 250 colonies per square kilometre). However, *A. dorsata* densities are probably much lower than this—perhaps ten colonies per square kilometre—allowing for a harvest cap of much fewer than 25 colonies, which is frequently exceeded. If this is the case, and assuming that the harvest rate stays the same, the population will be wiped out. According to hunters in Tamil Nadu, *A. dorsata* is becoming harder to find (Nath et al. 1994). We believe that overhunting is the main reason for this decline. The essential demographic characteristics of populations of hunted honeybees must be known immediately. To study a nesting site for a full reproductive season of giant bees, one must count the number of developed colonies at the beginning of the season, the number of migrants who join the nesting site, the amount of daughter colonies, and the survival of all of these (Rinderer et al. 2002; Wattanachaiyingcharoen et al. 2008).

17.5.2 Promoting the Sustainable Harvesting of Honey from Wild Colonies

Apis laboriosa and *A. dorsata* harvesting is commonly destructive, but it does not have to be. Bee hunters frequently also practise conservation, making them open to suggestions that could aid in bee conservation. They frequently converse out strongly in support of preserving forests. In Vietnam, Cambodia, and Kalimantan and some other Indonesian provinces, initiatives are being made to promote the non-destructive collection of honey from *A. dorsata* nests (Crane 1999). In order to harvest bees during the day rather than at night, bee smokers and safety clothing must be utilized to protect hunters from stings. Second, to draw migratory *A. dorsata* swarms, bee hunters can build "rafters" in the forest. Rafters are sturdy boards that are 2 m long and hung at a 45° angle in a clearing in the forest. Taking honey from a colony that is nesting 1 m off the ground on a rope is much simpler than taking honey from a wild colony that is nesting in a 20 m *Koompassia* tree. However, keep in mind that in many places, hunters lack the funds to buy smokers and bee veils, relying almost entirely on application that uses materials from the forest to create their basic hunting equipment. This is done to promote more environmentally friendly honey harvesting. (Oldroyd and Nanork 2009).

17.5.3 Utilization of Attraction Planks or Rafters

In order to prevent the bees and their young from being lost when honey was harvested, *Apis dorsata* combs with their bees and brood were held in place in the early 1960s using bamboo clips. A small amount of smoke was used. Many tribal people started to harvest honey two times from the same colonies by using clips, and the technique spread among the tribes of Madhya Pradesh's Zabua district (Mahindre 1968). Later, it was discovered that the scent of beeswax played a major role in the migration of *A. dorsata* colonies to the same location each year. Beeswax was applied to one side of wooden planks to draw swarming bees. Before the migratory swarms arrived, these were fastened to the lower side of the branch with the smeared side of the plank facing downward. During the following season, a few colonies found refuge on the planks fastened to the tree branches and on the foundation marks. Some swarms choose to congregate close to the planks, the underside of which is covered in beeswax. Similar to other *A. dorsata* colonies nearby, these colonies flourished. Later, the same procedure was followed to attract migratory bees in adjacent areas. Thus, attraction planks evolved into a crucial piece of gear for the effective management of *A. dorsata* colonies (Mahendra 1997).

REFERENCES

Abrol, D., and Bhat, A. 1990. Studies on 'Thai sac brood virus' affecting indigenous honeybee *Apis cerana indica* Fab. colonies-prospects and future strategies-I. *Journal of Animal Morphology and Physiology* 37(1–2): 101–108.

Akratanakul, P. 1990. *Beekeeping in Asia*. Food and Agriculture Organisation of the United Nations (FAO) Agricultural Services Bulletin 68/4. FAO, Rome, Italy, 73.

Allen, M.F. 1995. Bees and beekeeping in Nepal. *Bee World* 76(4): 185–194.

Allen, M.F., and Ball, B. 1996. The incidence and world distribution of honey bee viruses. *Bee World* 77(3): 141–162.

Anderson, D.L. 1994. Non-reproduction of *Varroa jacobsoni* in *Apis mellifera* colonies in Papua New Guinea and Indonesia. *Apidologie* 25(4): 412–421.

Anderson, D.L., and Morgan, M.J. 2007. Genetic and morphological variation of bee-parasitic Tropilaelaps mites (Acari: Laelapidae): New and re-defined species. *Experimental and Applied Acarology* 43(1): 1–24.

Buawangpong, N., Khongphinitbunjong, K., Chantawannakul, P., and Burgett, M. 2013. *Tropilaelaps mercedesae*: Does this honey bee brood mite parasite exhibit a sex preference when infesting brood of the adapted host *Apis dorsata*? *Journal of Apicultural Research* 52(3): 158–159.

Butani, D. 1950. An Apis dorsata colony in New Delhi. *Indian Bee Journal* 12: 115.

Caughley, G., and Sinclair, A.R.E. 1994. *Wildlife ecology and management*. Blackwell Science, London.

Corlett, R.T. 2004. Flower visitors and pollination in the Oriental (Indomalayan) region. *Biological Reviews* 79(3): 497–532.

Crane, E. 1999. *The world history of beekeeping and honey hunting*. Routledge, London.

Crane, E. 2004. A short history of knowledge about honey bees (Apis) up to 1800. *Bee World* 85(1): 6–11.

Delfinado-Baker, M., and Baker, E. 1982. Notes on honey bee mites of the genus *Acarapis* Hirst (Acari: Tarsonemidae). *International Journal of Acarology* 8(4): 211–226.

Koeniger, G., Koeniger, N., Anderson, D.L., Lekprayoon, C., and Tingek, S. 2002. Mites from debris and sealed brood cells of *Apis dorsata* colonies in Sabah (Borneo) Malaysia, including a new haplotype of Varroa jacobsoni. *Apidologie* 33(1): 15–24.

Koeniger, N., and Koeniger, G. 1980. Observations and experiments on migration and dance communication of *Apis dorsata* in Sri Lanka. *Journal of Apicultural Research* 19(1): 21–34.

Koeniger, N., Koeniger, G., and Tingek, S. 2010. *Honey bees of Borneo: Exploring the centre of Apis diversity*. Natural History Publications (Borneo), Kota Kinabalu, Malaysia.

Laigo, F., and Morse, R. 1968. The mite Tropilaelaps clareae in *Apis dorsata* colonies in the Philippines. *Bee World* 49(3): 116–118.

Mahendra, D. 1997. Improved methods of honey harvest from rock bee colonies. *Indian Bee J.* 59(2): 95–98.

Mahindre, D. 1968. Apis dorsata-giant bee of India. *Gleanings in Bee Culture* 96(2): 102.

Marshman, J., Blay-Palmer, A., and Landman, K. 2019. Anthropocene crisis: Climate change, pollinators, and food security. *Environments* 6(2): 22.

Mishra, R. 1995. *Honeybees and their management in India, publications and information division Indian council of agricultural research*. Krishi Anusandhan Bhavan, Pusa, New Delhi.

Nagaraja, D. 2016. Effect of insecticide poisoning on mortality of giant honeybee, *Apis dorsata* Colonies. *Pesqui Int J. Res.* 2(1): 96–100.

Nath, S., Roy, P., Leo, R., and John, M. 1994. *Honeyhunters and beekeepers of Tamil Nadu: A survey document*. Key Stone Foundation Kotagiri, Tamil Nadu, India, 138.

Nirmala, R., and Nehra, K. 2015. Toxic effect of insecticides on survival of giant honey bee (*Apis dorsata* F.). *Annals of Agri Bio Research* 20(1): 54–59.

Oldroyd, B.P., and Nanork, P. 2009. Conservation of Asian honey bees. *Apidologie* 40(3): 296–312.

Oldroyd, B.P., Osborne, K., and Mardan, M. 2000. Colony relatedness in aggregations of *Apis dorsata* Fabricius (Hymenoptera, Apidae). *Insectes Sociaux* 47(1): 94–95.

Oldroyd, B.P., and Wongsiri, S. 2009. *Asian honey bees: Biology, conservation, and human interactions*. Harvard University Press, Cambridge.

Otis, G.W. 1996. Distributions of recently recognized species of honey bees (Hymenoptera: Apidae; Apis) in Asia. *Journal of the Kansas Entomological Society* 311–333.

Perveen, N., and Ahmad, M. 2017. Toxicity of some insecticides to the haemocytes of giant honeybee, *Apis dorsata* F. under laboratory conditions. *Saudi Journal of Biological Sciences* 24(5): 1016–1022.

Ramchandra, T., Subashchandran, M., Joshi, N., and Balachandran, C. 2012. *ENVIS technical. report: 49*. Centre for Ecological Sciences, Indian Institute of Science, Bangalore.

Rinderer, T.E., Oldroyd, B.P., de Guzman, L.I., Wattanachaiyingchareon, W., and Wongsiri, S. 2002. Spatial distribution of the dwarf honey bees in an agroecosystem in southeastern Thailand. *Apidologie* 33(6): 539–543.

Sodhi, N.S., Koh, L.P., Brook, B.W., and Ng, P.K. 2004. Southeast Asian biodiversity: An impending disaster. *Trends in Ecology & Evolution* 19(12): 654–660.

Venkatesh, G. 2014. Study on population status in relation to urban development in few selected nesting site of rock bee colonies, *Apis dorsata* F. *International Journal of Scientific Research Publication* 4: 1–2.

Wattanachaiyingcharoen, W., Wongsiri, S., and Oldroyd, B.P. 2008. Aggregations of unrelated *Apis florea* colonies. *Apidologie* 39(5): 531–536.

18 Morphometric Analysis and Floral Resources of Giant Honeybee, *Apis Dorsata* F

A.J. Solomon Raju

18.1 INTRODUCTION

Among honeybees, *Apis mellifera* and *A. cerana* have been investigated for their morphometrics throughout their distribution range. The information ranges from the anecdotally descriptive accounts of earlier decades through univariate methods of analysis to the more recent application of complete multivariate analyses of morphometric characteristics for specific regions (Daly 1991; 1992; Ruttner 1988; Ruttner et al. 1978). There is hardly any information on the biology and morphoclusters/clines of undomesticated *A. dorsata*. Insuan et al. (2007) classified *A. dorsata* populations of Thailand into sub-species based on morphological characteristics. Makhmoor and Ahmad (1998) described 16 morphological characteristics of *A. dorsata* from the Jammu region of India. Parr et al. (2004) examined population differentiation in *A. dorsata* in northern India using molecular markers. Mujumdar and Kshirasagar (1986) documented clinal variation for *A. dorsata* in the sub-Himalayan region. This state of work on morpho- or biometric clusters of *A. dorsata* is not useful to understand the variation that exists in this bee species throughout its distribution range. Further, there are no studies on the morphometric aspects of *A. dorsata* occurring in South India. Morphometric characteristics of honeybees have been shown to be related to the nutrition levels and floral diversity that exists in their habitats (Tan et al. 2007). The honeybees concentrate mainly on those flowers that provide plenty of pollen and nectar by quality. Although it has been realized that the flower species in any ecological region play an important role in maintaining and multiplying the colonies of honeybees, there have been hardly any studies of floral calendars of honeybees in different regions, especially in India. Floral calendars are needed, specifically for each ecological region, for what may apply in one place may not apply in another—even sometimes quite nearby (Roubik 1995). The floral calendars are useful to grow plants that bloom during the dearth period for sustaining honeybees in general and *A. dorsata* in particular (Kevan 1995). With this backdrop, the work on morphometrics and floral calendar of *A. dorsata* was carried out in the erstwhile Andhra Pradesh state of India.

 DOI: 10.1201/9781003294078-18

18.2 MATERIALS AND METHODS

Erstwhile Andhra Pradesh is spread over an area of 2,76,754 km^2 (Latitude 12°41' and 22°00' N and Longitude 77°00' and 84°40' E) which is broadly divided into seven distinct agro-climatic zones: Krishna-Godavari Zone, North Coastal Zone, Southern Zone, North Telangana Zone, South Telangana Zone, Scarce Rainfall Zone, and High Altitudes and Tribal Areas Zone. Ten representative localities were selected in each agro-climatic zone based on the distance, topography and vegetation for the collection of samples of *Apis dorsata*. The distance between localities is 60–80 km. In each locality, 40 worker bees were collected, of which 20 were used for morphometrics and 20 others were deposited in the Department. They were collected using insect nets, killed with petroleum ether, air-dried for 2–3 days and placed in butter paper bags containing para-dichloro-benzene crystals to avoid fungal contamination. Latitude, longitude and altitude were recorded for each locality using Magellan Explorist 210 Model Digital GPS System.

Bees captured were dissected systematically into head, thorax and abdomen using a Magnus compound microscope. The measurements were done using a Leica M165C stereomicroscope with computer-aided measuring system and program. Twenty-one characteristics were measured: head length and width, proboscis length, distance between two dorsal ocelli, dorsal ocello-occular distance, antenna length, femur length, tibia length, metatarsus length and width, forewing length and width, cubital index, number of hamuli, third tergite length, third sternite length and width, wax mirror length and width, distance between two wax plates on third sternite and lateral width of fourth tomentum. These characteristics were reduced to 16 characteristics by taking the ratios of head length and width, metatarsus length and width, forewing length and width, third sternite length and width, and wax mirror length and width. All statistical calculations were performed using SPSS for Windows 16.0.0 and SYSTAT for Windows 13.0 statistical package. The data obtained were subjected to multivariate morphometric analysis, hierarchical cluster analysis and principal component analysis. Field observations were made on *A. dorsata* regarding the plant species used by them as forage throughout the year in all localities/agro-climatic zones to prepare a common floral calendar.

18.3 RESULTS

18.3.1 Results of Morphometric Analysis

Twenty-one morphometric characteristics of *Apis florea* were measured for each locality in each agro-climatic zone. The measured characteristics were then reduced to 16 characteristics, the average means and standard deviations of which are noted agro-climatic zone-wise in Table 18.1. The cluster analysis performed for the means of individual characteristics for each agro-climatic zone showed three clusters. The populations of Zones IV and V, followed by Zone VI, represented cluster 1, those of Zones I and III cluster 2, and those of Zones II and VII cluster 3 (Figure 18.1). The

TABLE 18.1
Means and Standard Deviation of Morphometric Characteristics of *Apis dorsata* from All Agro-Climatic Zones

	Krishna-Godavari		North Coastal		Southern		North Telangana		South Telangana		Scarce Rainfall		High Altitudes and Tribal Areas	
Character	Mean	SD	Mean	SD	Mean	SD	Mean	SD	Mean	SD	Mean	SD	Mean	SD
L:W	0.58	0.03	0.56	0.05	0.60	0.01	0.62	0.01	0.63	0.04	0.61	0.04	0.61	0.06
PL	3.78	0.09	3.85	0.14	4.04	0.16	4.05	0.19	4.05	0.12	4.04	0.12	4.07	0.28
DBO	0.35	0.01	0.34	0.02	0.36	0.01	0.34	0.02	0.35	0.02	0.35	0.02	0.34	0.02
DOOD	0.31	0.01	0.32	0.01	0.33	0.01	0.33	0.01	0.32	0.01	0.32	0.01	0.33	0.01
AL	3.68	0.05	3.63	0.12	3.83	0.18	3.76	0.09	3.73	0.10	3.78	0.10	3.74	0.07
FL	2.04	0.06	2.07	0.07	2.10	0.07	2.07	0.06	2.12	0.07	2.06	0.10	2.08	0.07
TL	2.83	0.11	2.83	0.19	2.98	0.08	2.89	0.08	2.99	0.06	2.93	0.14	2.91	0.18
ML:MW	2.22	0.09	2.24	0.07	2.07	0.11	2.07	0.06	2.04	0.06	2.05	0.07	2.15	0.15
FWL:FWW	2.85	0.06	2.83	0.12	2.55	0.12	2.46	0.03	2.49	0.08	2.46	0.03	2.66	0.20
CI a:b	6.80	0.63	6.89	0.75	6.54	0.46	7.77	0.51	7.60	0.90	7.33	0.44	6.40	0.48
HAM	25.19	0.46	24.53	0.56	25.10	0.47	25.2	0.48	25.22	0.54	25.35	0.83	24.32	0.48
3rd TL	2.13	0.04	2.13	0.06	2.13	0.04	2.09	0.03	2.13	0.06	2.10	0.04	2.14	0.04
3rd SL:SW	0.85	0.01	0.85	0.02	0.84	0.01	0.84	0.01	0.85	0.01	0.83	0.01	0.85	0.02
LWM:WWM	0.72	0.01	0.73	0.03	0.70	0.02	0.71	0.01	0.71	0.02	0.71	0.01	0.72	0.03
DBWP	0.29	0.02	0.27	0.03	0.23	0.04	0.20	0.02	0.21	0.03	0.22	0.02	0.24	0.03
4th TOMW	0.60	0.05	0.63	0.10	0.57	0.08	0.54	0.03	0.57	0.04	0.54	0.03	0.61	0.07

Notes: L:W = ratio of head length and width; PL = proboscis length; DBO = distance between two dorsal ocelli; DOOD = Dorsal ocello-ocular distance; AL = antennal length; FL = femur length; TL = tibia length; ML:MW = ratio of metatarsus length and width; FWL:FWW = ratio of fore-wing length and width; CI a:b = ratio of cubital index (a:b); HAM = number of hamuli; 3rd TL = tergite length; 3rd SL:SW = ratio of third sternite length and width; LWM:WWM = ratio of wax mirror length and width; DBWP = distance between two wax plates on third sternite; 4th TOMW = lateral width of fourth tomentum

cluster analysis performed for the means of individual characteristics for 70 localities at once showed 18 clusters (Figure 18.2).

The means and standard deviations of 16 morphometric characteristics of of *A. dorsata* from all localities were segregated into six latitudinal ranges: 13°12'–13°49' N (three localities in Zone III and one locality in VI), 14°06'–14°54' N (six localities in Zone III and two localities in Zone VI), 15°07'–15°52' N (two localities in Zone I, one locality in Zone III and six localities in Zone VI), 16°01'–16°56' N (five localities in Zone I, 1 locality in Zone IV, three localities in Zone V and one locality in Zone VI), 17°04'N–17°58' N (three localities in Zone I, five localities in Zone II, two localities in Zone IV, seven localities in Zone V and three localities in

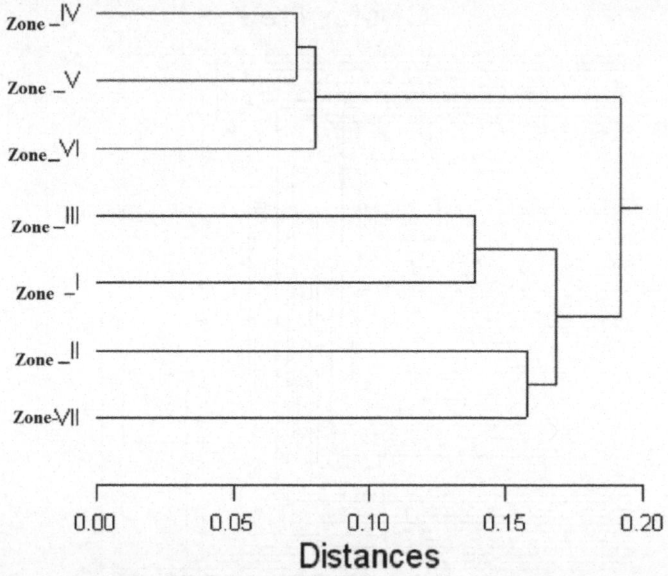

FIGURE 18.1 Characterization of *Apis dorsata* from all seven agro-climatic zones of Andhra Pradesh derived from cluster analysis using the single linkage method of squared Euclidean distance on means of 16 morphometric characteristics.

Zone VII) and 18°03'–19°40' N (five localities in Zone II, seven localities in Zone IV, seven localities in Zone VII). The data when subjected to hierarchical cluster analysis showed one cluster in 13°12'–13°49' N, two each in 14°06'–14°54' N and 16°01'–16°56' N, three in 15°07'–15°52' N and five each in 17°04'–17°58' N and 18°03'–19°40' N.

The means and standard deviations of 16 morphometric characteristics of of *A. dorsata* from all localities were segregated into three altitudinal ranges: 0.61–296 m (ten localities each in Zones I and II, six each in Zones III and VI, four in Zone IV, two in Zone V and eight in Zone VII), 316–595 m (two localities in Zone III, six in Zone IV, eight in Zone V and four in Zone VI) and 691–932 m (two localities each in Zones III and VII). When subjected to hierarchical cluster analysis for distinguishing morphometric clusters, the data showed 12 clusters in 0.61–296 m, five in 316–595 m and two in 691–932 m. The means and standard deviations for 16 morphometric characteristics for all 70 localities are presented in Table 18.2. In principal component analysis, the communalities are estimates of the variance in each variable accounted for by the components and are always equal to 1.0 for correlation analysis. The communalities obtained for the data fed indicate high values (>0.617) and hence the extracted components represent the variables well (Table 18.3). The data fed for Scree plot showed six components which are subsequently used for factor analysis (Figure 18.3).

Principal component analysis (PCA) was done using the means of 16 characteristics of worker bees of *A. dorsata* from all localities. Table 18.4 gives the data on

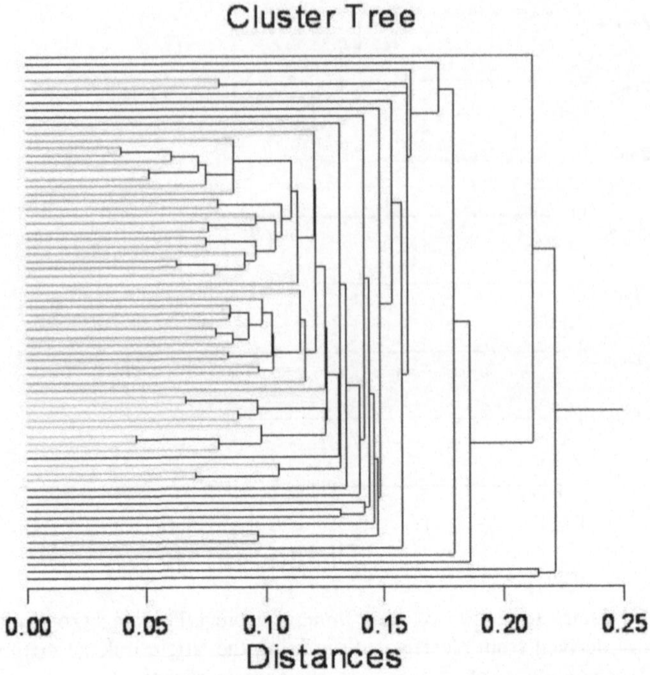

FIGURE 18.2 Characterization of *Apis dorsata* from 70 localities of seven agro-climatic zones derived from cluster analysis using the single linkage method of squared Euclidean distance on means of 16 morphometric characteristics. Localities serial wise: Rampachodavaram, Eleswaram, Tiruvur, Repalle, Tadepalligugdem, Araru, Jalantarakota, Kohlapur, Charakonda, Kunduluru, Nizamabad, Rayachoti, Proodatur, Medchal, Atmakur, Karimnagar, Kadiri, Medak, Aagadda, Mehubnagar, Bhainsa, Sangareddi, Vikarabad, Imrahimpatnam, Adilabad, Adoni, Warangal, Huzurnagar, Hindupur, Peddapalli, Annavaram, Tirupati, Adonki, Skota, Narsipatnam, Paderu, Khammam, Nellimarla, Suryapet, Chirala, Kandukur, Sattenepalli, Bobbil, Gannavaram, Srikakulum Kakinada, Vishakapatnam, Bhadrachalam, Venkatapurum, Komarada, Koyyuru, Patapatnam, Nellore, Nagarjunasagar, Chittoor, Gudur, Madanapalli, Parvatipuram, Secundrabad, Kurnool, Jangoon, Markapur, Tadipatri, Ananthapur, Hyderabad, Kamareddi, Giddalur, Udaygiri Addateegala Palakonda

variance accounted for by each component to the total variance in all of the variables. Table 18.5 gives the data of rotated component matrix (component loading values) that enables to identify the components for further analysis. The component score coefficient matrix data presented in Table 18.6 enables the selection of most important variables from 16 characteristics used for the study. The data used in the study were subjected to Kaiser-Meyer-Olkin measure of sampling adequacy and Bartlett's test of sphericity, and the results indicated the suitability of the data (Table 18.7).

The factor analysis yielded five factors with eigenvalues greater than 1. The first factor with loading values between 0.65 and 0.87 accounted for 28.50% of variation in the data, which was mainly associated with the proboscis length, ratio of metatarsus length and width, the ratio of forewing length and width, ratio of third sternite length

TABLE 18.2

Descriptive Statistics, Means and Standard Deviations of 16 Morphometric Characteristics of *Apis dorsata* from 70 Localities of All Agro-Climatic Zones

Descriptive Statistics

Character	Mean	Standard Deviation	Analysis (N)
L:W	0.60	0.04	70
PL	3.98	0.19	70
DBO	0.34	0.01	70
D0OD	0.32	0.01	70
AL	3.73	0.11	70
FL	2.07	0.07	70
TL	2.90	0.13	70
ML:MW	2.12	0.12	70
FWL:FWW	2.61	0.18	70
CI a:b	7.04	0.76	70
HAM	24.95	0.64	70
3rd TL	2.12	0.04	70
3rd SL:SW	0.84	0.01	70
LWM:WWM	0.71	0.02	70
DBWP	0.24	0.04	70
4th TOMW	0.58	0.06	70

TABLE 18.3

Apis dorsata Factor Analysis Communalities

Communalities

	Initial	Extraction
L:W	1.000	.670
PL	1.000	.617
DBO	1.000	.755
D0OD	1.000	.572
AL	1.000	.716
FL	1.000	.831
TL	1.000	.657
ML:MW	1.000	.777
FWL:FWW	1.000	.878
CI a:b	1.000	.752
HAM	1.000	.722
3rd TL	1.000	.747
3rd SL:SW	1.000	.640
LWM:WWM	1.000	.552
DBWP	1.000	.767
4th TOMW	1.000	.479

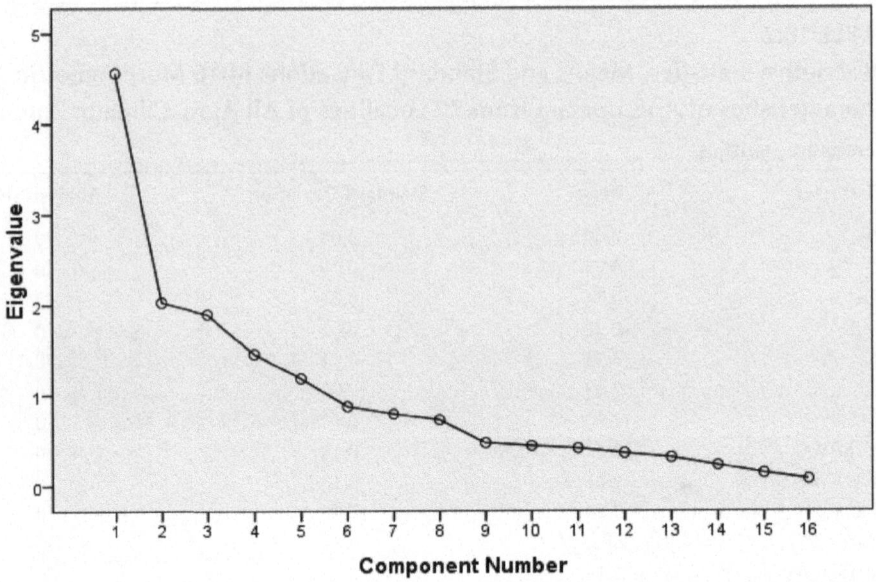

FIGURE 18.3 Factor analysis (principal component analysis) Scree plot determines the optimal number of components from 16 morphometric characteristics of *Apis dorsata* from 70 localities in seven agro-climatic zones of Andhra Pradesh. The eigenvalue of each component in the initial solution is plotted. The five extracted components are on the steep slope, while those on the shallow slope contribute little to the solution. The last big drop occurs between the fourth and fifth components; hence, the first five components are chosen for explanation.

and width, ratio of length and width of wax mirrors and finally distance between two wax plates. The second factor with loading values between 0.60 and 0.76 accounted for 12.66% of variation of the data, which was mainly associated with the distance between two dorsal ocelli, dorsal ocelli ocular distance and cubital index. The third factor with loading values between 0.56 and 0.88 accounted for 11.85% of variation of the data, which was mainly associated with the femur length, tibial length and third tergite length. The fourth factor with loading values between 0.59 and 0.81 accounted for 9.11% of variation of the data, which was mainly associated with the number of hamuli and lateral width of fourth tomentum. The fifth factor with loading values between 0.64 and 0.78 accounted for 7.42% of variation of the data, which was mainly associated with the ratio of head length and width and antennal length. The proboscis length varied from 3.57–4.54 mm, the hamuli number from 23.2–26.3 and the cubital index from 6.40–7.77.

18.3.2 Floral Calendar of *Apis dorsata*

In Table 18.8, the number of flowering trees, shrubs, herbs and creepers utilized by *A. dorsata* as forage sources in each month are provided. The data showed that the trees dominated the other life-forms during March–July, and they were hence

TABLE 18.4
Factor Analysis of Total Variance for 16 Morphometric Characteristics of *Apis dorsata* from Seventy Localities in All Agro-Climatic Zones

Total variance explained

	Initial Eigenvalues			Extraction Sums of Squared Loadings			Rotation Sums of Squared Loadings		
Component	Total	% of Variance	Cumulative %	Total	% of Variance	Cumulative %	Total	% of Variance	Cumulative %
1	4.561	28.508	28.508	4.561	28.508	28.508	4.155	25.972	25.972
2	2.027	12.666	41.174	2.027	12.666	41.174	1.903	11.895	37.867
3	1.897	11.857	53.031	1.897	11.857	53.031	1.875	11.717	49.584
4	1.458	9.112	62.143	1.458	9.112	62.143	1.640	10.250	59.834
5	1.188	7.422	69.565	1.188	7.422	69.565	1.557	9.731	69.565
6	.878	5.490	75.055						
7	.795	4.971	80.026						
8	.730	4.564	84.590						
9	.476	2.977	87.568						
10	.443	2.770	90.338						
11	.415	2.594	92.932						
12	.363	2.270	95.202						
13	.317	1.982	97.184						
14	.231	1.441	98.625						
15	.143	.894	99.518						
16	.077	.482	100.000						

Note: Extraction method used was principal component analysis.

TABLE 18.5
Apis dorsata Factor Analysis of the Rotated Component Matrix Indicating Five Components for Further Analysis for 16 Morphometric Characteristics of 70 Localities in Seven Agro-Climatic Zones

Rotated Component Matrix

	Component				
	1	2	3	4	5
L:W	−.087	−.363	.226	.262	.641
PL	−.658	−.043	.055	−.202	.373
DBO	−.072	.769	.005	.304	−.257
DOOD	−.280	.657	.193	−.064	.145
AL	−.214	.211	.132	.025	.780

(*Continued*)

TABLE 18.5 ***(Continued)***

Apis dorsata **Factor Analysis of the Rotated Component Matrix Indicating Five Components for Further Analysis for 16 Morphometric Characteristics of 70 Localities in Seven Agro-Climatic Zones**

Rotated Component Matrix

	Component				
	1	**2**	**3**	**4**	**5**
FL	−.143	−.118	.883	.094	.092
TL	−.459	.280	.565	.163	.149
ML:MW	.819	−.134	.120	−.225	−.153
FWL:FWW	.872	−.004	−.085	−.288	−.166
CI a:b	−.331	−.605	.250	.312	−.341
HAM	.157	.157	.111	.812	.036
3rd TL	.359	.125	.737	−.208	.129
3rd SL:SW	.740	−.132	−.119	.046	.242
LWM:WWM	.672	−.203	−.015	.231	−.075
DBWP	.759	.335	.024	−.237	−.146
4th TOMW	.294	.121	.115	−.594	−.108

TABLE 18.6

Factor Analysis Component Score Coefficient Matrix for 16 Morphometric Characteristics of *Apis dorsata* from 70 Localities in Seven Agro-Climatic Zones of Andhra Pradesh

Component Score Coefficient Matrix

	Component				
	1	**2**	**3**	**4**	**5**
L:W	.068	−.184	.037	.137	.422
PL	−.161	−.045	−.026	−.202	.192
DBO	.000	.419	.003	.229	−.189
D0OD	−.050	.335	.065	−.045	.054
AL	.036	.112	−.050	.000	.529
FL	−.008	−.090	.492	.011	−.079
TL	−.078	.129	.280	.055	−.021
ML:MW	.186	−.069	.118	−.077	−.041
FWL:FWW	.192	.004	.001	−.100	−.013
CI a:b	−.107	−.326	.197	.134	−.328
HAM	.133	.123	.028	.551	.028
3rd TL	.108	.041	.414	−.120	.027
3rd SL:SW	.230	−.045	−.088	.101	.274
LWM:WWM	.195	−.082	.013	.209	.019
DBWP	.178	.181	.045	−.064	−.023
4th TOMW	.015	.037	.104	−.358	−.062

TABLE 18.7
Factor Analysis of Kaiser–Meyer–Olkin Measure of Sampling Adequacy (KMO) and Bartlett's Test of Sphericity for 16 Morphometric Characteristics of *Apis dorsata* from 70 Localities in Seven Agro-Climatic Zones

KMO and Bartlett's Test

Kaiser–Meyer–Olkin Measure of Sampling Adequacy		.681
Bartlett's Test of Sphericity	Approx. chi-square	492.039
	df	120
	Sig.	.000

considered to be important in providing forage for *A. dorsata*. The total number of plants in flowering in each month ranged from 66–82. Most of the herbs occur as weeds and establish huge populations everywhere during rainy season; individual plants of some herbs provide copious pollen and nectar daily while some other herbs at population level provide adequate pollen and nectar for *A. dorsata*. Some mangrove flora also serve as pollen and nectar source for *A. dorsata* in estuarine areas of Andhra Pradesh.

18.4 DISCUSSION

In this study, 21 numerical morphological characteristics of *A. dorsata* have been used for identifying morphoclusters in the erstwhile Andhra Pradesh. These characteristics have been reduced to 16 by taking the ratios of certain characteristics such as head length and width, metatarsus length and width, forewing length and width, third sternite length and width, and wax mirror length and width. The other measured characteristics were taken as obtained. This consolidation has been done to enhance the accuracy in identifying morphoclusters of *A. dorsata*. The multivariate statistical analysis of the morphometric data of 16 morphological characteristics included hierarchical cluster analysis and factor analysis. The cluster analysis performed for the means of the 16 morphological characteristics of all seven agro-climatic zones indicated three distinct morphoclusters, the first cluster representing Zones IV, V and VI; the second representing Zones I and III; and the third representing Zones II and VII. These clusters appear to be a reflection of ecological and floral diversity that exists in these zones.

The analysis of the morphometric data indicated two distinct morphoclusters in factor scores 1–4 and one in factor score 5. The variation in factor score 1 is associated with proboscis length, ratio of metatarsus length and width, the ratio of forewing length and width, ratio of third sternite length and width, ratio of length and width of wax mirrors and distance between two wax plates; that in factor score 2 with the distance between two dorsal ocelli, dorsal ocelli ocular distance and cubital index; that in factor score 3 with the femur length, tibial length and third tergite length; that in factor score 4 with the number of hamuli and lateral

TABLE 18.8
Floral Calendar of *Apis dorsata* in Andhra Pradesh

January	February	March	April
Cocos nucifera (T)	*Mangifera indica* (T)	*Alangium salvifolium* (T)	*Alangium salvifolium* (T)
Phoenix sylvestris (T)	*Cocos nucifera* (T)	*Anacardium occidentale* (T)	*Anacardium occidentale* (T)
Ageratum conyzoides (H)	*Phoenix sylvestris* (T)	*Mangifera indica* (T)	*Mangifera indica* (T)
Cosmos bipinnatus (H)	*Cosmos bipinnatus* (H)	*Cocos nucifera* (T)	*Wrightia tinctoria* (T)
Cosmos sulphureus (H)	*Cosmos sulphureus* (H)	*Phoenix sylvestris* (T)	*Cocos nucifera* (T)
Wedelia chinensis (H)	*Wedelia chinensis* (H)	*Cosmos bipinnatus* (H)	*Phoenix sylvestris* (T)
Tridax procumbens (H)	*Tridax procumbens* (H)	*Cosmos sulphureus* (H)	*Cosmos bipinnatus* (H)
Tecoma stans (T)	*Tecoma stans* (T)	*Wedelia chinensis* (H)	*Cosmos sulphureus* (H)
Ceiba pentandra (T)	*Ceiba pentandra* (T)	*Tridax procumbens* (H)	*Wedelia chinensis* (H)
Caesalpinia pulcherrima (S)	*Caesalpinia pulcherrima* (S)	*Careya arborea* (T)	*Tridax procumbens* (H)
Carica papaya (T)	*Carica papaya* (T)	*Tecoma stans* (T)	*Careya arborea* (T)
Evolvulus alsinoides (H)	*Terminalia catappa* (T)	*Kigelia pinnata* (T)	*Tecoma stans* (T)
Ipomoea cairica (C)	*Evolvulus alsinoides* (H)	*Spathodea campanulata* (T)	*Kigelia pinnata* (T)
Ipomoea pes-caprae (C)	*Ipomoea cairica* (C)	*Boswellia ovalifoliolata* (T)	*Spathodea campanulata* (T)
Croton bonplandianum (H)	*Ipomoea pes-caprae* (C)	*Garuga pinnata* (T)	*Boswellia ovalifoliolata* (T)
Jatropha curcas (S)	*Croton bonplandianum* (H)	*Caesalpinia pulcherrima* (S)	*Garuga pinnata* (T)
Jatropha pandurifolia (S)	*Jatropha curcas* (S)	*Bauhinia racemosa* (T)	*Caesalpinia pulcherrima* (S)
Gliricidia sepium (T)	*Jatropha pandurifolia* (S)	*Peltophorum pterocarpum* (T)	*Bauhinia racemosa* (T)
Rhynchosia albiflora (C)	*Gliricidia sepium* (T)	*Carica papaya* (T)	*Peltophorum pterocarpum* (T)
Rhynchosia beddomei (S)	*Rhynchosia albiflora* (C)	*Suaeda maritima* (H)	*Tamarindus indica* (T)
Rhynchosia cana (S)	*Rhynchosia cana* (S)	*Suaeda nudiflora* (H)	*Crataeva magna* (T)
Rhynchosia densiflora (S)	*Rhynchosia densiflora* (S)	*Terminalia bellirica* (T)	*Carica papaya* (T)
Rhynchosia heynei (S)	*Rhynchosia heynei* (S)	*Terminalia catappa* (T)	*Suaeda maritima* (H)
Rhynchosia suaveolens (S)	*Rhynchosia suaveolens* (S)	*Evolvulus alsinoides* (H)	*Suaeda nudiflora* (H)
Rhynchosia viscosa (C)	*Rhynchosia viscosa* (C)	*Ipomoea cairica* (C)	*Terminalia bellirica* (T)
Tephrosia purpurea (H)	*Tephrosia purpurea* (H)	*Shorea roxburghii* (T)	*Terminalia catappa* (T)

Anisomeles indica (S)	*Anisomeles indica* (S)	*Diospyros chloroxylon* (T)	*Terminalia pallida* (T)
Anisomeles malabarica (S)	*Anisomeles malabarica* (S)	*Croton bonplandianum* (H)	*Evolvulus alsinoides* (H)
Hyptis suaveolens (S)	*Hyptis suaveolens* (S)	*Jatropha curcas* (S)	*Ipomoea cairica* (C)
Leucas aspera (H)	*Leucas aspera* (H)	*Jatropha pandurifolia* (S)	*Diospyros chloroxylon* (T)
Plectranthus mollis (H)	*Plectranthus mollis* (H)	*Gliricidia sepium* (T)	*Croton bonplandianum* (H)
Thymus vulgaris (S)	*Thymus vulgaris* (S)	*Pongamia pinnata* (T)	*Jatropha curcas* (S)
Couroupita guianensis (T)	*Couroupita guianensis* (T)	*Pterocarpus santalinus* (T)	*Jatropha pandurifolia* (S)
Malvastrum coromandelicanum (S)	*Malvastrum coromandelicanum* (S)	*Tephrosia purpurea* (H)	*Pongamia pinnata* (T)
Triumfetta pentandra (H)	*Triumfetta pentandra* (H)	*Anisomeles indica* (S)	*Pterocarpus santalinus* (T)
Acacia mellifera (T)	*Acacia mellifera* (T)	*Plectranthus mollis* (H)	*Tephrosia purpurea* (H)
Acacia nilotica (T)	*Acacia nilotica* (T)	*Thymus vulgaris* (S)	*Thymus vulgaris* (S)
Albizia lebbeck (T)	*Acacia sinuata* (T)	*Couroupita guianensis* (T)	*Couroupita guianensis* (T)
Pithecellobium dulce (T)	*Albizia lebbeck* (T)	*Malvastrum coromandelicanum* (S)	*Malvastrum coromandelicanum* (S)
Mimosa pudica (C)	*Pithecellobium dulce* (T)	*Azadirachta indica* (T)	*Azadirachta indica* (T)
Psidium guajava (T)	*Mimosa pudica* (C)	*Walsura trifoliata* (T)	*Walsura trifoliata* (T)
Argemone mexicana (H)	*Moringa oleifera* (T)	*Acacia mellifera* (T)	*Acacia mellifera* (T)
Antigonon leptopus (C)	*Psidium guajava* (T)	*Acacia nilotica* (T)	*Acacia nilotica* (T)
Portulaca oleracea (H)	*Argemone mexicana* (H)	*Acacia sinuata* (C)	*Acacia sinuata* (C)
Punica granatum (T)	*Antigonon leptopus* (C)	*Albizia lebbeck* (T)	*Albizia lebbeck* (T)
Zizyphus mauritiana (T)	*Portulaca oleracea* (H)	*Pithecellobium dulce* (T)	*Mimosa pudica* (C)
Zizyphus oenoplia (S)	*Punica granatum* (T)	*Mimosa pudica* (C)	*Moringa oleifera* (T)
Bruguiera gymnorrhiza (T)	*Zizyphus mauritiana* (T)	*Moringa oleifera* (T)	*Psidium guajava* (T)
Bruguiera cylindrica (T)	*Zizyphus oenoplia* (S)	*Psidium guajava* (T)	*Syzygium alternifolium* (T)
Ceriops decandra (T)	*Bruguiera gymnorrhiza* (T)	*Syzygium alternifolium* (T)	*Syzygium cuminii* (T)
Ceriops tagal (T)	*Bruguiera cylindrica* (T)	*Syzygium cuminii* (T)	*Argemone mexicana* (H)
Rosa centifolia (S)	*Ceriops decandra* (T)	*Argemone mexicana* (H)	*Antigonon leptopus* (C)
Rosa sp. (S)	*Ceriops tagal* (T)	*Antigonon leptopus* (C)	*Portulaca oleracea* (H)
Hamelia patens (S)	*Rosa centifolia* (S)	*Portulaca oleracea* (H)	*Punica granatum* (T)
Mitragyna parviflora (T)	*Rosa* sp. (S)	*Punica granatum* (T)	*Zizyphus mauritiana* (T)
Tarenna asiatica (S)	*Hamelia patens* (S)	*Zizyphus mauritiana* (T)	*Zizyphus oenoplia* (S)

(*Continued*)

TABLE 18.8 *(Continued)*
Floral Calendar of *Apis dorsata* in Andhra Pradesh

January	February	March	April
Santalum album (T)	*Mitragyna parviflora* (T)	*Zizyphus oenoplia* (S)	*Bruguiera gymnorrhiza* (T)
Sapindus emarginatus (T)	*Tarenna asiatica* (S)	*Bruguiera gymnorrhiza* (T)	*Bruguiera cylindrica* (T)
Solanum trilobatum (C)	*Santalum album* (T)	*Bruguiera cylindrica* (T)	*Ceriops decandra* (T)
Sterculia foetida (T)	*Madhuca longifolia* (T)	*Ceriops decandra* (T)	*Rosa centifolia* (S)
Sterculia urens (T)	*Mimusops elengi* (T)	*Rosa centifolia* (S)	*Rosa* sp. (S)
Clerodendrum inerme (S)	*Sterculia foetida* (T)	*Rosa* sp. (S)	*Hamelia patens* (S)
Lantana camara (S)	*Sterculia urens* (T)	*Hamelia patens* (S)	*Mitragyna parviflora* (T)
Duranta erecta (S)	*Clerodendrum inerme* (S)	*Mitragyna parviflora* (T)	*Morinda tomentosa* (T)
Vitex negundo (S)	*Lantana camara* (S)	*Tarenna asiatica* (S)	*Pavetta tomentosa* (S)
Achras zapota (T)	*Duranta erecta* (S)	*Wendlandia glabrata* (T)	*Tarenna asiatica* (S)
Dalbergia latifolia (T)	*Gmelina arborea* (T)	*Wendlandia tinctoria* (T)	*Wendlandia glabrata* (T)
Tecomaria capensis (T)	*Vitex negundo* (S)	*Madhuca longifolia* (T)	*Wendlandia tinctoria* (T)
Glycosmosis pentaphylla (T)	*Achras zapota* (T)	*Mimusops elengi* (T)	*Murraya koenigii* (T)
Tagetes erecta (S)	*Dalbergia latifolia* (T)	*Sterculia foetida* (T)	*Madhuca longifolia* (T)
Cestrum diurnum (S)	*Tecomaria capensis* (T)	*Clerodendrum inerme* (S)	*Mimusops elengi* (T)
Corallocarpus epigaeus (H)	*Prosopis juliflora* (T)	*Lantana camara* (S)	*Lantana camara* (S)
	Glycosmosis pentaphylla (T)	*Duranta erecta* (S)	*Clerodendrum inerme* (S)
	Cestrum diurnum (S)	*Gmelina arborea* (T)	*Gmelina arborea* (T)
	Corallocarpus epigaeus (H)	*Gmelina asiatica* (S)	*Gmelina asiatica* (S)
		Vitex negundo (S)	*Vitex negundo* (S)
		Achras zapota (T)	*Achras zapota* (T)
		Dalbergia latifolia (T)	*Dalbergia latifolia* (T)
		Tecomaria capensis (T)	*Tecomaria capensis* (T)
		Prosopis juliflora (T)	*Calliandra haematocephala* (S)
		Glycosmosis pentaphylla (T)	*Corallocarpus epigaeus* (H)
		Cestrum diurnum (S)	
		Corallocarpus epigaeus (H)	

May	June	July	August
Sesuvium portulacastrum (H)	*Agave americana* (S)	*Justicia procumbens* (H)	*Asystasia gangetica* (H)
Alangium salvifolium (T)	*Sesuvium portulacastrum* (H)	*Agave americana* (S)	*Justicia procumbens* (H)
Anacardium occidentale (T)	*Buchanania lanzan* (T)	*Sesuvium portulacastrum* (H)	*Agave americana* (S)
Mangifera indica (T)	*Buchanania axillaris* (T)	*Buchanania lanzan* (T)	*Buchanania lanzan* (T)
Wrightia tinctoria (T)	*Semecarpus anacardium* (T)	*Buchanania axillaris* (T)	*Buchanania axillaris* (T)
Cocos nucifera (T)	*Wrightia tinctoria* (T)	*Semecarpus anacardium* (T)	*Semecarpus anacardium* (T)
Phoenix sylvestris (T)	*Cocos nucifera* (T)	*Cocos nucifera* (T)	*Cocos nucifera* (T)
Cosmos bipinnatus (H)	*Cosmos bipinnatus* (H)	*Cosmos bipinnatus* (H)	*Ageratum conyzoides* (H)
Cosmos sulphureus (H)	*Cosmos sulphureus* (H)	*Cosmos sulphureus* (H)	*Cosmos bipinnatus* (H)
Wedelia chinensis (H)	*Wedelia chinensis* (H)	*Wedelia chinensis* (H)	*Cosmos sulphureus* (H)
Tridax procumbens (H)	*Tridax procumbens* (H)	*Tridax procumbens* (H)	*Wedelia chinensis* (H)
Careya arborea (T)	*Avicennia officinalis* (T)	*Avicennia officinalis* (T)	*Tridax procumbens* (H)
Tecoma stans (T)	*Avicennia marina* (T)	*Avicennia marina* (T)	*Avicennia officinalis* (T)
Kigelia pinnata (T)	*Avicennia alba* (T)	*Avicennia alba* (T)	*Avicennia marina* (T)
Spathodea campanulata (T)	*Tecoma stans* (T)	*Tecoma stans* (T)	*Avicennia alba* (T)
Caesalpinia pulcherrima (S)	*Kigelia pinnata* (T)	*Spathodea campanulata* (T)	*Tecoma stans* (T)
Bauhinia racemosa (T)	*Spathodea campanulata* (T)	*Caesalpinia coriaria* (T)	*Millingtonia hortensis* (T)
Peltophorum pterocarpum (T)	*Caesalpinia crista* (S)	*Caesalpinia crista* (S)	*Caesalpinia coriaria* (T)
Tamarindus indica (T)	*Caesalpinia pulcherrima* (S)	*Caesalpinia pulcherrima* (S)	*Caesalpinia crista* (S)
Crataeva magna (T)	*Bauhinia racemosa* (T)	*Tamarindus indica* (T)	*Caesalpinia pulcherrima* (S)
Carica papaya (T)	*Peltophorum pterocarpum* (T)	*Cleome chelidonii* (H)	*Tamarindus indica* (T)
Suaeda maritima (H)	*Tamarindus indica* (T)	*Cleome viscosa* (H)	*Cleome viscosa* (H)
Suaeda nudiflora (H)	*Cleome chelidonii* (H)	*Crataeva magna* (T)	*Carica papaya* (T)
Terminalia bellirica (T)	*Cleome viscosa* (H)	*Carica papaya* (T)	*Lumnitzera racemosa* (T)
Terminalia catappa (T)	*Crataeva magna* (T)	*Lumnitzera racemosa* (T)	*Terminalia catappa* (T)
Terminalia pallida (T)	*Carica papaya* (T)	*Terminalia catappa* (T)	*Evolvulus alsinoides* (H)
Evolvulus alsinoides (H)	*Terminalia bellirica* (T)	*Terminalia tomentosa* (T)	*Ipomoea cairica* (C)
Ipomoea cairica (C)	*Terminalia catappa* (T)	*Evolvulus alsinoides* (H)	*Ipomoea pes-caprae* (C)

(*Continued*)

TABLE 18.8 *(Continued)*
Floral Calendar of *Apis dorsata* in Andhra Pradesh

May	June	July	August
Croton bonplandianum (H)	*Terminalia pallida* (T)	*Ipomoea cairica* (C)	*Ipomoea tuba* (C)
Jatropha curcas (S)	*Terminalia tomentosa* (T)	*Croton bonplandianum* (H)	*Croton bonplandianum* (H)
Jatropha pandurifolia (S)	*Evolvulus alsinoides* (H)	*Excoecaria agallocha* (T)	*Excoecaria agallocha* (T)
Pongamia pinnata (T)	*Ipomoea cairica* (C)	*Jatropha curcas* (S)	*Jatropha curcas* (S)
Pterocarpus santalinus (T)	*Croton bonplandianum* (H)	*Jatropha gossypiifolia* (S)	*Jatropha gossypiifolia* (S)
Tephrosia purpurea (H)	*Excoecaria agallocha* (T)	*Jatropha pandurifolia* (S)	*Jatropha pandurifolia* (S)
Thymus vulgaris (S)	*Jatropha curcas* (S)	*Tephrosia purpurea* (H)	*Tephrosia purpurea* (H)
Couroupita guianensis (T)	*Jatropha pandurifolia* (S)	*Ocimum americanum* (H)	*Ocimum americanum* (H)
Malvastrum coromandelicanum (S)	*Pongamia pinnata* (T)	*Ocimum basilicum* (H)	*Ocimum basilicum* (H)
Walsura trifoliata (T)	*Tephrosia purpurea* (H)	*Thymus vulgaris* (S)	*Thymus vulgaris* (S)
Acacia mellifera (T)	*Thymus vulgaris* (S)	*Couroupita guianensis* (T)	*Couroupita guianensis* (T)
Acacia nilotica (T)	*Couroupita guianensis* (T)	*Hugonia mystax* (S)	*Hugonia mystax* (S)
Albizia lebbeck (T)	*Hugonia mystax* (S)	*Hibiscus tiliaceus* (T)	*Malvastrum coromandelicanum* (S)
Mimosa pudica (C)	*Hibiscus tiliaceus* (T)	*Malvastrum coromandelicanum* (S)	*Sida acuta* (S)
Moringa oleifera (T)	*Malvastrum coromandelicanum* (S)	*Sida acuta* (S)	*Sida cordifolia* (H)
Psidium guajava (T)	*Thespesia populneoides* (T)	*Sida cordifolia* (H)	*Triumfetta pentandra* (H)
Syzygium alternifolium (T)	*Acacia mellifera* (T)	*Thespesia populneoides* (T)	*Acacia caesia* (S)
Syzygium cuminii (T)	*Acacia nilotica* (T)	*Acacia caesia* (S)	*Acacia farnesiana* (S)
Argemone mexicana (H)	*Albizia lebbeck* (T)	*Acacia mellifera* (T)	*Acacia mellifera* (T)
Antigonon leptopus (C)	*Psidium guajava* (T)	*Acacia nilotica* (T)	*Acacia nilotica* (T)
Portulaca oleracea (H)	*Syzygium alternifolium* (T)	*Albizia lebbeck* (T)	*Albizia lebbeck* (T)
Punica granatum (H)	*Syzygium cuminii* (T)	*Aegiceras corniculatus* (T)	*Aegiceras corniculatus* (T)
Zizyphus mauritiana (T)	*Argemone mexicana* (H)	*Psidium guajava* (T)	*Psidium guajava* (T)
Zizyphus oenoplia (S)	*Antigonon leptopus* (C)	*Syzygium cuminii* (T)	*Syzygium cuminii* (T)
Bruguiera gymnorrhiza (T)	*Portulaca oleracea* (H)	*Argemone mexicana* (H)	*Argemone mexicana* (H)
Bruguiera cylindrica (T)	*Punica granatum* (T)	*Antigonon leptopus* (C)	*Pedalium murex* (H)
Ceriops decandra (T)	*Zizyphus mauritiana* (T)	*Portulaca oleracea* (H)	*Antigonon leptopus* (C)
Rosa centifolia (S)	*Zizyphus oenoplia* (S)	*Punica granatum* (T)	*Portulaca oleracea* (H)
Rosa sp. (S)	*Bruguiera gymnorrhiza* (T)	*Zizyphus mauritiana* (T)	*Punica granatum* (T)

Hamelia patens (S)	*Bruguiera cylindrica* (T)	*Zizyphus oenoplia* (S)	*Zizyphus mauritiana* (T)
Mitragyna parviflora (T)	*Ceriops decandra* (T)	*Bruguiera gymnorrhiza* (T)	*Zizyphus oenoplia* (S)
Morinda tomentosa (T)	*Rosa centifolia* (S)	*Bruguiera cylindrica* (T)	*Bruguiera gymnorrhiza* (T)
Pavetta tomentosa (S)	*Rosa* sp. (S)	*Ceriops decandra* (T)	*Bruguiera cylindrica* (T)
Tarenna asiatica (S)	*Hamelia patens* (S)	*Rosa centifolia* (S)	*Ceriops decandra* (T)
Wendlandia glabrata (T)	*Mitragyna parviflora* (T)	*Rosa* sp. (S)	*Rosa centifolia* (S)
Wendlandia tinctoria (T)	*Morinda tomentosa* (T)	*Hamelia patens* (S)	*Rosa* sp. (S)
Murraya koenigii (T)	*Pavetta tomentosa* (S)	*Mitragyna parviflora* (T)	*Borreria hispida* (H)
Santalum album (T)	*Tarenna asiatica* (S)	*Morinda tomentosa* (T)	*Hamelia patens* (S)
Madhuca longifolia (T)	*Murraya koenigii* (T)	*Pavetta tomentosa* (S)	*Mitragyna parviflora* (T)
Gmelina arborea (T)	*Santalum album* (T)	*Tarenna asiatica* (S)	*Morinda tomentosa* (T)
Gmelina asiatica (S)	*Duranta erecta* (S)	*Cardiospermum canescens* (C)	*Pavetta tomentosa* (S)
Lantana camara (S)	*Lantana camara* (S)	*Cardiospermum halicacabum* (C)	*Tarenna asiatica* (S)
Tectona grandis (T)	*Tectona grandis* (T)	*Solanum melongena* (H)	*Santalum album* (T)
Achras zapota (T)	*Vitex altissima* (T)	*Sonneratia apetala* (T)	*Allophylus serratus* (S)
Dalbergia latifolia (T)	*Tribulus terrestris* (H)	*Solanum trilobatum* (C)	*Cardiospermum canescens* (C)
Tecomaria capensis (T)	*Achras zapota* (T)	*Helicteres isora* (T)	*Cardiospermum halicacabum* (C)
Calliandra haematocephala (S)	*Dalbergia latifolia* (T)	*Duranta erecta* (S)	*Solanum melongena* (H)
	Tecomaria capensis (T)	*Lantana camara* (S)	*Sonneratia apetala* (T)
		Tectona grandis (T)	*Solanum trilobatum* (C)
		Vitex altissima (T)	*Helicteres isora* (T)
		Tribulus terrestris (H)	*Lantana camara* (S)
		Achras zapota (T)	*Duranta erecta* (S)
		Dalbergia latifolia (T)	*Vitex altissima* (T)
		Tecomaria capensis (T)	*Tribulus terrestris* (H)
			Achras zapota (T)
			Dalbergia latifolia (T)
			Tecomaria capensis (T)
			Prosopis juliflora (T)
			Glycosmosis pentaphylla (T)

(*Continued*)

TABLE 18.8 ***(Continued)***
Floral Calendar of *Apis dorsata* in Andhra Pradesh

September	October	November	December
Asystasia gangetica (H)	*Asystasia gangetica* (H)	*Asystasia gangetica* (H)	*Asystasia gangetica* (H)
Justicia procumbens (H)	*Justicia procumbens* (H)	*Justicia procumbens* (H)	*Cocos nucifera* (T)
Agave americana (S)	*Agave americana* (S)	*Cocos nucifera* (T)	*Ageratum conyzoides* (H)
Buchanania lanzan (T)	*Cocos nucifera* (T)	*Ageratum conyzoides* (H)	*Chromolaena odorata* (S)
Buchanania axillaris (T)	*Ageratum conyzoides* (H)	*Chromolaena odorata* (S)	*Cosmos bipinnatus* (H)
Semecarpus anacardium (T)	*Chromolaena odorata* (S)	*Cosmos bipinnatus* (H)	*Cosmos sulphureus* (H)
Cocos nucifera (T)	*Cosmos bipinnatus* (H)	*Cosmos sulphureus* (H)	*Wedelia chinensis* (H)
Cosmos bipinnatus (H)	*Cosmos sulphureus* (H)	*Wedelia chinensis* (H)	*Tridax procumbens* (H)
Cosmos sulphureus (H)	*Wedelia chinensis* (H)	*Tridax procumbens* (H)	*Tecoma stans* (T)
Wedelia chinensis (H)	*Tridax procumbens* (H)	*Tecoma stans* (T)	*Millingtonia hortensis* (T)
Tridax procumbens (H)	*Tecoma stans* (T)	*Millingtonia hortensis* (T)	*Ceiba pentandra* (T)
Tecoma stans (T)	*Millingtonia hortensis* (T)	*Caesalpinia crista* (S)	*Caesalpinia pulcherrima* (S)
Millingtonia hortensis (T)	*Caesalpinia crista* (S)	*Caesalpinia pulcherrima* (S)	*Bauhinia purpurea* (T)
Caesalpinia coriaria (T)	*Caesalpinia pulcherrima* (S)	*Bauhinia purpurea* (T)	*Peltophorum pterocarpum* (T)
Caesalpinia crista (S)	*Bauhinia purpurea* (T)	*Peltophorum pterocarpum* (T)	*Carica papaya* (T)
Caesalpinia pulcherrima (S)	*Peltophorum pterocarpum* (T)	*Carica papaya* (T)	*Evolvulus alsinoides* (H)
Peltophorum pterocarpum (T)	*Cleome viscosa* (H)	*Evolvulus alsinoides* (H)	*Ipomoea cairica* (C)
Cleome viscosa (H)	*Carica papaya* (T)	*Ipomoea cairica* (C)	*Ipomoea pes-caprae* (C)
Carica papaya (T)	*Evolvulus alsinoides* (H)	*Ipomoea pes-caprae* (C)	*Croton bonplandianum* (H)
Evolvulus alsinoides (H)	*Ipomoea cairica* (C)	*Croton bonplandianum* (H)	*Jatropha curcas* (S)
Ipomoea cairica (C)	*Ipomoea pes-caprae* (C)	*Jatropha curcas* (S)	*Jatropha pandurifolia* (S)
Ipomoea pes-caprae (C)	*Croton bonplandianum* (H)	*Jatropha pandurifolia* (S)	*Rhynchosia albiflora* (C)
Ipomoea tuba (C)	*Jatropha curcas* (S)	*Derris trifoliata* (S)	*Rhynchosia beddomei* (S)
Croton bonplandianum (H)	*Jatropha pandurifolia* (S)	*Rhynchosia beddomei* (S)	*Rhynchosia densiflora* (S)
Excoecaria agallocha (T)	*Derris trifoliata* (S)	*Rhynchosia heynei* (S)	*Rhynchosia heynei* (S)
Jatropha curcas (S)	*Rhynchosia heynei* (S)	*Rhynchosia rothii* (S)	*Rhynchosia rothii* (S)
Jatropha pandurifolia (S)	*Rhynchosia rothii* (S)	*Rhynchosia viscosa* (C)	*Rhynchosia viscose* (C)

Rhynchosia rothii (S)	*Tephrosia purpurea* (H)	*Tephrosia purpurea* (H)	*Tephrosia purpurea* (H)
Tephrosia purpurea (H)	*Anisomeles malabarica* (S)	*Anisomeles indica* (S)	*Anisomeles indica* (S)
Ocimum americanum (H)	*Hyptis suaveolens* (S)	*Anisomeles malabarica* (S)	*Anisomeles malabarica* (S)
Ocimum basilicum (H)	*Leucas aspera* (H)	*Hyptis suaveolens* (S)	*Hyptis suaveolens* (S)
Thymus vulgaris (S)	*Ocimum americanum* (H)	*Leucas aspera* (H)	*Leucas aspera* (H)
Couroupita guianensis (T)	*Ocimum basilicum* (H)	*Ocimum americanum* (H)	*Plectranthus mollis* (H)
Malachra capitata (H)	*Thymus vulgaris* (S)	*Ocimum basilicum* (H)	*Thymus vulgaris* (S)
Malvastrum coromandelicanum (S)	*Couroupita guianensis* (T)	*Plectranthus mollis* (H)	*Couroupita guianensis* (T)
Pavonia zeylanica (S)	*Malachra capitata* (H)	*Malvastrum coromandelicanum* (S)	*Malvastrum coromandelicanum* (S)
Sida acuta (S)	*Malvastrum coromandelicanum* (S)	*Pavonia zeylanica* (S)	*Pavonia zeylanica* (S)
Sida cordifolia (H)	*Pavonia zeylanica* (S)	*Sida acuta* (S)	*Sida cordifolia* (H)
Triumfetta pentandra (H)	*Sida acuta* (S)	*Sida cordifolia* (H)	*Triumfetta pentandra* (H)
Acacia caesia (S)	*Sida cordifolia* (H)	*Triumfetta pentandra* (H)	*Acacia mellifera* (T)
Acacia farnesiana (T)	*Triumfetta pentandra* (H)	*Acacia mellifera* (T)	*Acacia nilotica* (T)
Acacia mellifera (T)	*Acacia farnesiana* (S)	*Acacia nilotica* (T)	*Albizia lebbeck* (T)
Acacia nilotica (T)	*Acacia mellifera* (T)	*Albizia lebbeck* (T)	*Pithecellobium dulce* (T)
Albizia lebbeck (T)	*Acacia nilotica* (T)	*Pithecellobium dulce* (T)	*Mimosa pudica* (C)
Moringa oleifera (T)	*Albizia lebbeck* (T)	*Mimosa pudica* (C)	*Psidium guajava* (T)
Psidium guajava (T)	*Mimosa pudica* (C)	*Moringa oleifera* (T)	*Argemone mexicana* (H)
Argemone mexicana (H)	*Moringa oleifera* (T)	*Psidium guajava* (T)	*Antigonon leptopus* (C)
Pedalium murex (H)	*Psidium guajava* (T)	*Argemone mexicana* (H)	*Portulaca oleracea* (H)
Antigonon leptopus (C)	*Argemone mexicana* (H)	*Antigonon leptopus* (C)	*Punica granatum* (T)
Portulaca oleracea (H)	*Pedalium murex* (H)	*Portulaca oleracea* (H)	*Zizyphus mauritiana* (T)
Punica granatum (T)	*Antigonon leptopus* (C)	*Punica granatum* (T)	*Zizyphus oenoplia* (S)
Zizyphus mauritiana (T)	*Portulaca oleracea* (H)	*Zizyphus mauritiana* (T)	*Bruguiera gymnorrhiza* (T)
Zizyphus oenoplia (S)	*Punica granatum* (T)	*Zizyphus oenoplia* (S)	*Bruguiera cylindrica* (T)
Bruguiera gymnorrhiza (T)	*Zizyphus mauritiana* (T)	*Bruguiera gymnorrhiza* (T)	*Ceriops decandra* (T)
Bruguiera cylindrical (T)	*Zizyphus oenoplia* (S)	*Bruguiera cylindrica* (T)	*Ceriops tagal* (T)
Ceriops decandra (T)	*Bruguiera gymnorrhiza* (T)	*Ceriops decandra* (T)	*Rosa centifolia* (S)
Rosa centifolia (S)	*Bruguiera cylindrica* (T)	*Ceriops tagal* (T)	*Rosa* sp. (S)
Rosa sp. (S)	*Ceriops decandra* (T)	*Rosa centifolia* (S)	*Hamelia patens* (S)

(Continued)

TABLE 18.8 ***(Continued)***
Floral Calendar of *Apis dorsata* in Andhra Pradesh

September	October	November	December
Borreria hispida (H)	*Rosa centifolia* (S)	*Rosa* sp. (S)	*Mitragyna parviflora* (T)
Hamelia patens (S)	*Rosa* sp. (S)	*Borreria hispida* (H)	*Tarenna asiatica* (S)
Mitragyna parviflora (T)	*Borreria hispida* (H)	*Hamelia patens* (S)	*Santalum album* (T)
Tarenna asiatica (S)	*Hamelia patens* (S)	*Mitragyna parviflora* (T)	*Sapindus emarginatus* (T)
Santalum album (T)	*Mitragyna parviflora* (T)	*Tarenna asiatica* (S)	*Solanum trilobatum* (C)
Allophylus serratus (S)	*Tarenna asiatica* (S)	*Santalum album* (T)	*Sterculia urens* (T)
Cardiospermum canescens (C)	*Datura metel* (H)	*Sapindus emarginatus* (T)	*Lantana camara* (S)
Cardiospermum halicacabum (C)	*Solanum melongena* (H)	*Datura metel* (H)	*Achras zapota* (T)
Solanum melongena (H)	*Solanum trilobatum* (C)	*Solanum melongena* (H)	*Dalbergia latifolia* (T)
Sonneratia apetala (T)	*Helicteres isora* (T)	*Solanum trilobatum* (C)	*Tecomaria capensis* (T)
Solanum trilobatum (C)	*Duranta erecta* (S)	*Lantana camara* (S)	*Glycosmosis pentaphylla* (T)
Helicteres isora (T)	*Lantana camara* (S)	*Achras zapota* (T)	*Tagetes erecta* (S)
Duranta erecta (S)	*Vitex altissima* (T)	*Dalbergia latifolia* (T)	*Eucalyptus globulus* (T)
Lantana camara (S)	*Tribulus terrestris* (H)	*Tecomaria capensis* (T)	*Cestrum diurnum* (S)
Vitex altissima (T)	*Achras zapota* (T)	*Glycosmosis pentaphylla* (T)	*Corallocarpus epigaeus* (H)
Tribulus terrestris (H)	*Dalbergia latifolia* (T)	*Tagetes erecta* (S)	
Achras zapota (T)	*Tecomaria capensis* (T)	*Eucalyptus globulus* (T)	
Dalbergia latifolia (T)	*Glycosmosis pentaphylla* (T)	*Cestrum diurnum* (S)	
Tecomaria capensis (T)	*Eucalyptus globulus* (T)	*Corallocarpus epigaeus* (H)	
Prosopis juliflora (T)	*Cestrum diurnum* (S)		
Glycosmosis pentaphylla (T)			
Eucalyptus globulus (T)			

Notes: T = tree; H = herb; C = creeper/climber; S = shrub

width of fourth tomentum; that in factor score 5 with the ratio of head length and width and antennal length. The factor scores 1–4 show similar grouping with two morphoclusters while the factor score 5 shows a different group with one morphocluster. The morphological characteristics associated with factor score 1 and 2 are important variables, and the two morphoclusters formed in these scores have hence been considered as true representatives of populations of *A. dorsata*. The latitude-wise cluster analysis of morphometric data indicates that the factor score 1 represents mostly dry localities with sparse to moderate vegetation while factor score 2 represents mostly vegetation-rich localities. These localities in the respective zones represent different flora consisting of a combination of vegetable crops, fruit crops and wild flora. The observed values for morphological characteristics are relatable to the floral density, and the quantity and quality of floral rewards. Altitude-wise, most of the populations of *A. dorsata* in morphocluster 1 occur at 18–592 m while those of morphocluster 2 at 0.61–932 m. The study shows that in addition to seasonal floral diversity, latitude and altitude have a role in the occurrence of these morphoclusters and isolated populations. The ecological and floral diversity is also an important factor which brings about variation in morphological characteristics.

In *A. dorsata*, the proboscis length and the forewing length and width are variable characteristics and their variability depends on the extent of diversity in floral resources and flower types. These characteristics are part of five morphological characteristics which contributed to variation in factor score 1. Saini et al. (1983) reported that in honeybees, the hamuli are located near the proximal end of the anterior margin of each hind wing and couple it to a fold in the forewing to form a single, functional airfoil during flight. Richards (1949) reported that honeybees fly farther in dry conditions to get forage. and they hence need effective wings. In consequence, the bees develop longer forewings and produce a greater number of hamuli on the hind wings as adaptations to fly longer distances. Tan (2007) reported that the number of hamuli in *A. dorsata* is 25.7 ± 1.9 in Vietnam and 24.3 elsewhere. In this study, the number of hamuli in *A. dorsata* ranged from 23.2–26.3 The general trend is that there is an increase in hamuli number in dry localities where floral resources are not adequate. Therefore, the hamuli number in *A. dorsata* depends on the distance it flies regularly to collect the forage for its own needs, as well as for the needs of the colony. Ruttner (1988) used the cubital index in morphometric studies. Tan (2007) recorded that the cubital index of *A. dorsata* is 9.01 ± 3.32 in Vietnam and it is 7.25 elsewhere. In the present study, the cubital index of *A. dorsata* ranged from 5.65–8.84, and this variation is relatable to the variation in the length and width of forewing of *A. dorsata*.

The forage plants of *A. dorsata* indicated that 66–82 forage plants flower each month in all agro-climatic zones but trees serve as important forage sources year-round, especially during March–July when most herbaceous shrubs and herbs disappear. The floral calendar prepared for *A. dorsata* is of great help in managing their floral resources, especially during dearth period (Kevan 1984).

REFERENCES

Daly, H.V. 1991. Systematics and identification of Africanized honeybees. In: Spivak, M., Fletcher, D.J.C., and Breed, M.D. (Eds.), *The "African" honey bee*, Westview, Boulder, 13–44.

Daly, H.V. 1992. A statistical and empirical evaluation of some morphometric variables of honeybee classification. In: *Ordination in the study of morphology, evolution and systematics of insects: Applications and quantitative genetic rationales.* Elsevier, Amsterdam, 127–155.

Insuan, S., Deowanish, S., Klinbunga, S., Sittipraneed, S., Sylvester, H.A., and Wongsiri, S. 2007. Genetic differentiation of the giant honeybee (*Apis dorsata*) in Thailand analysed by mitochondrial genes and microsatellites. *Biochemical Genetics* 45: 345–361.

Kevan, P.G. 1984. *Insect pollination of economically important plants of tropical and Subtropical Asia*. Proc. FAO (United Nations) Expert consult on beekeeping with *Apis mellifera* in tropical and sub-tropical Asia, Bangkok, Thailand, 203–210.

Kevan, P.G. 1995. Bee botany, pollination, foraging and floral calendars. In: Kevan, P.G. (Ed.), *The Asiatic hive bee: Apiculture, biology and role in sustainable development in tropical and Subtropical Asia*. Enviroquest, Ontario, 113–116.

Makhmoor, H.D., and Ahmad, H. 1998. Biometric studies on four species of honeybees in Jammu region, India. *Indian Bee J.* 60: 141–142.

Mujumdar, S., and Kshirasagar, K.K. 1986. Morphometric characterization of *Apis dorsata* Fabr. workers. *Indian Bee J.* 48: 25–29.

Parr, J., Oldroyd, B.P., Huettinger, E., and Kastberger, G. 2004. Genetic structure of an *Apis dorsata* population: The significance of migration and colony aggregation. *J. Heredity* 95: 119–126.

Richards, O.W. 1949. The significance of the number of wing hooks in bees and wasps. *Proc. R. Ent. Soc. Lond.* 24: 7–9.

Roubik, D.W. 1995. Pollination of cultivated plants in the Tropics. *FAO Agricultural Services Bulletin* 118.

Ruttner, F. 1988. *Biogeography and taxonomy of honeybees.* Springer-Verlag, Berlin, Germany.

Ruttner, F., Tassencourt, L., and Louveaux, J. 1978. Biometrical statistical analysis of the geographic variability of *Apis mellifera* L. *Apidologie* 9: 363–381.

Saini, M.S., Aggarwal, R., and Dhillon, S.S. 1983. Functional morphology of the wing coupling apparatus and its bearing on the taxonomy of genus *Apis* Linnaeus. *Uttar Pradesh J. Zool.* 3: 10–14.

Tan, K., Radloff, S.E., Hepburn, H.R., Yang, M., Zhang, L., and Fan, X. 2007. Environmentally-induced developmental effects on morphometric characters of workers in *Apis cerana* colonies. *Apidologie* 38: 289–295.

Tan, N.Q. 2007. Biology of *Apis dorsata* in Vietnam. *Apidologie* 38: 221–229.

19 Biogeography of *Apis laboriosa* Smith and *Apis dorsata* Fabricius in Nepal

Ratna Thapa

19.1 INTRODUCTION

Nepal, located in the Himalayan region with an area of 47,181 square km with diverse topographical regions, is rich in honeybee diversity. In total, four native honeybee species—*Apis laboriosa*, *Apis dorsata*, *Apis florea*, and *Apis cerana*—and one exotic honeybee, *Apis mellifera*, exist at the base of the Himalayas in the Terai region. *Terai* is alluvial plains lying in the range of 60–200 m elevation. These areas have a subtropical climate with reasonably hotter summers and mild winters (mostly above freezing temperatures). Most rainfall is concentrated in the monsoon season. *Siwaliks* are foothills regions immediately north of the Terai region, reaching up to 1,800 m elevation. The climate in this region is also subtropical. *Hills* are characterized by a moderate elevation range of 1,500–2,500 m elevation. These areas have a sub-tropical climate at the bottom of the hills, but conditions become cooler with higher elevation. Some higher elevation experiences occasional snowfall during the winter season. *Middle Mountains*, topographically middle mountains, are characterized by steep slopes and narrow valleys lying north of the hills. The elevation could range from 2,000–4,000 m elevation. The climate is cold temperate and the temperature in some higher elevations often remains below the freezing point for five months of the year. Snowfall often covers the mountaintops during winter. *The High Mountains* are the northernmost part of the country. The elevation in this region is above 4,000 m, and reaches the highest point on the earth at 8,848 m above sea level. The climate is alpine to tundra, and most of the parts are under permafrost, snow, or glaciers throughout the year.

19.1.1 *Apis laboriosa*

Apis laboriosa Smith 1871, the largest Himalayan honeybee species in the world, is the least studied honeybee species. *A. laboriosa* was considered to be a subspecies of *A. dorsata* (Koeniger et al. 2011) until 1980 when Sakagami described *A. laboriosa* as a separate species (Engel 1999; Sakagami et al. 1980), which was supported

DOI: 10.1201/9781003294078-19

by genetic sequencing (Cao et al. 2012). *A. laboriosa* build a single comb under the rock ledges in deep vertical river valleys (Roubik et al. 1985). It is commonly known as the black Himalayan giant honeybee, and is found in high altitudes ranging from 1,200–3,700 m (Roubik et al. 1985; Thapa 2001; Woyke et al. 2012). The species is distributed almost continuously over a distance of 2,500 km along the Pan-Himalaya region from India (Uttarakhand, West Bengal), Bhutan, Nepal, Sikkim, Yunnan, Tibet, Laos, Myanmar, and Vietnam, along the Arakan Mountains in eastern Arunachal Pradesh, Nagaland, Manipur, the Shillong Hills of Meghalaya, and Mizoram (India) to Matupi in west-central Myanmar (Kitnya et al. 2020).

19.1.2 *APIS DORSATA*

A. dorsata Fabricius, 1793, a giant yellow honeybee species, is widely distributed in South and Southeast Asia. *A. dorsata* build a single large comb on the tallest tree, windows, balconies of buildings, and water tanks (Thapa and Wongsiri 2003). Colonies comprise up to 100,000 individuals (Morse and Laigo 1969). *A. dorsata* colonies prefer to nest on big branches of tall trees and human-built structures, particularly buildings, water tanks, and cliffs (Oldroyd and Wongsiri 2006; Thapa et al. 2000). Aggregation is very common on the same support in India (Reddy 1980), Nepal (Thapa 2001), the Philippines (Ruttner 1988), and Thailand (Wongsiri et al. 1996). Towards the end of the season, brood rearing is stopped (Ruttner 1988) and colonies eventually abandon permanent nesting sites for 3–4 months (Paar et al. 2004). At the end of the rainy season, honey and pollen are depleted (Ruttner 1988), brood rearing is stopped, and the colonies eventually abandon their permanent nesting sites. *A. dorsata* seasonally migrate 200 km from south to north, and vice-versa based on climatic rhythms, and seasonally return to their natal nesting (Koeniger and Koeniger 1980)

19.2 MATERIALS AND METHODS

19.2.1 IDENTIFICATION OF *A. LABORIOSA*

A. laboriosa has many morphological differences from *A. dorsata*. Sakagami et al. (1980) distinguished the two taxa based on thoracic hair color, which was "tawny musky brown" in *A. laboriosa* and "mostly dark" in *A. dorsata*. The workers of *A. laboriosa* phonologically have black abdominal segments with a white strip in each abdominal segment, whereas the workers of *A. dorsata* have yellow to pale yellow, white strips on each abdominal segment (Thapa and Wongsiri 2003).

19.2.2 DATA COLLECTION

Field data were collected personally by visiting the established nesting sites of *A. laboriosa* and *A. dorsata* in Kaski and Chitwan, respectively, between 2019 and 2022. The meteorological data from a network of Nepal and Meteoblue archiving were used to investigate the distribution and migration of *A. laboriosa* and *A. dorsata* in the Kaski and Chitwan districts.

19.3 RESULTS

19.3.1 Elevational Distribution of *A. laboriosa* and *A. dorsata*

A. laboriosa occurs between the mid-mountain area (862 m) and the upper Himalayas (4,300 m, Figure 19.2). *A. laboriosa* extends from Taplejung (eastern Nepal) to Rolpa (western Nepal, Table 19.1). Subsequently, *A. dorsata* is widely distributed in the southern Terai regions (60–200 m, Figure 19.2) from Jhapa (eastern Nepal) to Kanchanpur (far western Nepal, Table 19.2).

19.4 SEASONAL MIGRATION OF *A. LABORIOSA* AND *A. DORSATA*

19.4.1 *A. laboriosa*

A. laboriosa migrates seasonally to different altitudes to avoid harsh environmental conditions such as cold snaps, heat waves, heavy precipitation, and dwindling food

FIGURE 19.1 (A) *A. laboriosa* colonies on the rocky cliff at Ghachuk Cliff, Pokhara (49 nests). (B) *A. dorsata* nest at Chitwan National Park. (Photos by Ratna Thapa).

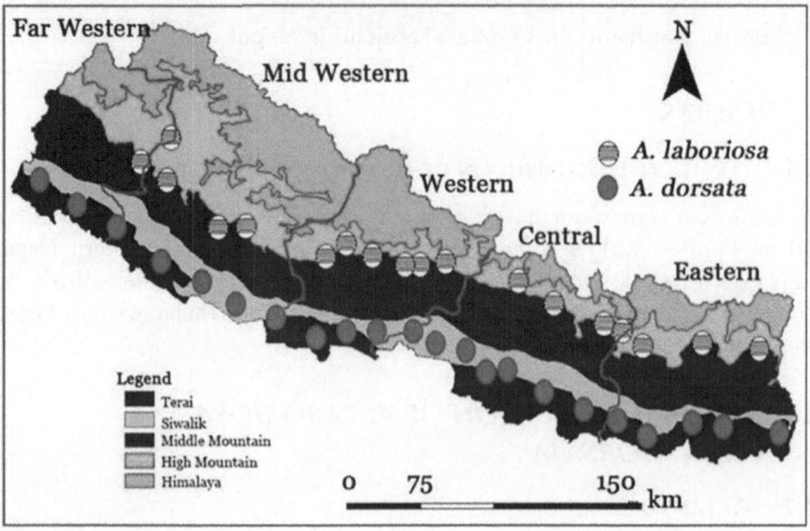

FIGURE 19.2 Distribution of *A. laboriosa* (black chocolate and green) and *A. dorsata* (blue and yellow) in Nepal.

TABLE 19.1
Location and Elevational Distribution of *A. laboriosa*

Location	Forest Types	Latitude (N)	Longitude (E)	Altitude (m)	Annual Mean Temp mp (°C) Min	Max	Annual Rainfall (mm)
Taplejung	Oak-rhododendron forest	27°21'14.48"	87°40'4.72"	1829	2	23	489
Sankhuwasaba	Tropical dipterocarp monsoon forest	27°36'51.09"	87°8'32.24"	2585	1	21	297
Solukumbu	Mixed hardwood forests	27°47'27.5"	86°39'39.99"	5121	13	21	312
Sindhupalchowk	Sal forest	27°57'4.33"	85°41'4.48"	1606	7	32	2,500
Gorkha	Rhododendron coniferous forests	27°59'3.06"	84°37'37.14"	862	23	38	467
Manang	Subalpine forest	28°38.5169'	84°11.1427'	3536	4	15	774
Lamjung	Broadleaf closed forest	28°16'35.58"	84°21'15.14"	1014	14	27	2,944
Kaski	Mixed broadleaved, rhododendron, pine	83°51' 57.78"	28°16.9416'	1122	23	30	347
Parbat	Sal forest	28°10'40.98	83°41'55.17"	1424	13	29	2,500
Magdi	Tropical deciduous riverine forest	28°36'10.42	83°20'10.28"	2828	22	28	676
Baglung	Temperate forest	28°16'38.76"	83°34'53.81"	1212	20	34	1,069
Rukum	Riverine forest	28°44.6065'	82°28.51.66'	3095	18	26	181
Rolpa	Oak forest	28°22'53.62"	82°38'54.04"	2067	−7	18	18
Lantang	Oak forest	28°15'0.00"	85°30'0.00"	4300	3	14	173

TABLE 19.2
Location and Elevational Distribution of *A. dorsata*

Location	Types of Forest	Latitude (N)	Longitude (E)	Altitude (m)	Annual Temp (°C)		Average Rainfall (mm)
					Min	Max	
Jhapa	Sal forest	26°38'49.33"	87°53'25.78"	113	11	33	2091
Morang	Sal forest	26°40'47.64"	87°27'37.43"	177	18	35	1312
Saptari	Tropical econdary forests	26°40'47.64"	26°40'47.64"	300	12	38	787
Siraha	Tropical forest	26°38'23.07"	86°11'7.25"	80	13	39	480
Sarlahi	Mixed riverine forest	26°57'45.77"	85°33'40.36"	105	8	41	292
Rautahat	Mixed hardwood and riverine forests	27°2'55.28"	85°18'49.04"	106	28	39	2,968
Parsa	Mixed riverine deciduous forests	27°10'24.92"	84°51'24.45"	105.59	9	37	1,402
Chitwan	Tropical and subtropical forests	27°31'44.87"	84°21'15.14"	197	15.2	43	2,407
Nawalparasi	Sal forest	27°39'25.88"	84°3'32.5"	236	7	39	2,145
Rupandehi	Sal forest	27°37'35.13"	83°22'44.18"	112	15	38	1,174
Kapilbastu	Tropical mixed hardwood	27°39'15.08"	82°59'35.38"	114	27	36	479
Dang	Subtropical forests	27°59'25.3"	82°18'6.38"	700	20	31	381
Banka	Sal forest	28°23'41.49"	84°7'26.43"	79	25	40	1,168
Bardiya	Riverine forests	28°18'36.84"	81°25'40.43"	138	24	37	1,900
Kanchanpur	Subtropical forest	28°50'13.92"	80°19'16.72"	1000	22	32	509

resources. As temperatures begin to drop below 20°C at the base of the Annapurna region, entire colonies of *A. laboriosa* migrate to the lowland nesting site (Gharchowk Cliff, 780 m) near Pokhara where they stay for two months (October–November). Beginning in mid-December, when the ambient temperature drops below 10°C, *A. laboriosa* again start to migrate to the hilly areas of lowlands (Butwal, 1,170 m) to hibernate for 2–3 months (December–February). As the spring season progresses from mid-February, *A. laboriosa* again reoccupy the Gharchowk Cliff for the second time and stay there for four months (February–May). When the temperature rises more than 30°C, intensely hitting the nests with tsunlight, entire colonies of *A. laboriosa* start to migrate upland (Korchan Cliff, 3,682 m) at Machhapuchhre, Annapurna for three months (July–September, the rainy season). *A. laboriosa* spend an entire rainy season in the mountains at the foot of Machhapuchhre, Annapurna (Figure 19.3).

19.4.2 *A. dorsata*

A. dorsata reoccupy the permanent human-built nesting sites, especially building windows, balconies, and water tanks. In Chitwan, colonies of *A. dorsata* arrive in

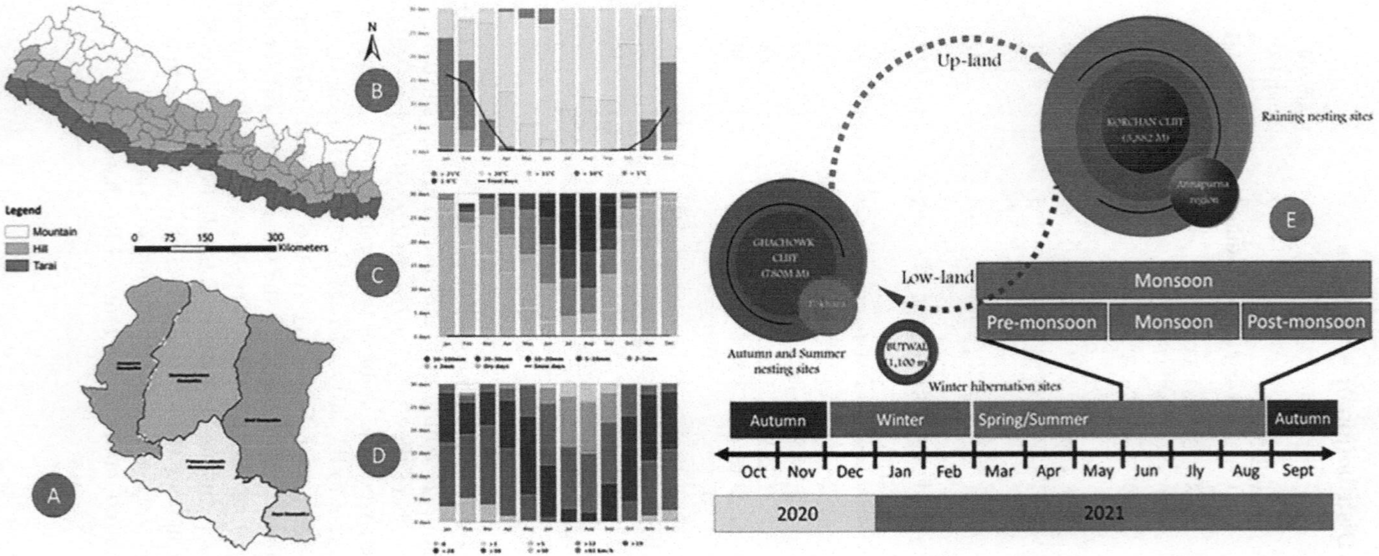

FIGURE 19.3 Simulated historical climate of Kaski, Pokhara: (A) study sites; (B) temperature; (C) Precipitation diagrams showing certain precipitation amounts; (D) diagram showing wind speeds—the monsoon creates steady strong winds from December–April, and calm winds from June–October; and (E) *A. laboriosa* seasonal migratory route to Kaski district. The climate prevailing in Nepal can be divided into four seasons, based on rainfall and temperature conditions. They are summer-monsoon (June–September, with rainy, hot and humid weather), December–February, with coldest ad driest weather, pre-monsoon (March–May, with hot weather and thunderstorms), and post-monsoon (October–November, with cool weather) (Source: www.meteoblue.com).

the post-monsoon period (October–November), stay there in the winter (December–February), spring (February–March), and early summer (April–May), and abandon the established nesting sites. During the extreme summer (39°C, in May), *A. dorsata* abandoned east-, south- and west-facing nesting sites. At the beginning of the rainy season (July), they built scattered nests on human-built structures. In July 2022 (monsoon), four established colonies were observed at Rhino River Side Hotel at Sauraha, Chitwan. The rest of the small colonies were observed on the branch of Bombax trees (*Bombax ceiba*) and mango trees (*Mangifera indica*) at Chitwan National Park (Figure 19.4).

19.5 DISCUSSION

A. laboriosa and *A. dorsata* colonies never overlap in their distribution and share the same ecological niches in Nepal. The most plausible explanation is that the climatic conditions of Nepal vary from one place to another according to the geographical features. In the north, summers are cool and winters severe, while in the south, summers are tropical and winters mild. Every 1,000 m of elevation in Nepal results in a 6°C drop in temperature. In the Terai, summer temperatures exceed 37°C higher in some areas, and winter temperatures range from 7°–23°C. The Terai is the northern part of the Indo-Gangetic plain, and the southern parts of Nepal, which extend nearly 800 km from east to west and about 30–40 km from south to north, and have an average elevation of below 59–750 m. The Outer Terai begins at the southernmost tip of the Siwalik Hills. The Inner Terai refers to elongated valleys lying between the Siwalik Range and the Mahabharata Range. The Siwalik Range (locally known as *Churia* hills, ranging from 700–1,500 m), consists of hills, steep land slopes, gorges, large spans of the river, and temporary streams, including 137 flowering plants; *Shorea robusta*, *Terminalia alata*, and *Semecarpus anacardium* are the most dominant plant species. The vegetation is much more diverse in Nepal.

In contrast to *A. dorsata*, which is found in Nepal between 79 m and 700 m above sea level, *A. laboriosa* is found in both middle and high mountains, with an elevation range of 780–3,662 meters. The far western regions of Nepal are noticeably lacking of *A. laboriosa*. This may be due to the far western regions of Nepal's unusually dry environment and habitat destruction. The amount of precipitation reduces from east to west as distance from the Bay of Bengal increases. According to field investigations conducted in eastern and mid-mountainous regions, *A. laboriosa* colonies were found along the Langtang River at elevations ranging from 2,460–4,100 m on the southern slope of the Eastern Himalayas (Roubik et al. 1985).

Arunachal Pradesh's western Denling Forest in northeastern India had the fewest *A. laboriosa* colonies (Sakagami et al. 1980). In Arunachal Pradesh, the Indian team found five locations where *A. laboriosa* foragers foraged with its sister species *A. dorsata*. The sympatric presence of these two species in Arunachal Pradesh was confirmed by Kitnya et al. (2020). However, I did not see *A. laboriosa* and *A. dorsata* workers foraging together in Nepal during my field visits from Pokhara to Machhapuchhre, the Annapurna Himalayas base, and to Chitwan. This might be because *A. laboriosa* and *A. dorsata* are not very common in Nepal.

Underwood (1986) reported that *A. laboriosa* colonies abandon most of the permanent nesting cliffs in the fall and spend the winter in combless swarms at lower

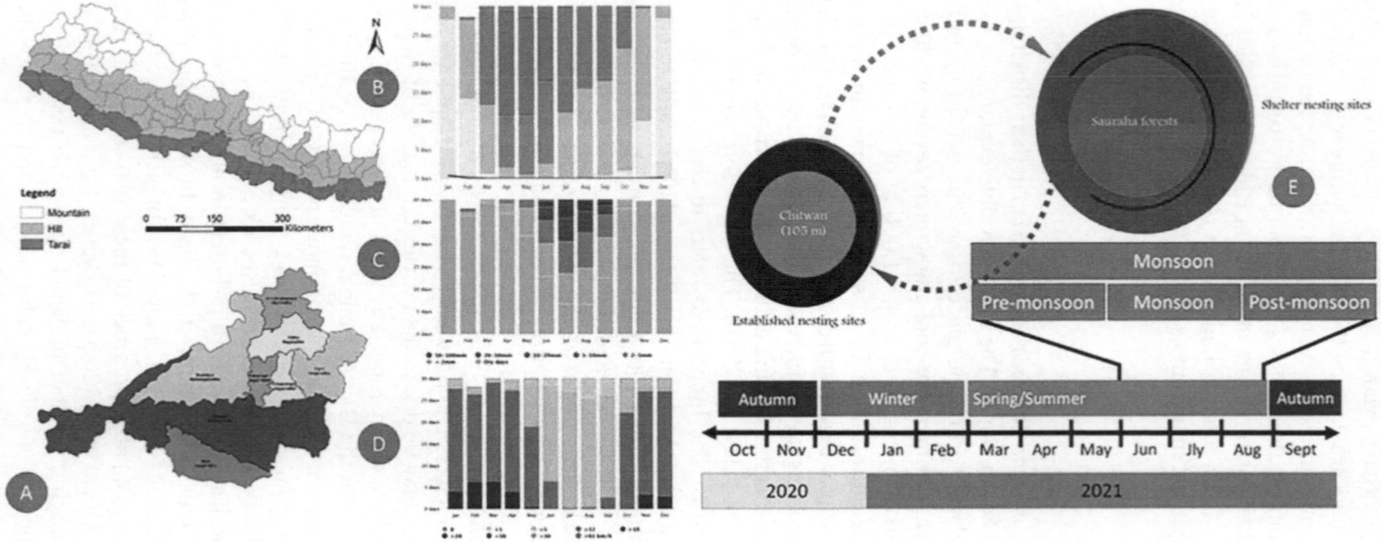

FIGURE 19.4 Simulated historical climate of Bharatpur, Chitwan: (A) study sites; (B) temperature; (C) precipitation diagrams showing certain precipitation amounts; (D) diagram showing the wind reaches a certain speed. The monsoon creates steady strong winds from December–April, and calm winds from June–October; and (E) seasonal migratory route of *A. dorsata* in Chitwan district (Source: www.meteoblue.com).

elevations. By April, they recolonized lower-elevation nesting cliffs. As the season progresses, colonies continue to move higher in elevations along river valleys during the summer before they retreat downhill for the winter. During my field survey, *A. laboriosa* colonies abandoned their permanent nesting sites in the rainy season (May–June), and they migrate to the upland (Machhapuchhre, high-land nesting sites), where they spend a whole rainy season, and reoccupied the permanent nesting sites (low-land nesting sites) by October/November. *A. laboriosa* again abandon their low-land nesting sites to escape from the winter. During the winter, they spend three months in the mountainous area of lowland with a small comb, and reoccupied the permanent nesting sites by February for the proliferation of colonies. *A. laboriosa*'s migratory route implies that they have regular seasonal up-and-down migrations based on the ambient temperature.

A. dorsata is present all around the year between Terai and Siwalik regions. The Siwalik regions are seasonal. I observed several colonies of *A. dorsata* all around the year at building sites of Chitwan National Park. The survey suggests that the migration schedule of *A. dorsata* does not seem to be due to differences in the local weather and climatic conditions but is governed by temperature and available flora resources. In Terai, a lowland area, cold waves have been occurring annually since first being recorded in 1990, between mid-December and mid-January; when the temperature drops up to 5°C for a few hours during the night, *A. dorsata* did not abandon their nesting sites. During this time of the year, mustard (*Brassica nigra*) bloom in the Terai region. In May, when the ambient temperature exceeded 38°C, *A. dorsata* started to abscond. Adult workers are not well survived from about 36°C, and the combs started to melt. This behavior suggests that *A. dorsata* is highly sensitive to high temperature and somewhat less sensitive to low temperature. Similar observations were made in Malaysia (Mardan and Kevin 2002).

My observation was not consistent with previous reports of *A. dorsata* migration routes by Koeniger and Koeniger (1980) and Paar et al. (2000). They state that *A. dorsata* migrate to the foothills of the Himalayas (north), and after 7–8 months at the end of the season, they reoccupied the old nesting sites in India. Similarly, Koeniger and Koeniger (1980) reported that *A. dorsata* seasonally migrates 200 km to alternative nesting sites from south to north and vice-versa in Sri Lanka. I did not observe *A. dorsata* migrate such long distances in Nepal. The average width of Nepal from south to north is 193 km. However, *A. dorsata* migrate within 50 km between Terai and Siwalik regions. Similar observations were made in Malaysia. Swarms reportedly cross the 50 km-wide Strait of Malacca between Sumatra and Malaysia (Mardan 1989). *A. dorsata* swarms have been observed to travel between habitats with different blooming seasons (Crane et al. 1993; Koeniger and Koeniger 1980; Mahindre 2000).

19.6 CONCLUSION

The timing of the giant honeybee's migration seemed to be influenced by two main factors: ambient temperature coupled with availability of bee forage. *A. laboriosa* does not overlap in distribution with *A. dorasta* in Nepal. *A. laboriosa* has two established nesting sites; first, rainy established nesting sites at upland and the second, fall and spring established nesting sites at lowland. The nesting sites of *A. laboriosa* are observed at the lowest elevation of 780 m and highest established nesting sites at 3,682 m, whereas *A. dorsata* is found between 59 m and 700 m in Nepal. The distribution of these two giant honeybee species is separated from the range of Siwalik (Churia hills) in Nepal. *A. laboriosa* migrate from Pokhara to Machhapuchhre (26 km), whereas *A. dorsata* migrate between Terai and Siwalik (50 km). These giant honeybees are key pollinators in the central Himalayan region, and the continued decline in their numbers is resulting in a severe loss of pollination services in the Himalayan ecosystem regions.

ACKNOWLEDGMENTS

The author wishes to thank Mr. Man Bdr. Pun for field assistance and information of migratory routes and nesting sites of *A. laboriosa* in Machhapuchre, Kaski District. The author also states that there were no conflicts of interest during the study or in the publication of this research.

REFERENCES

Cao, L.F., Zheng, H.Q., Hu, C.Y., He, S.Y., Kuang, H.O., and Hu, F.L. 2012. Phylogeography of *Apis dorsata* (Hymenoptera: Apidae) from China and neighboring Asian areas. *Annals of the Entomological Society of America* 105(2): 298–304.

Crane, E., van Luyen, V., Mulder, V., and Ta, T.C. 1993. Traditional management system for *Apis dorsata* in submerged forests in southern Vietnam and central Kalimantan. *Bee World* 74: 27–40.

Engel, M.S. 1999. The taxonomy of recent and fossil honey bees (Hymenoptera: Apidae; *Apis*). *Jr. Hymen. Res.* 8(2): 165–196.

Kitnya, N., Prabhudev, M.V., Bhatta, C.P., Pham, T.H., Nidup, T., Megu, K., Chakravorty, J., Brockmann, A., and Otis, G.W. 2020. Geographical distribution of the giant honey bee Apis laboriosa Smith, 1871 (Hymenoptera, Apidae). *ZooKeys* 951: 67.

Koeniger, G., Koeniger, N., and Phiancharoen, M. 2011. Comparative reproductive biology of honeybees. In: Hepburn, H., and Radloff, S. (Eds.), *Honeybees of Asia*. Springer, Berlin, Heidelberg. https://doi.org/10.1007/978-3-642-16422-4_8.

Koeniger, N., and Koeniger, G. 1980. Observations and experiments on migration and dance communication of *Apis dorsata* in Sri Lanka. *Jr. Apic. Res.* 19: 21–34.

Mardan, M.B. 1989. *Thermoregulation in the Asiatic giant honeybee Apis dorsata (Hymenoptera: Apidae)*. Thesis, University of Guelph, Guelph.

Mardan, M., and Kevan, P.G. 2002. Critical temperature for survival of brood and adult workers of the giant honey bee, *Apis dorsata* (Hymenoptera: Apidae). *Apidologie* 33: 295–301. https://doi.org/10.1051/apido:2002017.

Mahindre, D.B. 2000. Developments in the management of *Apis dorsata* colonies. *Bee World* 81: 155–163.

Morse, R., and Laigo, F.M. 1969. *Apis dorsata in the Philippines*. Philippines Assoc. of Entomologists, Laguanna.

Oldroyd, B., and Wongsiri, S. 2006. *Asian honeybees: Biology, conservation and human interactions*. Harvard University Press, Cambridge, 340.

Paar, J., Oldroyd, B.P., Huettinger, E., and Kastberger, G. 2004. Genetic strcuture of an *Apis dorsata* population; The significance of migration and colony aggregation. *J. of Heredity* 95(2): 119–126.

Paar, J., Oldroyd, B.P., and Kastberger, G. 2000. Giant honeybees return to their nest sites. *Nature* 406, 475.

Reddy, C.C. 1980. Studies on the nesting behavior of Apis dorsata F. *Intl. Conf. Apic. Trop. Climate* 2: 391–397.

Ruttner, F. 1988. *Biogeography and taxonomy of honeybee*. Springer-Verlag, Berlin.

Roubik, D.W., Sakagami, S.F., and Kudo, I. 1985. A note on distribution and nesting of the Himalayan honey bee Apis laboriosa Smith (Hymenoptera: Apidae). *Jr. of the Kansas Entomological Society* 58(4): 746–749.

Sakagami, S.F., Matsumura, T., and Ito, K. 1980. *Apis laboriosa* in Himalaya, the little known world largest honeybee (Hymenoptera, Apidae). *Insecta Matsumurana* 19: 47–78.

Thapa, R. 2001. The Himalayan giant honey bee and its role in ecotourism development in Nepal. *Bee World* 82(3): 139–141.

Thapa, R., Shrestha, R., Manandhar, D.N., and Kafle, B. 2000. *Beekeeping in Nepal*. 7th International Bee Research and 4th Asian Apiculture Association Conference Proceeding of 7th International Conference, Tropical Bee Management and Diversity, 409–413. https://link.springer.com/content/pdf/10.1007/978-3-642-16422-4.pdf.

Thapa, R., and Wongsiri, S. 2003. The flying predators of honeybee: Apis dorsata and A laboriosa in Nepal. *American Bee Jr.* 143(7): 540–542.

Underwood, B.A. 1986. *The natural history of Apis laboriosa Smith in Nepal*. Master's thesis. Cornell University, Ithaca, NY, 88.

Wongsiri, S., Thapa, R., Oldroyd, B.P., and Burgett, D.M. 1996. A magic bee tree. Home to *Apis dorsata* Fab. *Am. Bee J.* 136(11): 796–799.

Woyke, J., Wilde, J., and Wilde, M. 2012. Swarming and migration of *Apis dorsata and Apis laboriosa* honey bees in India, Nepal and Bhutan. *Journal of Apicultural Science* 56: 81–91. https://doi.org/10.2478/v10289-012-0009-7.

20 Diversity, Recent Distribution, and Nesting Behavior of Giant Honeybees in Indonesia and Their Role in Natural and Agricultural Ecosystems

S. Kahono, D. Peggie, S. Subiyakto, J.S.A. Lamerkabel and M.S. Engel

20.1 INTRODUCTION

With its complex geological history, Indonesia is home to the greatest diversity of advanced eusocial bees in Asia (Kahono et al. 2018). Two lineages of advanced eusocial corbiculate bees are abundant across Indonesia: stingless bees (Meliponini) and honeybees (Apini). While stingless bees and meliponiculture is blessed with a considerable diversity of species, honeybees and apiculture has comparatively few species with which to contend (Michener 2007; Engel and Rasmussen 2021). Nonetheless, of the seven species of honeybees (genus *Apis* L.), more species and subspecies may be found natively in Indonesia than in any neighboring region (Maa 1953; Kahono et al. 2018). This unique faunal composition for honeybees makes Indonesia an ideal region in which to explore the complexities of their biology, ecology, interspecific interactions, and intricate evolutionary history and biogeography.

Honeybees, like all advanced eusocial corbiculate bees, live in large perennial colonies with fixed worker and queen castes (Michener 1974, 2007). Species of *Apis* construct a double layered comb of hexagonal waxen cells in which the bees store reserves and rear their brood (Michener 2007). The comb is covered by workers, which comprise the majority of a colony's population, and which, in contrast to their stingless bee relatives, have functional stings for defense, although stinging results in autotomy and the defending worker's demise (Seeley et al. 1982; Winston 1991).

Although Indonesia is rich in native species of honeybees, the most familiar species has been *Apis mellifera* L., a species introduced from Australia (Mashudi and

 DOI: 10.1201/9781003294078-20

Suwanda 1988; Sihombing 1997), where it was also not native (Ruttner 1988). In addition to this introduced species, there are several other species truly native to Indonesia. Two species of dwarf honeybees are found in Indonesia, *A. andreniformis* Smith and *A. florea* Fabricius. The origin and status of *A. florea* has been uncertain (Hadisoesilo 2001), and Museum Zoologicum Bogoriense (MZB) holds collections of *A. florea* sampled from Jakarta in 1920, 1931, 1953, and from the small island of Pulau Dua, near Jakarta, in 1972. Unfortunately, there is no additional information on the species from these locations, and modern surveys need to be conducted to explore the occurrence as it has been believed that the presence of *A. florea* in Indonesia was the result of recent introductions (Silva et al. 2020). Indeed, it may be that *A. florea* has been introduced multiple times over the last century, but whether these populations have been able to persist until recently is in need of detailed study. The giant honeybee, *A. dorsata* Fabricius, is found across most of western Indonesia and is similar to the dwarf honeybees in that it plesiomorphically nests in the open, with an exposed comb and colony, although often hanging high in trees, on cliff overhangs, or even on the exposed rafters or roof lines of homes (the dwarf honeybees often nest in shrubs or smaller trees) (Ruttner 1988). Among cavity-nesting honeybees, *A. cerana* Fabricius, *A. koschevnikovi* Enderlein, and *A. nigrocincta* Smith are native to Indonesia (Ruttner 1988; Otis 1991, 1996; Radloff et al. 2011; Hepburn and Radloff 2011; Smith 2021; Kahono et al. 2021), as is the subspecies *A. cerana nuluensis*, which is considered specifically distinct by some authors or as a synonym by others (Tingek et al. 1996; Hadisoesilo 2001; Radloff et al. 2011; Engel 2012).

Apis dorsata is quite abundant across tropical Asia (Batra 1977; Radloff et al. 2011; Hepburn and Radloff 2011; Smith 2021). The species is a focus of interest owing to its economic importance as a primary source for national honey production (Hadisoesilo and Kahono 2011, unpublished data; Muslim 2014), and may be severely impacted by underway human-induced environmental changes. The species has the largest body size, greatest number of colony members, and strongest sting of any species in the genus *Apis*. In nature, their combs hang on high and massive tree branches or high cliffs, well outside of the reach of most animal predators, except humans. Accordingly, their natural predators—and honey hunters—must go to extreme measures to collect the honey or attack hives. Since the species nests in the open and can be migratory, efforts to manage the species in a controlled hive context have not been successful. However, in Indonesia, Cambodia, and Thailand, colonies of *A. dorsata* have been successfully attracted to nest on human-built rafters set up at about head-height so as to make it easier to harvest honey (Kahono 2002; Muslim 2014; Waring and Jump 2015; Chuttong et al. 2019; Kahono et al. 2021). Quite recently, local honey hunters in the city of Bogor (West Java) have successfully prepared semi-natural nesting trees, similar to the examples using low rafters, by bending down branches from larger trees or by attracting the bees to shorter trees (recent unpublished data).

As a migratory species, *A. dorsata* is a well-known seasonal long-distance traveler, conducting long-distance flights between geographically discrete ecosystems and islands (Dingle 1980; Koeniger and Koeniger 1980; Dyer and Seeley 1991, 1994). Colonies tend to have some degree of site fidelity and return to places where they have previously nested (Kahono et al. 1999). This behavior is dependent on the stability of local environmental conditions. While the bees can fly far, there is no evidence of endemic subspecies moving across Wallace's Line on their own and between the islands of Kalimantan and Sulawesi (Hill 2009).

20.2 DIVERSITY AND CURRENT DISTRIBUTIONS

There are four subspecific varieties of giant honeybee, although some of these are at times considered specifically distinct (Maa 1953; Ruttner 1988; Engel 1999; Radloff et al. 2011; Engel and Rasmussen 2021; Smith 2021). There is also at least one extinct species of giant honeybee found in southern Japan's Iki Island of the Tsushima Strait, well outside of present-day giant honeybee distribution (Engel 2006). *Apis dorsata laboriosa* is endemic to the Indian highlands (Otis 1996), *A. d. breviligula* is endemic to the Philippines, while in Indonesia *A. d. dorsata* and *A. d. binghami* are distributed West and East of Wallace's Line, respectively, although the former has made transgressions into the central Moluccas. The two subspecies can be easily distinguished by the coloration of workers, as well as their distinct geographic isolation. The phylogenetic placement of giant honeybees has been extensively explored on the basis of morphological, ethological, and molecular data (Alexander 1991; Smith 1991; Smith and Hagen 1996; Engel and Schultz 1997; Engel 1999; Raffiudin and Crozier 2007; Koeniger et al. 2011; Raffiudin and Shullia 2020).

Workers of *A. d. dorsata* have abdomens that are largely brownish with orange bands (Hadisoesilo 2001), while progressing along the eastern islands the coloration tends to become progressively darker (Kahono unpublished data). This subspecies is the most widespread throughout India and Southeast Asia (Otis 1991; Smith 2021). In Indonesia, the distribution of this form includes Sumatra, Borneo, Java, and Bali, across Wallace's Line from Lombok to Timor and the Kei Islands (Otis 1991; Smith 2021). Although *A. d. dorsata* can fly long distances, there is no record of its movement to Sulawesi and adjacent islands. Their spread has been recorded in the islands of Sermata, Romang, Wetar, and Moa (southwest Moluccas), and more recently colonies have occupied the islands of Buru, Ambon, and Seram (Moluccas) (recent unpublished data) (Figure 20.1). This subspecies is the only one that today occupies areas on either side of Wallace's Line, and the degree to which human activity has made this possible needs detailed study.

Sulawesi was formed from three different regions around 50 million years ago (Mya) (Hill 2009). It has a large number of endemic species, including a unique, monotypic highland genus of stingless bees, *Wallacetrigona incisa* (Rasmussen et al. 2017; Kahono et al. 2018), as well as a distinctive subspecies of giant honeybee, *A. d. binghami* (Maa 1953). Workers of *A. d. binghami* have a more predominantly black abdomen with white to yellow pubescent bands, and are found on Sulawesi (Otis 1991; Hadisoesilo 2001; Nagir et al. 2016), as well as the surrounding islands of Peleng, Banggai Laut, Taliabu, and Sula (Labi Mapok personal comm.; recent unpublished data).

Local people in Sulawesi conduct annual honey harvests at about the same period across the island from January–April (Nagir et al. 2016; recent unpublished data). There is no report on the bees' migration among islands of the Wallacean region. From available data on the diversity and the distribution of flowering forest

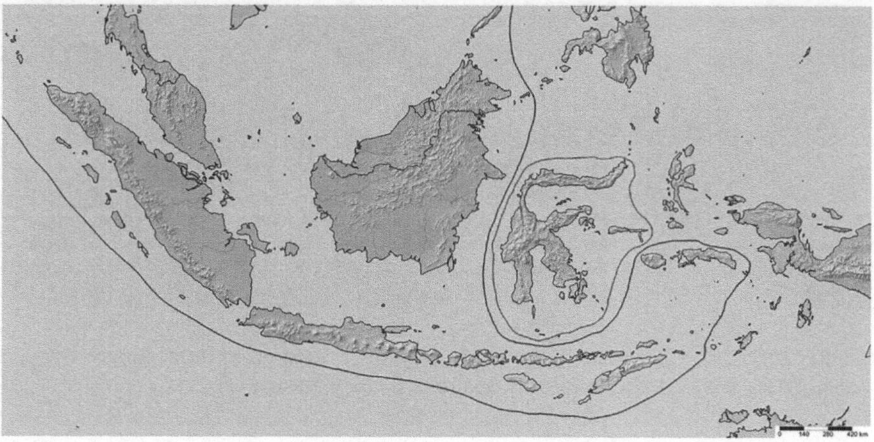

FIGURE 20.1 Recent distributions of giant honeybee subspecies *A. d. dorsata* (blue line) and *A. d. binghami* (red line) across Indonesia.

plants on Sulawesi and surrounding islands (Lam 1945; van Steenis 1978), there is no trend toward phenological groups that could trigger bee migration. According to the belief of local honey hunters, the main trigger for the bees' movements is the difference in plant phenology within and among the forests and agricultural plants (Kahono et al. 2021).

20.3 NESTING BEHAVIOR

Apis dorsata constructs a large, exposed, single vertical comb. However, in locations with abundant populations of its natural raptor predator, the bees tend to nest on a branch covered with climbing plants or small tight branches. Incipient colonies have a rather round to conical form since the comb is not yet fully completed. Later, as the comb continues to expand, the colony takes on the more familiar flattened and expanded shape, and finally when fully matured it is of an overall elliptical shape, although sometimes there are subtle deviations (e.g., Figure 20.2). The bilateral comb of hexagonal cells is constructed exclusively from self-produced wax (Ruttner 1988), and is affixed to the lower surface of a flat horizontal or inclined object. Within a comb, honey is located in the uppermost corner, bee bread sticks to the honeycomb, and broods are reared below. Unlike observations from Nepal, where *A. dorsata* tends to nest in orientation such that the comb faces easterly to capture the sun's rays (Neupane et al. 2013), there is no pattern of cardinal coordinate orientation in populations across Indonesia.

In nature, *A. dorsata* nests under branches of large, tall trees of the forests, or in some places, it nests under limestone cliffs. The bark selected for attaching the nest is usually exposed fine bark that does not peel easily, and with fewer branches such as *Ficus* sp. (Moraceae), *Intsia* sp. (Leguminosae), *Dryobalanops* sp. and *Shorea* sp. (Dipterocarpaceae), *Canarium* sp. (Burseraceae), *Chisocheton* sp. (Meliaceae),

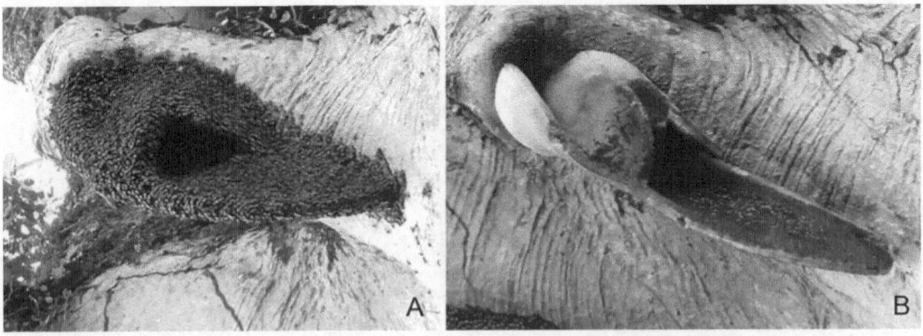

FIGURE 20.2 An uncommon shape of a comb of *A. d. dorsata* in the Bogor Botanical Garden, West Java: A. living colony; B. abandoned nest exposing comb.

Artocarpus sp. (Moraceae) (Kahono et al. 1999), and "kapuk randu" (*Ceiba pentandra*) (Bombacaceae). *Apis dorsata* prefers to nest on trees close to water sources, and on Konaweha (East Kolaka, Southeast Sulawesi), Rongkong (North Luwu, South Sulawesi), and Semongkat (Batu Dulang, Sumbawa, Nusa Tenggara Barat), the bees prefer to nest along river basins (Hadisoesilo and Kahono 2011, unpublished data; Indonesian Forest Honey Association [JMHI], personal communication). The bees need more water to cool their hives (Muslim 2014; recent unpublished data).

There is a difference between the two subspecies as to the number of colonies in a single tree. On a nesting tree, in Sumatra and Kalimantan, *A. d. dorsata* may congregate in the tens to hundreds of colonies (Hadisoesilo 2001; recent unpublished data), and even though the nests may be close together on the same tree there are no reports of inter-colony agonistic interactions (Seely et al. 1982; recent unpublished data). In Sumatra, the trees in which bees may aggregate are called "sialang" trees and may belong to several different forest-plant species. The strategy of clustered nesting in tall trees in Sumatra and Kalimantan may be a strategy to defend bees from attacks by their natural enemies, such as the protected endemic sun bear (*Heliarctos malayanus*) and orangutans (*Pongo pygmaeus*). By contrast, *A. d. binghami* tends to form aggregations of up to three colonies, and more exceptionally with a maximum of ten colonies on a tree (Nagir et al. 2016). Both subspecies nest at about similar heights, from 1–32 m above the ground, and both have similar comb length, reaching about 2 m (Nagir et al. 2016).

The periods of stay and annual honey harvests of *A. d. dorsata* on the islands of Sumatra, Kalimantan, Java, and Sumbawa are from June–November (Kahono et al. 1999; JMHI, personal communication; recent unpublished data), whereas those of *A. d. binghami* on Sulawesi and surrounding smaller islands are from February–April (Hadisoesilo and Kahono 2011, unpublished data; Nagir et al. 2016; JMHI and Labi Mapok, personal communication; recent unpublished data). Considering the danger involved in undertaking honey harvesting from colonies of *A. dorsata* in the canopies of high trees, people in Indonesia have practiced rafter beekeeping, i.e., "sunggau" in Bengkulu (Bengkulu province) and Belitung (Bangka Belitung province), "tingku" in Poso (West Sulawesi province), "tikung" in Danau Sentarum National Park (West

Kalimantan province) (Hadisoesilo and Kuntadi 2007; Koeniger et al. 2010), "sapangan" in Peleng (Central Sulawesi province), and bowing tree trunks and branches in a variety of locations (recent unpublished data).

Colonies of *A. dorsata* need healthy ecosystems in which to reproduce, produce healthy workers, and, at times, multiply colonies (Kahono et al. 1999; Koeniger et al. 2010), and they prefer environments rich in floral species and of high abundance. As a migratory species, the bees usually move from one ecosystem to another and tend to revisit sites where they have previously nested (Kahono et al. 1999); however, if their original nesting trees are cut down, some colonies use neighboring trees, if still present. With the increase of deforestation and the large-scale conversion of land to monoculture plantations, associated with an increase in forest fires, colonies are increasingly forced to nest in new locations, often within these monoculture plantations such as "akasia" (*Acacia* spp.), oil palm (*Elaeis guineensis*), rubber (*Hevea brasiliensis*), coconut (*Cocos nucifera*), "mahoni" (*Swietenia mahagoni*), and "meranti" (*Shorea* spp.); in public gardens and around human settlements, for example, the buildings of local houses, the campuses of the Universitas Indonesia (Jakarta), Pattimura University (Ambon, Moluccas), and even office buildings such as banks (Bank Mandiri, Bank Rakyat Indonesia, and Bank Negara Indonesia in Padang, West Sumatra) (Figure 20.3). Even if an old nesting site remains, it is often abandoned as the trees become weakened.

After forest fires in Kalimantan and Sumatra, there is ample evidence that—not surprisingly—there is an increase in new nesting sites for colonies in temporary residence. These data indicate that forest fires are one of the triggers that disrupt the

FIGURE 20.3 Colonies of *A. d. dorsata*. A: nests in an oil palm tree (*Elaeis guineensis*) at Riau, Sumatra. B: nests on a local house.

migratory patterns of *A. dorsata*. The relocation of such colonies to new sites, owing to destruction of their original nesting site, often results in colonies placing themselves in closer proximity to human settlements, and these can instill fear of stings in residents, who at times burn and forcibly expel the bees. Indeed, against growing bee colonies, some people provoke attacks by predatory raptors, and the bees' defense of raptor attacks leads to much misdirection and ill-advised activities in local human populations (Kahono et al. 2020).

Interestingly, *A. dorsata* is the only honeybee that consistently maintains some nocturnal activity, including foraging after dusk (Dyer 1985). Recent data confirms that about 10%–20% of workers captured at light traps were carrying pollen in their metatibial corbiculae (recent unpublished data). Additionally, local honey hunters attest to the night-foraging activity based on hunters who stayed overnight in the forests under large flowering trees and heard the sound of masses of workers visiting the flowers (unpublished data). *Apis dorsata* has comparatively enlarged ocelli relative to other species of *Apis*, allowing it to orient in dim light (Somanathan et al. 2009). This presumably allows the bees to send larger numbers of workers to forage without diminishing the collective defense of the hive.

20.4 ROLE IN NATURAL AND AGRICULTURAL ECOSYSTEMS

Apis dorsata plays a vital role in the local food chain, for wildlife as well as for humans. The bees are a source of food for a large array of predatory animals, e.g., birds (bee-eater *Merops* spp., and "srigunting" *Dicrurus macrocercus*), spiders (crab spiders that mimic flowers, the golden silkweed spider, *Nephila* spp.) (Koeniger et al. 2010; recent unpublished data), lowland hornets (*Vespa analis*, *V. affinis*, and *V. tropica*), a highland hornet (*V. velutina*) (recent unpublished data), praying mantises (*Hierodula* spp.), asilid robber flies, weaver ants (*Oecophylla smaragdina*), and other diverse reptiles, amphibians, birds, and small mammals (Koeninger et al. 2010; recent unpublished data). These predators can tell whether the bees' comb contains brood (larvae, pupae), honey, pollen, and/or wax. The comb is a nutritious target of animal predators, including protected endemic mammals such as the honey bear (*H. malayanus*) and orangutans (*P. pygmaeus*), as well as many others (Koeniger et al. 2010; recent unpublished data). An important flying predator of the giant honeybee is a migrant raptor, the oriental honey-buzzard (*Pernis ptilorhynchus orientalis*) (Oldroyd and Wongisiri 2006; Koeniger et al. 2010; Kahono et al. 2020). These raptors migrate annually from Siberia and Japan to Indonesia (Ash 1993; Nijman 2004), and this migration coincides with the activity of *A. dorsata* (Kahono et al. 2020).

Indonesia has diverse natural ecosystems (Kartawinata 2013), with a large number of flowering plants: 6,888 species in Java, 5,888 species in Kalimantan, 5,296 species in Sumatra, 4,296 species in Papua, 3,688 species in Sulawesi, 2,421 species in the Moluccas, and 1,680 species in the Lesser Sunda Islands (Retnowati et al. 2019). As an agricultural country, many flowering plants grow in various agricultural ecosystems (Sumarno et al. 2014). These floral species, of course, encompass a staggering diversity of floral's biology and morphology (Backer and Brink 1963; van Steenis 1978; Hatta and Darnaedi 2005), each with their unique pollination requirements. Bees are among the most efficient insect pollinators due to their diversity in species, behavior, and morphology, and their interdependency on flowers (Faegri and van der Pijl 1971; Barth 1991; Rasmussen et al. 2020). Mass-flowering patterns in Indonesia are influenced by rainfall patterns (Bawa 1990; Inoue and Nakamura 1990; Whitmore 1990; Nakamura et al. 1994). Pollination success during mass flowerings in tropical forests requires the migratory giant honeybee *A. dorsata*, as well as many other diverse local pollinators. The occurrence of colonies of *A. dorsata* in any region depends on seasonal flowering sequence and available floral diversity. Therefore, plant diversity must be maintained in each region in order to sustain populations of *A. dorsata* and promote the concomitant success of these same regional plant species (Raju and Rao 2004).

As noted previously, *A. dorsata* undertake long-distance migration and also have comparatively large foraging areas, extending quite some distance from their nesting sites. During the life of a colony at a specific site, workers will visit agricultural areas, wild forests, and other unmanaged ecosystems surrounding their nesting location (Kahono et al. 1999; Raffiudin et al. 2018; JMHI and Sumbawa Forest Honey Association [JMHS], personal communication; recent unpublished data). The importance of the bees' role in pollination has been established from studies on their patterns of occurrence and visitation to flowers in natural and agricultural ecosystems with diverse floral assemblages (Kahono et al. 1999; Kahono 2000, 2001, 2002, 2018).

The migration of *A. dorsata* in many areas shows almost identical patterns, with bees arriving at the end of the rainy season or the beginning of the dry season, and leaving at the end of the dry season or the beginning of the rainy season the following year (Kahono et al. 1999; JMHI, personal communication; recent unpublished data). The bees annually visit almost all protected forests and many other green environments such as Betung Kerihun National Park, Danau Sentarum National Park, Ujung Kulon National Park, Meratus Mountains (South Kalimantan), Tesso Nilo National Park (Kampar, Riau), Meru Betiri National Park and Alas Purwo National Park (East Java), and the forests of Mutis Mountains, Adonara and Lembata Islands (Nusa Tenggara Timur), mountains in East Flores, East Kolaka (Southeast Sulawesi), North Luwu (South Sulawesi), and Batu Dulang Sumbawa Nusa Tenggara Barat. *Apis dorsata* has been recorded pollinating many mass-flowering plants in monocultural plantation forests such as "mahoni" (*Swietenia mahagoni*), "akasia" (*Acacia mangium*), "kayu putih" (*Eucalyptus globulus*); wild flowers of "putri malu" (*Mimosa* spp.), "kersen" (*Muntingia calabura*); agricultural plants and crops such as corn (*Zea mays*), *Coffea* spp., melon (*Cucumis melo*), strawberry (*Fragaria daltoniana*), watermelon (*Citrullus* spp.), coconut (*Cocos nucifera*), guava (*Psidium guajava*), "jambu air" (*Syzygium aqueum*), "jambu biji" (*Syzygium* spp.), apple (*Malus sylvestris*), mango (*Mangifera indica*), "kweni" (*Mangifera odorata*), mangosteen (*Garcinia mangostana*), orange (*Citrus* spp.), and rambutan (*Nephelium lappaceum*). Additionally, along with bats, *A. dorsata* provides nocturnal pollination for durian (*Durio zibethinus*), dragonfruit (*Hylocereus undatus*), and "petai" (*Parkia speciosa*).

Apis dorsata supplies much of Indonesia's national honey production (Hadisoesilo and Kuntadi 2007; Muslim 2014), and honey from *A. dorsata* is the most important product of both high nutritional and medicinal value (Moniruzzaman et al. 2013; Pribadi and Wiratmoko 2019). In the lowland forests of Ujung Kulon National Park (Banten), the color and taste of the honey and bee pollen varies at the beginning, middle, and end of the harvesting season, which roughly reflects the different species of flowers visited and the flowering phenology in the region (recent unpublished data). The bees produce high quality natural honey used to supplement drinks and medicines, and the wax is the best material for batik printing and other purposes. These products are a source of livelihood for the local people (JMHI, JMHS, personal communication; recent unpublished data). The period of honey harvesting is usually in the dry season, from June to October, and the amount of honey produced correlates with the amount of annual rainfall. Strong drought (e.g., La Niña) causes the number of flowers to decrease so that the honey yield decreases (Hadisoesilo and Kahono 2011, unpublished data; Kahono 2011).

Recently, the number of bee nesting trees has been gradually reduced, and some of them have been lost entirely due to deforestation and fires. In the provinces of Riau and Jambi (Sumatra), there has been a decline in forest honey production due to changes in bee habitat, diminishing number of "sialang" trees, and occurrence of forest fires. The number of colonies in each nesting site is gradually on the decline, a recent trend that is of great concern. Forest fires and deforestation are the main reason for the loss of nesting trees, and these are both exacerbated by global climate change. There are colonies nesting in new sites which may remain for an unusually long period of time after a fire occurs but under normal circumstance would consist of only a shorter period. Many local people are against colonies persisting for extended periods around areas of habitation, leading to the destruction of these colonies—a double tragedy for the bees whose original nesting sites more distant from humans were destroyed and then, when they are forced in a situation closer to human settlements, they are aggressively attacked and often killed. Thus, these critical pollinators are being driven down in their abundance from multiple angles. With the diminishing numbers of bees, in few natural ecosystems that are not slashed and burned through human action, humans will find pollinators to be a limiting factor to their sustainability.

Pollination is vital to natural and agricultural ecosystems (Heithaus 1974; IPBES 2017), and in Indonesia, pollination is utterly dependent on our diverse wild pollinators. Continued research into the role of *A. dorsata* in the pollination of wild and cultivated plants is critically needed, particularly with rising concerns about global pollinator declines (Kovács-Hostyánszki et al. 2018). With the growing loss of nesting sites for *A. dorsata* and the associated decline in the number of colonies in almost all parts of Indonesia, this issue must be addressed properly and as soon as possible—before declines reach a critical point of no return. Fortunately, some minor steps are being made. There is a government regulation on forest conservation that automatically protects nesting trees inside the forest. Additionally, some indigenous peoples, such as the Dayak tribes in Kalimantan and those in Riau Province (Sumatra), promote traditional wisdom for protecting nesting trees and "sialang" outside of protected forests. Recently, the Indonesian government, JMHI, JMHS, and other non-governmental organizations (NGOs) have raised concerns about how to harvest and obtain bee products safely, hygienically, and sustainably. These are certainly positive actions that will help to promote honeybees across Indonesia. Nonetheless, much more is needed to save these and other critical pollinators, particularly in ceasing the relentless burning of natural forests and the vital biodiversity they harbor. Our very lives are dependent on these forests, as they protect us from emerging diseases by providing natural buffers between human populations and wildlife harboring pathogens; they provide food, medicines, and natural products; they produce the very oxygen we breathe; and they are restorative to the body, mind, and soul (e.g., "forest bathing" and "healing forests"). Our security is dependent on diverse forests and their myriad pollinators, of which *A. dorsata* is one conspicuous and important member.

REFERENCES

Alexander, B.A. 1991. A cladistic analysis of the genus *Apis*. In: Smith, D.R. (Ed.), *Diversity in the genus Apis*. Westview Press, Boulder, 1–28.

Ash, J.S. 1993. Raptor migration on Bali, Indonesia. *Forktail* 9: 3–11.

Backer, C.A., and Brink Jr, R.C.B.V.D. 1963. *Flora of java*. Vol. 1 (Spermatophytes only). NVP Noordhof, Groningen, The Netherlands.

Barth, F.G. 1991. *Insects and flowers: The biology of a partnership*. Translated by Biederman-Thorson, M.A. Princeton University Press, Princeton, NJ, 408.

Batra, S.W.T. 1977. Bees of India (Apoidea) their behaviour, management and a key to the genera. *Oriental Insects* 11: 289–324.

Bawa, K.S. 1990. Plant-pollinator interactions in tropical rain forests. *Annual Review of Ecology and Systematics* 21: 399–422.

Chuttong, B., Somana, W., and Burgett, M. 2019. Giant honey bee (*Apis dorsata* F.) rafter beekeeping in Southern Thailand. *Bee World* 96: 66–68.

Dingle, H. 1980. Ecology and evolution of migration. In: Gautheaux Jr, S.A. (Ed.), *Animal migration, orientation, and navigation*. Academic Press, New York, 1–101.

Dyer, F.C. 1985. Nocturnal orientation by the Asian honey bee, *Apis dorsata*. *Anim. Behav.* 33: 769–774.

Dyer, F.C., and Seeley, T.D. 1991. Nesting behavior and the evolution of worker tempo in four honey bee species. *Ecology* 72: 156–170.

Dyer, F.C., and Seeley, T.D. 1994. Colony migration in the tropical honey bee *Apis dorsata* F (Hymenoptera: Apidae). *Insectes Soc.* 41: 129–140.

Engel, M.S. 1999. The taxonomy of recent and fossil honey bees (Hymenoptera: Apidae; *Apis*). *Journal of Hymenoptera Research* 8: 165–196.

Engel, M.S. 2006. A giant honey bee from the middle Miocene of Japan (Hymenoptera: Apidae). *American Museum Novitates* 3504: 1–12.

Engel, M.S. 2012. The honey bees of Indonesia (Hymenoptera: Apidae). *Treubia* 39: 41–49.

Engel, M.S., and Rasmussen, C. 2021. Corbiculate bees. In: Starr, C.K. (Ed.), *Encyclopedia of social insects*. Springer, Cham, 302–310.

Engel, M.S., and Schultz, T.R. 1997. Phylogeny and behavior in honey bees (Hymenoptera: Apidae). *Annals of the Entomological Society of America* 90: 43–53.

Faegri, K., and van der Pijl, L. 1971. *The principles of pollination ecology*. Pergamon Press, Oxford, New York, Toronto, Sydney and Braunschweig, 291.

Hadisoesilo, S. 2001. Review: Diversity of Indonesian honey bees. *Biodiversitas* 2(1): 123–128 (translated).

Hadisoesilo, S., and Kuntadi, S. 2007. *Traditional wisdom in aquaculture forest bee (Apisdorsata)*. Department of Forestry Research and Development. Agency of Forestry, Research Centre and Development for Forest and Nature Conservation. Bogor Ministry of Forestry of Republic of Indonesia Press, 40 (translated).

Hatta, H., and Darnaedi, D. (Eds.). 2005. *Phenology and growth habits of tropical trees*. National Science Museum Monographs No. 30. National Science Museum, Tokyo, 441.

Heithaus, E.R. 1974. The role of plant-pollinator interactions in determining community structure. *Annals of the Missouri Botanical Garden* 61(3): 675–691.

Hepburn, H.R., and Radloff, S.E. 2011. Biogeography. In: Hepburn, H.R., and Radloff, S.E. (Eds.), *Honeybees of Asia*. Springer, Berlin, 51–67.

Hill, R. 2009. Southeast Asia's changing palaeogeography. *Blumea* 54: 148–161.

Inoue, T., and Nakamura, K. 1990. Physical and biological background for insect studies in Sumatra. In: Sakagami, S.F., Ohgushi, R., and Roubik, D.W. (Eds.), *Natural history of social wasps and bees in equatorial sumatra*. Hokkaido University Press, Sapporo, 1–12.

IPBES. 2017. The assessment report on pollinators, pollination and food production. Summary for policy makers. In: *Annex to document IPBES/4/INF/1/Rev.1 thematic assessment on pollinators, pollination and food production (deliverable 3 (a)): Individual chapters and their executive summaries the intergovernmental science-policy platform on biodiversity and ecosystem services (IPBES) expert on pollinators and pollinations*. UNESCO, Geneva, 11–51.

Kahono, S. 2000. *Bees and wasps as pollinators at tropical rain forest of Gunung Halimun National Park and their distributions in Indonesia*. National seminar on utilization of natural bioresources and its management. Faculty of Biology-Christian University of Satyawacana, Salatiga (translated).

Kahono, S. 2001. Roles and problems of insect pollinators in Indonesia. *Fauna Indonesia* 5(2): 9–16 (translated).

Kahono, S. 2002. *Distribution of nesting areas of wild honeybee Apis dorsata F. (Hymenoptera: Apidae) in the wet mountain forest of Halimun National Park and surrounding areas*. Technical Report of Research Center for Biology, Indonesian Institute for Science (LIPI) Press (translation), Jakarta.

Kahono, S. 2011. *Effect of extreme wet climate to the number of immigrant colonies of the giant honeybee Apis dorsata F.* International Conference on Conservation and Management of Pollinators, Kuching, Sarawak, Malaysia, September 14–17.

Kahono, S. 2018. *Diversity and the potency of Indonesian native bees*. 14th Conference of Asian Apiculture Association, Jakarta, Indonesia, October 22–25.

Kahono, S., Chantawannakul, P., and Engel, M.S. 2018. Social bees and current status of beekeeping in Indonesia. In: Chantawannakul, P., Williams, G., and Neumann, P. (Eds.), *Asian beekeeping in the 21th century*. Springer, Cham, 287–306.

Kahono, S., Nakamura, K., and Amir, M. 1999. Seasonal migration and colony behavior of the tropical honeybee *Apis dorsata* F (Hymenoptera: Apidae). *Treubia* 31(3): 283–297.

Kahono, S., Peggie, D., and Sulistyadi, E. 2021. Diversity of the closed-nested honey bees (Apidae: *Apis* spp.) and the traditional honey collecting and beekeeping in four islands of Indonesia. *Treubia* 48(2): 141–152.

Kahono, S., Prawiradilaga, D.M., Erniwati, D.P., and Sulistyadi, E. 2020. First report on hunting behavior of migratory oriental honey-buzzard (*Pernis ptilorhynchus orientalis*) toward migratory giant honeybee (*Apis dorsata dorsata*) (Hymenoptera: Apidae) on Java Island, Indonesia. *Treubia* 47(2): 123–132.

Kartawinata, K. 2013. *Diversity of Indonesian natural ecosystems*. Yayasan Pustaka Obor Indonesia, Jakarta, 124 (translated).

Koeniger, N., and Koeniger, G. 1980. Observations and experiments on migration and dance communication of *Apis dorsata* in Sri Lanka. *J. Apic. Res.* 19: 21–34.

Koeniger, N., Koeniger, G., and Smith, D.R. 2011. Phylogeny of the genus *Apis*. In: Hepburn, H.R., and Radloff, S.E. (Eds.), *Honeybees of Asia*. Springer, Berlin, 23–50.

Koeniger, N., Koeniger, G., and Tingek, S. 2010. *Honeybees of Borneo: Exploring the centre of apis diversity*. Natural History Publications (Borneo), Kota Kinabalu, 262.

Kovács-Hostyánszki, A., Li, J., Pettis, J., Settele, J. et al. 2018. Chapter 2. Drivers of change of pollinators, pollination networks and pollination. In: *Annex to document IPBES/4/INF/1/Rev.1 thematic assessment on pollinators, pollination and food production (deliverable 3 (a)): Individual chapters and their executive summaries. The intergovernmental science-policy platform on biodiversity and ecosystem services (IPBES) expert on pollinators and pollinations.* UNESCO, Geneva, 44–242.

Lam, H.J. 1945. Contribution of our knowledge of the flora of Celebes (collections of C. Monod de Froidville) and of some other Malaysian islands. *Blumea* 5(3): 554–599.

Maa, T.C. 1953. An inquiry into the systematics of the tribus Apidini or honeybees (Hymenoptera). *Treubia* 21: 525–640.

Mashudi, K.P., and Suwanda, O. 1988. *Honey bees, bees honey in Indonesia year 2000.* PAP Press, Jakarta, 268 (translated).

Michener, C.D. 1974. *The social behavior of the bees: A comparative study.* Harvard University Press, Cambridge.

Michener, C.D. 2007. *The bees of the world.* 2nd edn. Johns Hopkins University Press, Baltimore.

Moniruzzaman, M., Khalil, M.I., Sulaiman, S.A., and Gan, S.H. 2013. Physicochemical and antioxidant properties of Malaysian honeys produced by *Apis cerana*, *Apis dorsata*, and *Apis mellifera*. *BMC Complementary and Alternative Medicine* 13: 43.

Muslim, T. 2014. *Potential of forest honey and its management in Indonesia.* Proceeding of the Seminar of Balitek KSDA, Balikpapan, 67–82 (translated).

Nagir, M.T., Atmowidi, T., and Kahono, S. 2016. The distribution and nest-site preference of *Apis dorsata binghami* at Maros Forest, South Sulawesi, Indonesia. *Journal of Insect Biodiversity* 4(23): 1–14.

Nakamura, K., Noerdjito, W.A., and Hasyim, A. 1994. Regional difference and seasonality of rainfall in Java, with special reference to Bogor. *Tropics* 4(1): 93–103.

Neupane, K.R., Woyke, J., and Poudel, S.M. 2013. *Nesting site preference and behavior of giant honey bee Apis dorsata.* Paper presented on Apimondia, Kyiv, Ukraine, September 29–October 4.

Nijman, V. 2004. Magnitude and timing of migrant raptors in Central Java, Indonesia. *Ardea* 92: 161–168.

Oldroyd, B.P., and Wongisiri, S. 2006. *Asian honey bees: Biology conservation and human interaction.* Harvard University Press, Cambridge, MA and London, 204–207.

Otis, G.W. 1991. A review of the diversity of species within Apis. In: Smith, R.E.D. (Ed.), *Diversity in the Genus of Apis.* Westview Press, Boulder, San Francisco and Oxford, 29–50.

Otis, G.W. 1996. Distributions of recently recognized species of honey bees (Hymenoptera: Apidae: Apis) in Asia. *Journal of the Kansas Entomological Society* 69(4), Supplement: 311–333.

Pribadi, A., and Wiratmoko, M.E. 2019. Characteristics of forest bee honey (Apis dorsata Fabr.) From various bioregions in Riau. *Journal of Forest Product Research* 37(3): 185–200 (translated).

Radloff, S.E., Hepburn, H.R., and Engel, M.S. 2011. The Asian species of *Apis*. In: Hepburn, H.R., and Radloff, S.E. (Eds.), *Honeybees of Asia*. Springer, Berlin, 1–22.

Raffiudin, R., and Crozier, R.H. 2007. Phylogenetic analysis of honey bee behavioral evolution. *Molecular Phylogenetics and Evolution* 43: 543–552.

Raffiudin, R., Dyahastuti, M., Nugraha, R., Djuita, N.R., Suwananda, E., Alluioningruni, V., Mardhony, R., Biagioni, S., Saad, A., and Behling, H. 2018. *The effect of land-use change on the foraging behavior and pollen diversity in honey of the Giant Bee Apis dorsata in Sumatra, Indonesia.* 14th Asian Apiculture Association Conference, Jakarta, Indonesia, October 22–25.

Raffiudin, R., and Shullia, N.I. 2020. Phylogenies of Asian honey bees. In: Ilyasov, R.A., and Kwon, H.W. (Eds.), *Phylogenetics of bees*. CRC Press, Boca Raton, 28–57.

Raju, A.J.S., and Rao, S.P. 2004. The rock honey bee, *Apis dorsata* and its pollination potential in some plants. *Journal for Nature Conservation* 16(1): 59–65.

Rasmussen, C., Engel, M.S., and Vereecken, N.J. 2020. A primer of host-plant specialization in bees. *Emerging Topics in Life Sciences* 4: 7–17.

Rasmussen, C., Thomas, J.C., and Engel, M.S. 2017. A new genus of Eastern Hemisphere stingless bees (Hymenoptera: Apidae), with a key to the supraspecific groups of Indomalayan and Australasian Meliponini. *American Museum Novitates* 3888: 1–33.

Retnowati, A., Rugayah, J.S.R., and Arifiani, D. 2019. *Status of Indonesia's biodiversity: The wealth of Indonesian plants and fungi*. LIPI Press, Jakarta, 139 (translated).

Ruttner, F. 1988. *Biogeography and taxonomy of honeybees*. Springer-Verlag, Berlin, 284.

Seeley, T.D., Seeley, R.H., and Akratanakul, P. 1982. Colony defense strategies of the honeybees in Thailand. *Ecological Monographs* 52(1): 43–63.

Sihombing, D. 1997. *The honey bee science*. Gadjah Mada Press, Yogyakarta (translated).

Silva, D.P., Castro, A.C.F., Vilela, B., Onr, X.R., Thomas, J.C., Alqarni, A.S., Engel, M.S., and Ascher, J.S. 2020. Colonizing the East and the West: Distribution and niche properties of a dwarf honey bee invading Africa, the Middle East, the Malay Peninsula, and Taiwan. *Apidologie* 51: 75–87.

Smith, D.R. 1991. Mitochondrial DNA and honey bee biogeography. In: Smith, D.R. (Ed.), *Diversity in the genus Apis*. Westview Press, Boulder, 131–176.

Smith, D.R. 2021. Biogeography of honey bees. In: Starr, C.K. (Ed.), *Encyclopedia of social insects*. Spinger, Cham, 117–131.

Smith, D.R., and Hagen, R.H. 1996. The biogeography of *Apis cerana* as revealed by mitochondrial sequence data. *J. Kans. Entomol. Soc.* 69: 294–310.

Somanathan, H., Warrant, E.J., Borges, R.M., Wallén, R., and Kelber, A. 2009. Resolution and sensitivity of the eyes of the Asian honeybees *Apis florea*, *Apis cerana* and *Apis dorsata*. *The Journal of Experimental Biology* 212: 2448–2453.

Sumarno, H., Mustika, I., and Bahagiawati (Eds.). 2014. *Indonesian agricultural genetic resources: Food crops—horticultural plantations*. Center for Agricultural Research and Development, Ministry of Agriculture of the Republic of Indonesia, IAARD Press, Bogor, 496 (translated).

Tingek, S., Koeniger, G., and Koeniger, N. 1996. Description of a new cavity dwelling species of Apis (*Apis nuluensis*) from Sabah, Borneo with note on its occurrence and reproductive biology (Hymenoptera, Apoidea, Apini). *Senckenbergiana Biologica* 76: 115–119.

van Steenis, C.G.G.J. (Ed.). 1978. *Flora Malesiana*. Vol. 8. Noordhoff-Kolff N.V., Djakarta, 1–577.

Waring, C., and Jump, D.R. 2015. Rafter beekeeping in Cambodia with *Apis dorsata*. *Bee World* 85(1): 14–18.

Whitmore, T.C. 1990. *An introduction to tropical rain forest*. Oxford University Press, New York, 226.

Winston, M.L. 1991. *The biology of the honey bee*. Revised edn. Harvard University Press, Cambridge.

21 Distribution, Nesting Biology, and Floral Preference of Giant Honeybee (*Apis dorsata* Fabricius) in Southern West Bengal, India

Ujjwal Layek, Nandita Das, Rajib Mondal and Prakash Karmakar

21.1 INTRODUCTION

Honeybees are highly eusocial insects that belong to the genus *Apis* of the family Apidae under the order Hymenoptera. They are beneficial to humans as they provide bee products (honey, pollen, wax, and bee-venom), and they are also involved in increasing crop yield through pollination (Rader et al. 2016; Sáez et al. 2020). According to the U.S. Department of Agriculture (USDA), the worth of a honeybee colony is 100 times more to the community than the beekeeper, which means the value they deliver extends well beyond their actual price (McGregor 1976). In India, honey-producing bees are *Apis cerana* Fabricius (Indian hive bee), *Apis dorsata* Fabricius (rock bee), *Apis florea* Fabricius (red dwarf bee), *Apis laboriosa* Smith (Himalayan giant honeybee), *Apis mellifera* Linnaeus (European or Italian bee), and *Tetragonula iridipennis* Smith (Indian stingless bee or dammar bee). All these bee species differ from each other morphometrically, as well as in their nesting biology, activity patterns, floral preferences, and effectiveness as pollinators (Corlett 2011). According to phylogeny, these different species form separate monophyletic groups, with the small species in basal and honeybees being the sister of the stingless bees (Arias and Sheppard 2005). Among the true honeybees, dwarf honeybees are the most ancestral in the genus *Apis*, as is evident from the morphological and molecular data (Arias and Sheppard 2005).

Apis dorsata is also called Asian giant honeybees due to its large body size (approximately 17 mm in length). Giant honeybee colonies exhibit a migratory lifestyle, with several colony movements during an annual cycle (Koeniger and Koeniger 1980; Oldroyd et al. 2000). Relocation of each colony requires the construction of a

DOI: 10.1201/9781003294078-21

new comb, which represents one of the significant investments of energy and wax material because of the large size of individual combs. *Apis dorsata* is widely distributed in tropical and subtropical Asia, and it is an important source of honey and wax for local communities. Giant honeybees pollinate both wild and cultivated plants. Therefore, understanding its distribution, nesting biology, and floral resources will help manage and conserve the bee species.

This chapter contains detailed information about the distribution, nesting biology, and floral resources of giant honeybees (*Apis dorsata*) in southern West Bengal, India.

21.2 DISTRIBUTION OF *APIS DORSATA*

The Asian giant honeybee (*Apis dorsata*) extends from the Indian subcontinent to Southeast Asia, including the Philippines (Ruttner 1988; Hepburn and Radloff 2011). The more significant numbers of *A. dorsata* colonies are found in China, India, Indonesia, Pakistan, and Sri Lanka. The species migrates seasonally between summer and winter nesting sites, and queens appear to return to their previously occupied site (Paar et al. 2004). Only a few colonies remain within the same place throughout the year. There is a strong tendency for colonies of this species to aggregate at perennially utilized nesting sites that may be shared with several colonies (Figure 21.1). Their abundance significantly varied among the seasons in southern West Bengal. More wild colonies were recorded during the winter, spring, and summer seasons (Table 21.1). After that, they migrate to other regions due to unfavorable weather conditions and a shortage of floral resources. Therefore, the number of colonies drastically reduced in monsoon and autumn. Colony density was relatively lower in agricultural field areas than in forest areas and human settlement areas. Unlike red dwarf honeybees (Layek and Karmakar 2020), the number of colonies does not significantly differ in between forest and human settlement areas.

21.3 NESTING BIOLOGY OF *APIS DORSATA*

Nest site selection is critical for wild honeybee species because poor choice can heighten predation risks, be placed in an adverse environment, and result in reproductive failure (Franks et al. 2002). The giant honeybee (*Apis dorsata)* often nests in single or in aggregations, and it is located in sites inaccessible to most predators (Seeley et al. 1982). The bee species utilize tall trees, shrubs, building structures, and water tanks as nesting substrates in West Bengal (Figures 21.1–21.2). The most common nesting substrates are limbs of trees and building structures. As nesting substrate, frequently used trees were *Ficus benghalensis* L., *Ficus religiosa* L., *Mangifera indica* L., and *Tamarindus indica* L. (Table 21.2). The diameter of the limbs ranged from 4.1–14.8 cm, with an average value of 8.11 ± 2.61 cm. Most of the nesting trees have a rough bark surface texture. Nests are in the open that hang freely from the substratum. The nests of *Apis dorsata* were found to vary in height from ground level. The majority of the nests were at an altitude of >10–15 m (Figure 21.3), which is comparatively higher than the elevation of the red dwarf honeybee (Layek and Karmakar 2020), though a few nests are much closer to ground level when

FIGURE 21.1 Aggregated colonies of *Apis dorsata*; a. tree (*Ficus benghalensis*); b. water tank.

TABLE 21.1
Number of *Apis dorsata* Colonies per km² Area in Southern West Bengal

Season	Human Settlement Area	Forest Area	Agricultural Field Area
Summer	2.2 ± 1.14	2.0 ± 1.15	0.6 ± 0.70
Monsoon	0.7 ± 0.82	0.3 ± 0.48	0
Autumn	0.6 ± 0.70	0.2 ± 0.42	0
Late autumn	0.9 ± 0.88	1.2 ± 1.03	0
Winter	1.9 ± 0.99	2.3 ± 0.67	0.9 ± 0.99
Spring	2.3 ± 1.16	2.4 ± 1.26	0.8 ± 0.79
Mean	$1.43^{a} \pm 1.17$	$1.40^{a} \pm 1.25$	$0.38^{b} \pm 0.69$

Notes: Values given as mean ± standard deviation; means in the row followed by same superscript letters do not differ significantly (Duncan's multiple range test [DMRT] at 5%)

FIGURE 21.2 Nesting substrates of *Apis dorsata*: a. house structure; b–f. live plant limbs (a. *Azadirachta indica*; b. *Butea monosperma*; c. *Ficus religiosa*; d. *Senna siamea*; e. *Spondias pinnata*).

found in the bushy jungle having thick canopy and covered by certain liana species (Figure 21.2c). The attachment region of a hive with a substratum is not always in a straight line. Depending on the substrate, it may be curved (Figure 21.2f). To represent the nest orientation of the rock bee, we considered the honey storage end of the colonies, which their swollen nature can identify from the other end. In an absconded hive, remnant honey storage cells are much whiter than the other cells of the comb. The colonies found on buildings oriented their nests towards northeast (22.92%), east (18.75%), and southeast (16.67%) direction in more significant percentages. In contrast, colonies on trees showed nest orientation almost equally towards all directions. A similar nest orientation was also observed in Karnataka, India (Nagaraja and Yathisha 2015). Individual nests bear a single wax comb made up of hexagonal cells, with the comb structure assuming a semicircular shape under most circumstances. The length of the hives (which was measured at the attachment point with the substrates) was ranged from 42.6–1,14.3 cm, with a mean value of 74.80 ± 16.51 cm. The height of the hives at the maximum point was ranged from 36.2–52.8 cm, with a mean value of 44.02 ± 4.65 cm. The comb cells are used for food storage and to rear brood. The one end of the horizontal plane to the substratum is utilized to store

TABLE 21.2

Plant Species Utilized as Nesting Substrate by *Apis dorsata* in Southern West Bengal

Plant Species	Occurrence (%)	Limb Diameter (cm)	Bark Surface
***Acacia auriculiformis* A. Cunn. ex Benth.**	1.32	5.9	rough
***Alangium salviifolium* (L. f.) Wangerin**	1.32	5.3	deeply furrowed
***Albezia lebbeck* (L.) Benth.**	1.32	6.7	rough
***Alstonia scholaris* (L.) R. Br.**	3.95	5.2–6.1 (5.6)	rough
***Azadirechta indica* A. Juss.**	2.63	16.4–7.0 (11.7)	rough
***Bombax ceiba* L.**	2.63	6.6–7.3 (6.95)	thorny
***Bridelia retusa* (L.) A. Juss.**	2.63	6.5–6.8 (6.65)	rough
***Butea monosperma* (Lam.) Taub.**	1.32	4.1	deeply furrowed
***Diospyros Montana* Roxb.**	1.32	6.2	rough
***Ficus benghalensis* L.**	35.53	4.9–14.8 (9.09)	rough
***Ficus religiosa* L.**	11.84	6.1–11.4 (8.76)	rough
***Holoptelea integrifolia* (Roxb.) Planch.**	3.95	7.4–8.5 (8.03)	rough
***Mangifera indica* L.**	14.47	5.7–10.3 (7.41)	rough
***Neolamarckia cadamba* (Roxb.) Bosser**	1.32	7.3	rough
***Senna siamea* (Lam.) Irwin et Barneby**	2.63	9.4–10.2 (9.8)	rough
***Spondias pinnata* (L. f.) Kurz**	1.32	7.1	smooth
***Syzygium cumini* (L.) Skeels.**	2.63	6.9–7.4 (7.15)	rough
***Tamarindus indica* L.**	6.58	6.3–8.6 (7.12)	rough
***Terminalia arjuna* (Roxb.) Wight & Arn.**	1.32	7.2	smooth

Note: Mean values of limb diameter are given within the parentheses.

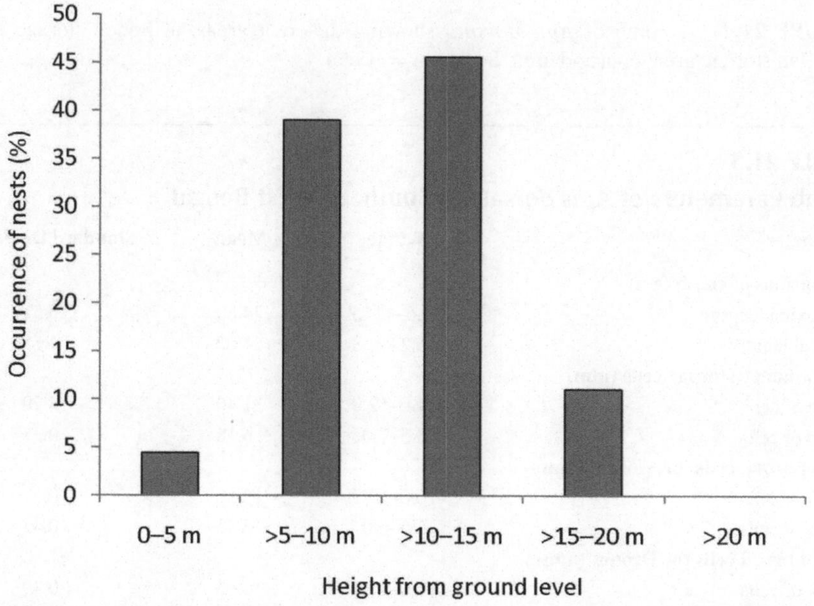

FIGURE 21.3 Height of the nests from ground level.

honey (Figure 21.4). Pollen loads are stored surrounding this zone, and the remaining part of the hive serves as brood area. On average, 81.36% of the total comb area was used for brood production. The brood cells are almost uniform in both diameter and cell depth. Unlike the red dwarf honeybee (*Apis florea*), *Apis dorsata* drones' brood cells are identical to workers' brood cells, though Tan (2007) reported a significant difference in size between the drone and worker cells. The average diameter and depth of worker brood cells are 5.45 ± 0.43 mm and 16.58 ± 0.52 mm, respectively (Table 21.3). Drone cells are 5.60 ± 0.42 mm in diameter and 16.65 ± 0.49 mm in

FIGURE 21.4 A comb of *Apis dorsata* showing different areas: a. honey storage area; b. pollen storage area; c. brood area. Scale bar = 10 cm.

TABLE 21.3
Comb Parameters of *Apis dorsata* in Southern West Bengal

Parameter	Range	Mean	Standard Deviation
Dimensions of comb (cm)			
Horizontal length	42.6–114.3	74.80	16.51
Vertical length	36.2–52.8	44.02	4.65
Size of honey storage cells (mm)			
Depth of cells	11.0–62.0	31.40	14.29
Width of cells	5.5–7.0	6.18	0.37
Size of brood cells for Workers (mm)			
Depth of cells	16.0–17.5	16.58	0.52
Width of cells	5.5–6.0	5.45	0.43
Size of brood cells for Drones (mm)			
Depth of cells	16.0–17.5	16.65	0.49
Width of cells	5.5–6.0	5.60	0.42

FIGURE 21.5 Different comb cells: a–b. honey storage cells; c. pollen storage cells; d. ruptured pollen storage cells; e. arrow showing bee bread; f. brood cells and arrow showing a queen cell. Scale bars = 10 mm.

depth. Queen cells are large, nut-like and protrude vertically from the lower edges of the nest (Figure 21.5f). The diameters of honey storage cells are almost uniform, but the depth of the cells varies significantly. Peripheral honey storage cells are smaller than the cells located at the middle zone of the honey storage area. On average, the depth and width of honey storage cells of a comb are significantly greater than the brood cells for workers and drones.

21.4 FOOD SOURCES

Apis dorsata depends on mainly floral resources (nectar and pollen) as food resources. Workers can also collect fruit sap, sugar solution, or other bioproducts like turmeric flour. Food sources may determine by direct observation of the foragers or by melissopalynological analysis of honey and pollen samples (Layek et al. 2016, 2020a, 2020c; Layek and Karmakar 2018). Microphotographs of a few pollen types obtained from palynological analysis of honey and pollen loads of *Apis dorsata* are given in Figure 21.6. To collect nectar and pollen grains, they utilize diverse plant species (Table 21.4). However, workers have vigorously foraged on only a few numbers of plants. The foraged plants (i.e., bee flora) are much similar to the bee flora of other

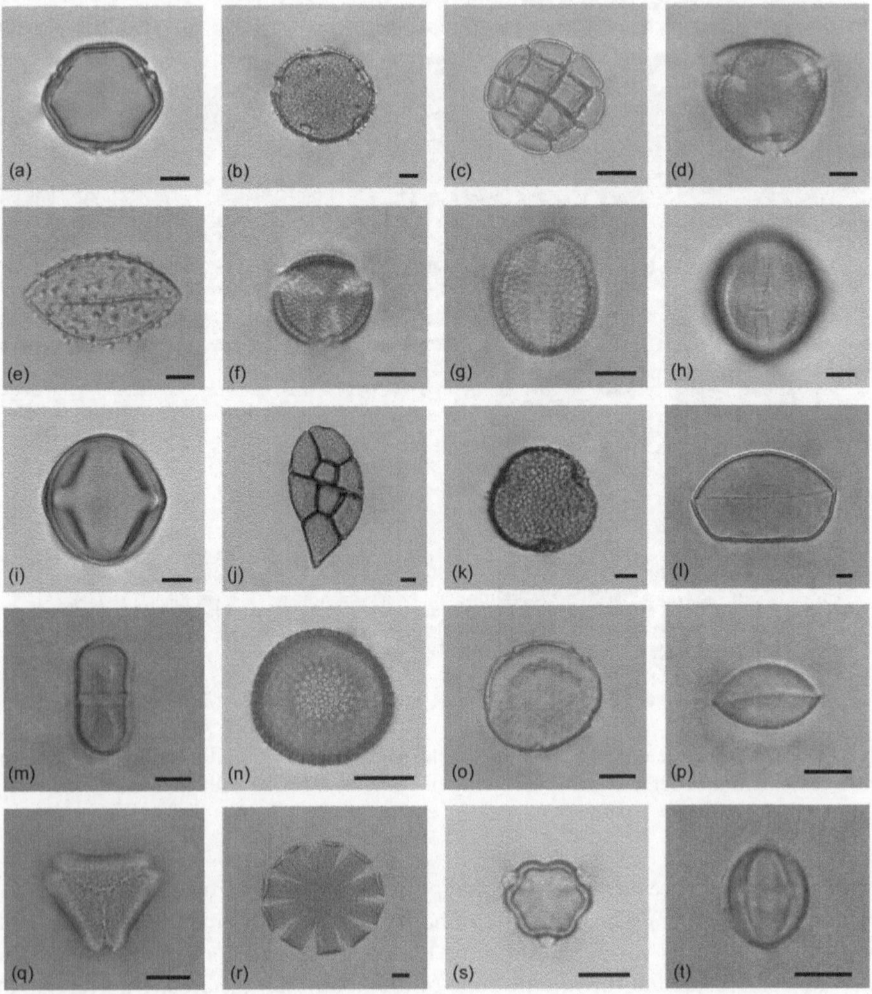

FIGURE 21.6 Microphotographs of a few pollen types obtained from honey and pollen samples of *Apis dorsata*: a. *Ailanthus excelsa*; b. *Alangium salviifolium*; c. *Albizia lebbeck*; d. *Anacardium occidentale*; e. *Borassus flabellifer*; f–g. *Brassica* type; h–i. *Butea monosperma*; j. *Calliandra umbrosa*; k. *Ceiba pentandra*; l. *Cocos nucifera*; m. *Coriandrum sativum*; n. *Croton bonplandianum*; o. *Holoptelea integrifolia*; p. *Phoenix sylvestris*; q. *Psidium guajava*; r. *Sesamum indicum*; s–t. *Terminalia arjuna*. Scale bars = 10 μm.

honeybees (Layek and Karmakar 2016; Layek et al. 2020b). Regarding the palynological analysis of honey samples, predominant pollen types were *Borassus flabellifer* type, *Brassica* type, *Bridelia retusa* type, *Butea monosperma* type, *Eucalyptus* type, *Holoptelia integrifolia* type, *Lannea coromandelica* type, and *Sesamum indicum* type. From pollen loads analysis, very frequent pollen types were *Acacia auriculiformis* type, *Alangium salviifolium* type, *Borassus flabellifer* type, *Brassica* type, *Bridelia retusa* type, *Coriandrum sativum* type, *Eucalyptus* type, *Lannea coromandelica* type, *Phoenix sylvestris* type, *Shorea robusta* type, and *Terminalia arjuna* type. Within the foraging spectrum of the bee species, highly represented families were Fabaceae (17 plant species), Asteraceae (nine plant species), Cucurbitaceae (seven plant species), Myrtaceae (seven plant species), Malvaceae (six plant species), and Rutaceae (six plant species). Among the listed plants (Table 21.4), the bee species utilized most taxa (64.06%) for both nectar and pollen sources.

The number of blooming plants varies according to the seasons in the study areas. The numbers of entomophilous blooming taxa were higher in monsoon season and summer (Table 21.5). However, the giant honeybee utilized a higher number of plant species during spring (61 plant species), summer (59 plant species) and monsoon (55 plant species). The bee species utilized 41.29% of local angiosperm flora and 51.90% of the entomophilous plant species throughout the year. Their floral host selection criteria depend on different parameters like flower availability index of a plant species (Layek et al. 2020b), quality and quantity of available floral resources, and floral attractants like floral volatile organic compounds (Layek et al. 2020a). In contrast to the number of utilized plant taxa during late autumn and winter, giant honeybee colonies stored a greater amount of honey and pollen due to the blooming of the bee-loving plants, viz. *Brassica juncea*, *Brassica nigra*, *Eucalyptus globulus*, and *Eucalyptus tereticornis*. In monsoon season and autumn, bee-loving blooming plants are scarce. Workers selected many plant species to forage during monsoon season to mitigate the colony's demand.

The rock bee–visited plants are much diversified concerning plant habit, floral characteristics, and pollen morphology. Within the bee flora, dominating plant life forms are trees (43.75%), herbs (34.38%), shrubs (13.28%), and climbers (8.59%). Most of the bee-visited plants show the synchronous type of flowering with 2–3 days longevity of individual flowers. The size, shape, and color of bee-visited flowers also varied. We classified the flowers into five size groups, based on the length of floral parts, as very small (<5 mm), small (5–15 mm), medium (>15–30 mm), large (>30–50 mm), and very large (>50 mm). With concern, 81 bee-visited plant species were with small flowers, 27 were with medium-sized flowers, seven were with large flowers, seven were with very large flowers, and six plant species were with very small flowers. The most common shapes of flowers were brush and dish. Considering flower colors as perceived by humans, yellow, cream, and white flowers were dominant within the bee flora, while brown, orange, green, blue, and lilac were less represented within the bee-visited flora of the region. When we look at the pollen presentation of the polleniferous flora (meaning that the plants supply pollen grains to the bee species), the exposed type was dominant (74.19%) over the partially hidden and well-hidden types. The morphologies of bee-collected pollen grains were also varied. Most of the bee-visited plants (contributing pollen grains) have

TABLE 21.4
List of Plant Species Foraged by *Apis dorsata* in Southern West Bengal

Family	Plant Species	Resource	Season					
			Summer	Monsoon	Autumn	Late Autumn	Winter	Spring
Acanthaceae	*Hemigraphis hirta*	N+P	+	+	–	–	–	–
	Hygrophila auriculata	N+P	+	–	–	+	+	+
	Justicia adhatoda	N	+	+	–	–	–	–
	Justicia gendarussa	N	–	–	–	–	–	+
Aizoaceae	*Trianthema portulacastrum*	N+P	+	+	+	–	–	–
Amaranthaceae	*Alternanthera paronychioides*	N	+	+	–	–	–	–
	Amaranthus spinosus	P	+	+	–	–	–	–
	Celosia argentea	N	–	–	–	–	–	+
	Spinacia oleracea	P	–	–	–	–	+	+
Amaryllidaceae	*Allium cepa*	N+P	–	–	–	–	–	+
Anacardiaceae	*Anacardium occidentale*	N	+	–	–	–	–	–
	Lannea coromandelica	N+P	–	–	–	–	–	+++
	Mangifera indica	N	–	–	–	–	++	+
	Semecarpus anacardium	N+P	+	–	–	–	–	–
	Spondias pinnata	N+P	–	–	–	–	++	++
Apiaceae	*Coriandrum sativum*	N+P	–	–	–	–	+++	+++
	Foeniculum vulgare	N+P	–	–	–	–	+	+
Apocynaceae	*Calotropis gigantea*	N	+	+	–	–	–	–
	Gymnema sylvestris	N	+	+	–	–	–	–
Arecaceae	*Borassus flabellifer*	N+P	+++	–	–	–	–	+++
	Cocos nucifera	N+P	+	++	++	+	+	+
	Phoenix sylvestris	N+P	–	–	–	–	+++	+++
Asparagaceae	*Agave amica*	P	+	+	–	–	–	–
Asteraceae	*Ageratum conyzoides*	N+P	–	+	+	–	–	–
	Chromolaena odoratum	N+P	–	–	–	–	+	+

	Chrysanthemum indicum	N+P	–	–	–	–	+	+
	Cyanthillium cinereum	N+P	+	–	–	–	–	+
	Helianthus annuus	N+P	–	–	–	–	+	+
	Mikania scandens	N+P	–	–	–	+	+	+
	Tagetes erecta	N+P	–	–	–	–	+	+
	Taraxacum officinale	N+P	+	–	–	–	–	+
	Tridax procumbens	N+P	+	+	+	+	–	–
	Xanthium strumarium	P	–	–	+	+	–	–
Brassicaceae	*Brassica juncea*	N+P	–	–	–	–	+++	+++
	Brassica nigra	N+P	–	–	–	–	+++	+++
	Raphanus sativus	N+P	–	–	–	–	+	+
Cactaceae	*Cereus pterogonus*	N+P	–	+	–	–	–	–
Cannabaceae	*Trema orientalis*	N+P	+	+	–	–	–	–
Capparaceae	*Capparis zeylanica*	P	–	–	–	–	–	+
Cleomaceae	*Cleome viscosa*	N+P	+	+	–	–	–	–
Combretaceae	*Terminalia arjuna*	N+P	+++	–	–	–	–	–
Commelinaceae	*Commelina benghalensis*	N	–	+	+	+	–	–
	Cyanotis axillaris	N+P	–	+	+	+	–	–
Convolvulaceae	*Evolvulus nummularius*	N+P	–	+	–	–	–	–
Cornaceae	*Alangium salviifolium*	N+P	–	–	–	–	–	++
Cucurbitaceae	*Citrullus lanatus*	N+P	+	–	–	–	–	+
	Coccinia grandis	N+P	+	+	–	–	–	–
	Cucumis sativus	N	–	+	–	–	–	–
	Cucurbita maxima	N	++	++	–	–	–	–
	Luffa aegyptiaca	N+P	+	+	+	–	–	–
	Momordica charantia	N+P	+	+	–	–	–	+
	Trichosanthes dioica	N	–	+	–	–	–	–
Dipterocarpaceae	*Shorea robusta*	N+P	–	–	–	–	–	+
Euphorbiaceae	*Croton bonplandianum*	N+P	+	+	+	–	–	–
	Jatropha curcas	N	+	+	–	–	–	–
	Ricinus communis	N+P	–	+	+	+	+	–

(*Continued*)

TABLE 21.4 *(Continued)*
List of Plant Species Foraged by *Apis dorsata* in Southern West Bengal

Family	Plant Species	Resource	Season					
			Summer	Monsoon	Autumn	Late Autumn	Winter	Spring
Fabaceae	*Acacia auriculiformis*	N+P	–	+	+++	+	–	–
	Albizia lebbeck	N+P	+	–	–	–	–	+
	Butea monosperma	N	–	–	–	–	–	+++
	Cassia fistula	N+P	–	–	–	–	–	+
	Calliandra umbrosa	N	–	+	+	–	–	–
	Dalbergia sissoo	N+P	+	+	–	–	–	–
	Lablab purpureus	N	–	–	–	–	+	+
	Leucaena leucocephala	N+P	+	+	+	–	–	–
	Millettia pinnata	N	++	–	–	–	–	++
	Mimosa pudica	N+P	+	+	–	–	–	–
	Peltophorum pterocarpum	N+P	++	–	–	–	–	+
	Pithecellobium dulce	N	–	–	–	–	+	+
	Senna occidentale	N	–	+	+	–	–	–
	Tamarindus indica	N	++	++	–	–	–	–
	Tephrosia purpurea	N	–	+	–	–	–	–
	Vachellia nilotica	N+P	–	+	+	–	–	–
	Vigna radiata	N	+	+	–	–	–	–
Lamiaceae	*Gmelina arborea*	N+P	+	–	–	–	–	+
	Leucas aspera	N	–	–	–	–	+	+
	Ocimum basilicum	N+P	–	+	–	–	–	–
	Vitex negundo	N	+	+	–	–	–	–
Lecythidaceae	*Barringtonia acutangula*	N+P	+	–	–	–	–	–
Lythraceae	*Lagerstroemia speciosa*	N+P	+	+	–	–	–	–
	Punica granatum	P	+	+	–	–	–	–

Malvaceae	*Bombax ceiba*	N+P	–	–	–	–	–	+
	Ceiba pentandra	N+P	–	–	–	–	–	+
	Corchorus capsularis	N+P	+	–	–	–	–	–
	Corchorus olitorius	N+P	+	–	–	–	–	–
	Grewia asiatica	N+P	–	–	–	+	+	–
	Sida cordifolia	N+P	–	–	–	+	+	–
Meliaceae	*Azadirachta indica*	N	+	–	–	–	–	+
	Melia azedarach	N	–	–	–	–	–	+
Moringaceae	*Moringa oleifera*	N	–	–	–	–	++	++
Myrtaceae	*Eucalyptus globulus*	N+P	–	–	–	+++	+++	++
	Eucalyptus tereticornis	N+P	–	–	–	+++	+++	++
	Melaleuca citrina	N	–	–	–	–	–	+
	Psidium guajava	N+P	+	–	–	–	+	+
	Syzygium cumini	N+P	–	–	–	–	–	++
	Syzygium jambos	N+P	–	–	–	–	+	+
	Syzygium reticulatum	N+P	++	–	–	–	–	–
Oxalidaceae	*Averrhoa carambola*	N	+	+	–	–	–	–
Papaveraceae	*Papaver somniferum*	P	–	–	–	–	+	+
Pedaliaceae	*Sesamum indicum*	N+P	++	–	–	–	–	–
Phyllanthaceae	*Bridelia retusa*	N+P	–	++	–	–	–	–
	Phyllanthus emblica	N+P	+	+	–	–	–	–
Plantaginaceae	*Bacopa monnieri*	N	+	+	–	–	–	–
Poaceae	*Echinochloa frumentacea*	P	–	+	–	–	–	–
	Zea mays	P	+	+	–	–	–	–
Polygonaceae	*Antigonon leptopus*	N	+	+	+	–	–	–
Ranunculaceae	*Nigella sativa*	N+P	–	–	–	–	+	+
Rhamnaceae	*Ziziphus mauritiana*	N+P	–	+	+	–	–	–
Rubiaceae	*Haldina cordifolia*	N+P	–	+	+	–	–	–
	Meyna spinosa	N	+	+	–	–	–	–
	Neolamarckia cadamba	N+P	–	+	–	–	–	–

(*Continued*)

TABLE 21.4 (Continued)
List of Plant Species Foraged by *Apis dorsata* in Southern West Bengal

			Season					
Family	**Plant Species**	**Resource**	**Summer**	**Monsoon**	**Autumn**	**Late Autumn**	**Winter**	**Spring**
Rutaceae	*Aegle marmelos*	N	+	–	–	–	–	–
	Citrus × aurantiifolia	N+P	+	++	–	–	–	–
	Citrus × limon	N+P	–	–	–	–	+	+
	Citrus maxima	N+P	–	–	–	–	–	+
	Murraya koenigii	N	–	–	–	–	–	+
	Murraya paniculata	N+P	+	+	–	–	–	+
Salicaceae	*Flacourtia jangomas*	N+P	–	–	–	–	–	+
Sapindaceae	*Litchi chinensis*	N+P	–	–	–	–	–	+
	Sapindus mukorossi	N+P	–	–	–	–	+	–
	Schleichera oleosa	N+P	+	–	–	–	–	+
Sapotaceae	*Madhuca longifolia*	N+P	–	–	–	–	–	++
	Mimusops elengi	N+P	+	+	–	–	–	–
Simaroubaceae	*Ailanthus excelsa*	N+P	–	–	–	–	+	+
Solanaceae	*Solanum lycopersicum*	N	–	–	–	–	+	+
	Solanum melongena	P	+	–	–	–	–	–
	Solanum sisymbriifolium	P	+	–	–	–	–	–
Verbenaceae	*Lippia alba*	N+P	+	+	–	–	–	–
Ulmaceae	*Holoptelea integrifolia*	N+P	–	–	–	–	–	++

Notes: N = nectar, P = pollen; frequency: high (+++), medium (++) and low (+)

TABLE 21.5
Utilization of Local Angiosperm Flora as Floral Resources (Nectar and Pollen Grains) by *Apis dorsata* in Southern West Bengal, India

	Blooming Angiosperm Flora				
Season	**Anemophilous**	**Entomophilous**	**Amphiphilous**	**Others**	**Total**
Summer					
No. of blooming taxa	35	93	6	4	138
No. of used plant taxa	3 (8.57%)	51 (54.84%)	5 (83.33%)	0	59 (42.75%)
Monsoon					
No. of blooming taxa	50	118	5	3	176
No. of used plant taxa	4 (8%)	47 (39.83%)	4 (80%)	0	55 (31.25%)
Autumn					
No. of blooming taxa	38	57	4	7	106
No. of used plant taxa	0	16 (28.07%)	3 (75%)	0	19 (17.92%)
Late autumn					
No. of blooming taxa	34	47	4	7	92
No. of used plant taxa	0	10 (21.28%)	2 (50%)	0	12 (13.04%)
Winter					
No. of blooming taxa	26	54	5	6	91
No. of used plant taxa	2 (7.69%)	25 (46.30%)	3 (60%)	0	30 (32.97%)
Spring					
No. of blooming taxa	12	63	6	4	85
No. of used plant taxa	1 (8.33%)	52 (82.54%)	6 (100%)	3 (75%)	61 (71.76%)
Total no. of plant taxa	76	210	11	13	310
Total no. of used taxa	6 (7.89%)	109 (51.90%)	10 (90.91%)	3 (23.08%)	128 (41.29%)

pollen grains in monads, while three plant species (*Acacia auriculiformis*, *Albizia lebbeck* and *Calliandra umbrosa*) are with polyad types of pollen grains and a single species (*Mimosa pudica*) is of tetrad type of pollen grains. Pollen types belonging to dicot plant families showed radial symmetry, whereas all monocot pollen types (except members of Poaceae) were of bilateral symmetry. Based on the multiplication value of polar axis (PA) and equatorial diameter (ED), we classified the pollen grains into five size groups viz. very small (PA × ED <100 μm^2), small (PA × ED = 100–500 μm^2), medium (PA × ED = 500–1500 μm^2), large (PA × ED = 1500–2500 μm^2), and gigantic (PA × ED >2500 μm^2). Most of the collected pollen types were small to medium-sized. The shape of pollen grains ranged from per-oblate to per-prolate with the dominance of prolate-spheroidal, spheroidal, and oblate-spheroidal. Most of the pollen grains were with circular amb, followed by triangular, oval, and trilobite outlines. The apertural patterns were different types, with trizonocolporate as the most common. The members of Acanthaceae (*Hemigraphis hirta* and *Hygrophila auriculata*) and Combretaceae (*Terminalia arjuna*) showed heterocolpate aperture type. All the bilaterally symmetrical pollen grains of monocot plant taxa show sulcate type of aperture. Two plant species, *Croton bonplandianum* and *Jatropha curcas*, exhibit inaperturate pollen grains. Exine ornamentations of the observed pollen types were also greatly varied. Though, the most common sculpturing types were reticulate, echinate and striate.

Like other honeybees, giant honeybees also prefer to forage several cultivated plant species (Figure 21.7). While visiting the flowers of a crop plant, if the foragers worked from the top of the flower (i.e., they came in contact with the reproductive

FIGURE 21.7 *Apis dorsata* foraging on some agri-horticultural plants to collect floral rewards: a. *Brassica nigra*; b. *Citrullus lanatus*; c. *Citrus* × *aurantiifolia*; d. *Coccinia grandis*; e. *Corchorus olitorius*; f. *Halianthus annuus*; g. *Momordica charantia*; h. *Nigella sativa*; i. *Papaver somniferum*; j. *Psidium guajava*; k. *Sesamum indicum*; l. *Trichosanthes dioica*. Scale bars = 20 mm.

organ of the flowers), these foragers were characterized as pollinators of the reference crop. If the foragers worked from the side of the flower to steal nectar only and did not come in contact with the flower's reproductive organs, then this forager was characterized as a non-pollinator (or nectar thief) of the reference crop. In some plants bearing tubular flowers like *Sesamum indicum*, foragers can act as either legitimate visitors or nectar robbers. In general, foragers visit several flowers of a plant species in each bout and maintain considerable floral constancy. Additionally, hairy body surface, a large number of colony workers, and greater flower visitation rate (number of flowers visited per minute) compared to the other insect orders (viz., Lepidoptera, Coleoptera) make them effective pollinators of several plant species. Thereby, they play an essential role to conserve the local flora and to increase the yield (fruit quality, fruit set, and seed set) of some agricultural and horticultural crop plants like *Brassica campestris*, *Brassica juncea*, *Fragaria* × *ananassa* (Sihag 1986; Abrol et al. 2019).

21.5 CONCLUSION

The giant honeybee *(Apis dorsata)* is widely distributed in tropical and subtropical Asia. They prefer for nest construction water tanks, house structures, and limbs of tall trees having rough bark texture. The abundance of rock bee colonies was comparatively higher during winter to summer (due to favorable weather conditions and greater availability of floral resources) and reduced during monsoon season to autumn by their migratory nature. Their nests are in single or in aggregations with a single large and semicircular comb of hexagonal cells. The honey storage cells are located at one end of the comb, variable in depth and considerably larger than the brood cells. During honey flow season (i.e., blooming time of important nectariferous plants), a colony can store a large amount of honey and become much more attractive to honey hunters. Though they vigorously foraged on only a few plants like *Borassus flabellifer*, *Brassica* spp., *Eucalyptus* spp., and *Sesamum indicum*, their forage spectrum constitutes a large numbers number of plant species. Due to diversified plant habits, flowering phenology, and floral characteristics of the bee-visited plants, it may conclude that *Apis dorsata* is an obligatory generalist flower visitor. Additionally, *Apis dorsata*'s high flower visitation rate, hairy body surface, and floral constancy make them effective pollinators of several wild and cultivated crop plants.

ACKNOWLEDGEMENTS

The authors express their gratitude to the authorities of Vidyasagar University (VU) for utilizing laboratories. They would also like to thank Mr. Dipankar Mandal (staff, USIC section of VU) for microscopy. They are grateful to Prof. Dharam P. Abrol (faculty of agriculture, SKUAST, India) for his invitation and lovely cooperation as an editor.

REFERENCES

Abrol, D.P., Gorka, A.K., Ansari, M.J., Al-Ghamdi, A., and Al-Kahtani, S. 2019. Impact of insect pollinators on yield and fruit quality of strawberry. *Saudi Journal of Biological Sciences* 26(3): 524–530. https://doi.org/10.1016/j.sjbs.2017.08.003.

Arias, C.A., and Sheppard, W.S. 2005. Phylogenetic relationships of honey bees (Hymenoptera: Apinae: Apini) inferred from nuclear and mitochondrial DNA sequence data. *Molecular Phylogenetics and Evolution* 37: 25–35. https://doi.org/10.1016/j.ympev.2005.02.017.

Corlett, R.T. 2011. Honeybees in natural ecosystems. In: Hepburn, R.H., and Radloff, S.E. (Eds.), *Honeybees of Asia*. Springer, Berlin, Germany, 215–225.

Franks, N.R., Pratt, S.C., Mallon, E.B., Britton, N.F., and Stumper, D.J.T. 2002. Information flow, opinion polling and collective intelligence in house-hunting social insects. *Philosophical Transactions of the Royal Society of London B* 357: 1567–1583. https://doi.org/10.1098/rstb.2002.1066.

Hepburn, R., and Radloff, S.E. 2011. *Honey bees of Asia*. Springer-Verlag, New York.

Koeniger, N., and Koeniger, G. 1980. Observations and experiments on migration and dance communication of *Apis dorsata* in Sri Lanka. *Journal of Apicultural Research* 19: 21–34.

Layek, U., Bisui, S., Mondal, R., Das, N., De, S.K., and Karmakar, P. 2020a. Floral traits and chemical cues associated with rock bee (*Apis dorsata* Fabricius) for the host selection in West Bengal, India. *Grana* 60: 310–323. https://doi.org/10.1080/00173134.2020.1823466.

Layek, U., and Karmakar, P. 2018. Pollen analysis of *Apis dorsata* Fabricius honeys in Bankura and Paschim Medinipur districts, West Bengal. *Grana* 57: 298–310. https://doi.org/10.1080/00173134.2017.1390604.

Layek, U., and Karmakar, P. 2020. Distribution, nesting biology and floral resources of red dwarf honey bee (*Apis florea* Fabricius) in West Bengal, India. In: Abrol, D.P. (Ed.), *The future role of dwarf honeybees in natural and agricultural systems*. CRC Press, Taylor and Francis, Boca Raton, 301–309.

Layek, U., Manna, S.S., and Karmakar, P. 2020b. Pollen foraging behaviour of honey bee (*Apis mellifera* L.) in Southern West Bengal, India. *Palynology* 44(1): 114–126. https://doi.org/10.1080/01916122.2018.1533898.

Layek, U., Mondal, R., and Karmakar, P. 2020c. Honey sample collection methods influence pollen composition in determining true nectar-foraging bee plants. *Acta Botanica Brasilica* 34(3): 1–9. https://doi.org/10.1590/0102-33062020abb0086.

Layek, U., Nandi, T., and Karmakar, P. 2016. Pollen forage and storage pattern of *Apis dorsata* Fabricius in Bankura and Paschim Medinipur districts, West Bengal. *International Journal of Pure and Applied Bioscience* 4: 59–71. https://doi.org/10.18782/2320-7051.2384.

McGregor, S.E. 1976. *Insect pollination of cultivated crop plants*. United States Department of Agriculture, University of Virginia, USA 411 pages.

Nagaraja, N., and Yathisha, V. 2015. Nest orientation of Asian giant honeybee, *Apisdorsata* in plains of Karnataka, India. *Journal of Entomological Research* 39(3): 197–201. https://doi.org/10.5958/0974-4576.2015.00019.5.

Oldroyd, B.P., Osborne, K.E., and Mardan, M. 2000. Colony relatedness in aggregations of *Apis dorsata* Fabricius (Hymenoptera, Apidae). *Insectes Sociaux* 47: 94–95. https://doi.org/10.1007/s000400050015.

Paar, J., Oldroyd, B.P., Huettinger, E., and Kastberger, G. 2004. Genetic structure of an *Apis dorsata* population: The significance of migration and colony aggregation. *Journal of Heredity* 95(2): 119–126. https://doi.org/10.1093/jhered/esh026.

Rader, R., Bartomeus, I., Garibaldi, L.A. et al. 2016. Non-bee insects are important contributors to global crop pollination. *Proceedings of the National Academy of Science of the USA* 113: 146–151. https://doi.org/10.1073/pnas.1517092112.

Ruttner, F. 1988. *Biogeography and taxonomy of honey bees*. Springer-Verlag, Heidelberg, Germany.

Sáez, A., Aizen, M.A., Medici, S., Viel, M., Villalobos, E., and Negri, P. 2020. Bees increase crop yield in an alleged pollinator-independent almond variety. *Scientific Reports* 10(1): 1–7. https://doi.org/10.1038/s41598-020-59995-0.

Seeley, T.D., Seeley, R.H., and Akratanakul, P. 1982. Colony defence strategies of the honeybees in Thailand. *Ecological Monographs* 52: 43–63.

Sihag, R.C. 1986. Insect pollination increases seed production in cruciferous and umbelliferous crops. *Journal of Apicultural Research* 25(2): 121–126.

Tan, N.Q. 2007. Biology of *Apis dorsata* in Vietnam. *Apidologie* 38: 221–229. https://doi.org/10.1051/apido:2007002.

Index

A